Towns and Temples Along the Mississippi

Towns and Temples Along the Mississippi

Edited by
David H. Dye
and
Cheryl Anne Cox

A Dan Josselyn Memorial Publication

The University of Alabama Press
Tuscaloosa • London

Manufactured in the United States of America

Library of Congress Cataloging-in-Publication Data

Towns and temples along the Mississippi / edited by David H. Dye
and Cheryl A. Cox.
p. cm.
Bibliography: p.
Includes index.
ISBN 0-8173-0455-X (alk. paper)
1. Mississippian culture. 2. Indians of North America—
Mississippi River Valley—Architecture. I. Dye, David H.
II. Cox, Cheryl Anne. 1953–
E99.M6815T68 1990
977'.00497—dc20 89–32994
CIP

British Library Cataloguing-in-Publication Data Available

Contents

Figures vii
Tables ix
Preface xi
Introduction 1

1
James B. Griffin
Comments on the Late Prehistoric Societies in the Southeast 5

2
George J. Armelagos and M. Cassandra Hill
An Evaluation of the Biocultural Consequences of the Mississippian Transformation 16

3
R. Barry Lewis
The Late Prehistory of the Ohio–Mississippi Rivers Confluence Region, Kentucky and Missouri 38

4
James E. Price and Cynthia R. Price
Protohistoric/Early Historic Manifestations in Southeastern Missouri 59

5
Dan F. Morse
The Nodena Phase 69

6
Mary Lucas Powell
Health and Disease at Nodena: A Late Mississippian Community in Northeastern Arkansas 98

7
Phyllis A. Morse
The Parkin Site and the Parkin Phase 118

8
Gerald P. Smith
The Walls Phase and Its Neighbors 135

9
Stephen Williams
The Vacant Quarter and Other Late Events in the Lower Valley 170

10
Charles Hudson, Marvin T. Smith, and Chester B. DePratter
The Hernando de Soto Expedition: From Mabila to the Mississippi River 181

11
Michael P. Hoffman
The Terminal Mississippian Period in the Arkansas River Valley and Quapaw Ethnogenesis 208

12
Ian W. Brown
Historic Indians of the Lower Mississippi Valley: An Archaeologist's View 227

13
George F. Fielder, Jr.
Comprehensive Planning for the Protection and Preservation of Mississippian Sites in Tennessee 239

References 246

Contributors 280

Index 284

Figures

2-1. Model for the Interpretation of Stress Indicators in Paleoepidemiological Research 17

2-2. Mississippian Culture Area ca. A.D. 1400 24

2-3. Time Spans of Analyses of Populations from the Eastern United States 26

3-1. Mississippi Period Sites in the Ohio–Mississippi Rivers Confluence Region 39

3-2. The Late Prehistoric Chronological Sequence in the Ohio–Mississippi Rivers Confluence Region 41

3-3. Topographic Map of the Adams Site 46

3-4. Reconstruction of the Adams Site Mound Group 47

3-5. Vertical Profile of Test Unit 1 at the Adams Site 49

3-6. Late Mississippi Period Ceramics 50

3-7. Rim Profiles 51

3-8. Plan of Subfloor Features, Structure 4, Hess Site 52

5-1. Upper Nodena Site 71

5-2. Nodena Phase Sites 79

5-3. Nodena Phase Vessel Forms 89

5-4. Nodena Phase Vessel Forms 91

5-5. Nodena Phase Vessel Forms 92

5-6. Nodena Phase Chipped- and Ground-Stone Tools 93

6-1. Comparison of Demographic Profiles of Three Amerindian Population Samples 103

6-2. Healing Caries Sicca in Adult Female Cranium from Upper Nodena 109

6-3. Percentage of Enamel Hypoplasis Lesions Formed at Given Postnatal Developmental Ages for Upper and Middle Nodena 115

7-1. Contour Map of the Parkin Site 122

7-2. Distribution of Parkin Phase Sites 124

8-1. Basic Vessel Forms, Northern Delta Regional Tradition 136

8-2. Basic Vessel Forms, Northern Delta Regional Tradition 137

8-3. Distinguishing Walls Phase Ceramic Types 138

8-4. Walls and Adjacent Phases 139

8-5. Configuration of Sherd Type Frequencies by Phases 141

8-6. The Walls Phase 142

8-7. The Boxtown Phase 148

8-8. Vessel Forms Characteristic of Specific Phases 149

8-9. The Commerce Phase 150

8-10. The Tipton Phase 152

8-11. The Jones Bayou Phase 154

8-12. The Early and Late Kent Phases 156

8-13. The Horseshoe Lake Phase 158

8-14. The Cramor Phase 160

8-15. The Nodena Phase 162

8-16. The Pemiscot Bayou and Campbell Phases 163

9-1. The Vacant Quarter: The Southeastern United States ca. A.D. 1500 174

10-1. The de Soto Route from Athahachi to Quizquiz 182

10-2. De Soto in the Black Warrior River Drainage 185

11-1. Quapaw Indian ca. 1700 210

11-2. Protohistoric Phases Along the Arkansas River 217

12-1. Various Historic Indian Sites in the Southern Portion of the Lower Mississippi Valley 228

Tables

2-1. Cross-cultural Secular Trends of Selected Stress Indicators 25

3-1. Ohio–Mississippi Rivers Confluence Region Late Mississippi Period Radiocarbon Dates 45

4-1. Sherd Percentages in Surface Collections from Three Protohistoric-Historic Sites in Southeastern Missouri 64

6-1. Percentages of Nodena Individuals Represented by Cranial and Postcranial Remains 100

6-2. Demographic Profile of the Nodena Skeletal Series 102

6-3. Prevalence of Periostitis in Long Bones 108

6-4. Molar Occlusal Wear 112

6-5. Prevalence of Dental Caries and Antemortem Tooth Loss in Four Prehistoric Amerindian Populations 113

6-6. Caries Prevalence by Tooth Type 113

6-7. Caries Prevalence by Loci 114

7-1. Diagnostic Types Used by Brain et al. 1974 129

8-1. Species Composition of Potential Meat Supply at Chucalissa 146

Preface

This volume presents an overview of the most advanced Native American cultures north of Mexico at the time of initial European contact. The story of their way of life is now unfolding as new information accumulates from the efforts of long-term archaeological research programs. These Mississippian people, organized into complex chiefdoms, lived in the Central Mississippi Valley from approximately A.D. 1350 to A.D. 1650. They occupied the fertile natural levees bordering cutoff lakes in the broad Mississippi Alluvial Valley. Their cultural adaptation to the Mississippi floodplain environment presents an interesting case study of societies that have not yet become fully urbanized but yet have developed a sophisticated political and military organization. Such studies are an important source of information for understanding the growth and development of cultures in other parts of the world and in outlining the rise of civilization in the New World and Old World.

The idea of producing a volume on the Late Prehistoric and Early Historic Mississippian inhabitants of the Memphis area began in 1981 when the senior editor taught a course entitled "Indians of the Mid-South" at Memphis State University. In teaching this course it became apparent that a comprehensive publication covering the last expression of the Mississippian florescence in the Mid-South was needed for students, professional archaeologists, ethnohistorians, and the interested public. The Morses' excellent overview of the Central Mississippi Valley, *Archaeology of the Central Mississippi Valley*, had not been published and few sources were available to students interested in local archaeology.

In 1983 we decided to host an Archaeological Institute of America symposium on the Native American cultures in the Memphis area in conjunction with an exhibit being prepared on artifacts from the Central Mississippi Valley in the Memphis State University Art Gallery. We applied for a Regional Symposium Grant through the Archaeological Institute of America and were encouraged by Delano Black, Assistant Dean of the College of Arts and Sciences at Memphis State, to apply for further financial support from the President's Academic Enrichment Trust within the State University Foundation.

Archaeologists, ethnohistorians, and anthropologists who were actively involved in archaeological research in the Memphis area were invited to participate in the symposium and share their thoughts on the nature of Late Mississippian society. The afternoon prior to the meeting, an old-fashioned Memphis barbecue was held at Chucalissa Indian village. The symposium was held at Memphis State University in the Faulkner Lounge on October 18, 1985. We were delighted with the enthusiastic response and, at the urging of Delano Black, made plans for the symposium's eventual publication.

A number of individuals were instrumental in the production of the gallery exhibit and the symposium from which this publication evolved. Anne I. Lockhart, MSU Art Gallery Director, Richard R. Ranta, Dean of the College of Communication and Fine Arts, Carol J. Crown, Chair of the Art Department, and Teresa Hays, gallery registrar, all contributed their time and energy to making the gallery exhibit, "Towns and Temples Along the Mississippi: Art of the Mississippian Peoples," a successful exhibition of Native American artifacts.

The College of Arts and Sciences symposium, "Towns and Temples Along the Mississippi: Mississippian in the Memphis Area," was a direct result of the efforts of a number of individuals and departments on the Memphis State campus. Support from Thomas W. Collins, Chair of the Anthropology Department, Sharon Harwood, Chair of the Foreign Languages and Literatures Department, and Delano Black, Assistant Dean of the College of Arts and Sciences, were instrumental in their support and encouragement. The Archaeological Institute of America and the Mid-South Association of Professional Anthropologists generously helped defray costs in the production of the symposium. Special thanks go to James Wiseman, President of the AIA, Peter Kuniholm, Chair of the Subcommittee on Regional Symposia, and Martha Richardson, formerly assistant director of the AIA, for their encouragement and support. Gerald P. Smith, Director of the C. H. Nash Museum, graciously allowed the use of the Chucalissa Indian Village for the barbecue on Friday afternoon.

We offer this volume to the professional and amateur archaeologists in the Central Mississippi Valley whose dedication and perseverance provided the information that made this work possible. In particular, we wish to thank the participants of the Towns and Temples symposium for their interest and enthusiasm. Without their support and patience this book would not have been possible.

David H. Dye
Cheryl A. Cox

Towns and Temples Along the Mississippi

Introduction

Towns and Temples Along the Mississippi is a collection of scholarly essays that outline the Late Prehistoric (A.D. 1350–1541), Protohistoric (1541–1700), and Early Historic (1700–1800) periods in the Central Mississippi Valley. The volume focuses upon the life-style of the native inhabitants, the Mississippian Indians, at a critical juncture in their history, that period during which European political, economic, and biological forces came in contact with native populations and institutions. This period of change and cultural reorientation represents a complex mosaic of rapidly changing social patterns. The nature of these cultural changes, in increasing and devastating contact with the European societies, is revealed in the documentation of the changing life-style of the Central Mississippi Valley inhabitants from the early sixteenth century until their forced removal in the early nineteenth century. That we know the broad pattern of native biological and cultural adjustment to the rapidly changing world system is due to the tireless efforts of a number of archaeologists, ethnohistorians, and cultural and physical anthropologists who have been engaged in answering basic questions concerning the nature of Mississippian life-style for the past 100 years. While there is still much to learn, the broad outline of Mississippian cultural development has been established as a solid record to be tested and expanded upon by future archaeological research in the Memphis area.

Most of the following essays were presented at Memphis State University on October 18, 1985, at a symposium entitled "Towns and Temples Along the Mississippi: Mississippian in the Memphis Area." The authors of the various chapters are concerned with a period of time that is virtually unknown to all but a small number of historians and anthropologists. Beginning in southeastern Missouri and continuing southward to the Natchez Bluffs, archaeologically and ethnohistorically known societies are described that once flourished along the banks of the Mississippi and its associated oxbow lakes. In order to cover areas not dealt with in the symposium, several scholars with interests in the Central Mississippi Valley were asked to contribute essays.

The lead chapter by the acknowledged dean of eastern United States prehistory, James B. Griffin, presents a comprehensive overview of Native American populations in the Southeast during the Late Prehistoric cultural expression. This succinct review outlines features of Mississippian cultures including the growth of pre-Mississippian societies and sixteenth-century European imperialism. This scholarly background sets the stage for appreciating the way of life of these Mississippian peoples by emphasizing their games, warfare, trade and exchange, subsistence, art, religious beliefs, and possible historic tribal affiliations.

George J. Armelagos and M. Cassandra Hill place one critical aspect of archaeological research in the domain of the physical anthropologist: the nature of a prehistoric population's biological adaptation and adjustment. The authors present a scholarly approach to such a study and suggest several hypotheses concerning the Late Prehistoric, Protohistoric, and Early Historic cultures in the Central Mississippi Valley. Reevaluating our methodological approach to data from the past is imperative if we are to advance beyond our present state of knowledge.

R. Barry Lewis summarizes current knowledge of the Late Prehistoric and Early Historic sequence in western Kentucky and eastern Missouri with special emphasis on the results of the University of Illinois's Western Kentucky Research Project. Knowledge of the Ohio–Mississippi River confluence region in western Kentucky has developed rather slowly and Lewis's contribution sheds new light on this important area of the Mid-South between A.D. 1300 and 1700.

James E. Price and Cynthia R. Price draw our attention to the Protohistoric period in southeastern Missouri when the resident Mississippian populations were concentrated in a few villages along Pemiscot Bayou between A.D. 1550 and 1650. Their discussion of these sites and of the artifacts collected from them suggests several new ideas concerning the human occupation of southeastern Missouri. The Campbell site is particularly significant because of its importance throughout the Protohistoric period. Their work with the Leo Anderson collection is a testament to the value of amateur and professional archaeologists working together when their combined values are based upon scientific research.

The development and aftermath of the Nodena phase, A.D. 1400 to 1650, is evaluated by Dan F. Morse. As a result of contact with European diseases from the Spanish expedition, which was resident in the Nodena phase (Pacaha) in June 1541, the native inhabitants of eastern Arkansas underwent rapid depopulation. Morse provides a superb description of the precontact life-style of the Mississippian people in the Memphis area and the resulting social changes in the late sixteenth century. Special emphasis is placed on the Upper and Middle Nodena sites and the settlement and subsistence patterns, material culture, society, and environmental setting of the Nodena phase.

Mary Lucas Powell provides an important dimension to Mississippian studies through her detailed analysis of 159 skeletal individuals from the Upper Nodena site and 69 individuals from the Middle Nodena site. Her analysis stresses dental pathology, demography, and skeletal morphology. She outlines the risk factors and biological stress to which these people were exposed. Her paper is an integral aspect of any discussion of a prehistoric population's adaptation to its social and biophysical environment.

Phyllis A. Morse brings earlier discussions of the Parkin phase up to date in a well-balanced and anthropological approach to the archaeology of the St. Francis River area. Her interpretation of the de Soto entrada augments well the discussions by Dan Morse, Charles Hudson, Martin T. Smith, and Chester B. DePratter.

In his discussion of the Memphis area, Gerald P. Smith proposes several new geographical divisions of native populations along the Mississippi River during Late Prehistoric and Protohistoric times. His formulation undoubtedly will be tested against future archaeological fieldwork and represents an important stage in the development of archaeological cultures in a spatial and temporal framework.

Stephen Williams outlines his perspective of the archaeology of the Memphis area from some 35 years of research on the prehistoric and historic cultures of the Lower Mississippi Valley. This overview of archaeological research orients the reader with an appreciation of the accumulation of knowledge that archaeolgists have garnered in the recent past.

Charles Hudson, Marvin T. Smith, and Chester B. DePratter bring Hernando de Soto and his entrada of foot soldiers, horsemen, priests, attendants, bureaucrats, and Indian slaves to the Memphis area from the chiefdom of Tascaluza. Hudson, Smith, and DePratter's day-by-day description of the sixteenth-century Spanish conquerors in the Memphis area provides the most recent interpretation of the route and exemplifies the use of ethnographic and archaeological approaches to ethnohistorical research.

Michael P. Hoffman's extensive research with Quapaw ethnohistory and the Protohistoric period is brought to bear in his chapter on Quapaw ethnogenesis. The nature of the Protohistoric and Early Historic Quapaw life-style is not well understood; even the origins of the Quapaw currently are being debated. Hoffman presents a lucid description of the eighteenth-century Quapaw and their adjustment to the predations of European culture.

The Natchez have held a fascination for European explorers since the earliest French descriptions in the early eighteenth century. Ian W. Brown documents his involvement with Natchez archaeology and ethnohistory and their development over the past 20 years. Natchezean archaeology is particularly important to archaeology in the Memphis area. The Natchez embody the best example of a chiefly society still functioning in the Mississippi Valley and, as such, the dynamic ethnohistoric accounts provide archaeologists with a rich corpus of ideas to be tested against the archaeological record.

George F. Fielder, Jr., discusses several means by which archaeological site preservation might be effected in the Mid-South. Comprehensive planning is vital and essential for integrating scientific knowledge with the impending destruction of our limited and nonrenewable cultural resources. If archaeological sites are to be preserved for future use and enjoyment by all citizens, it is incumbent upon cultural resource managers to develop and carry forth comprehensive management plans and ascertain the importance and significance of specific cultural resources.

The papers in this volume cover the 300-year period from ca. A.D. 1350 to 1650 from a variety of theoretical and methodological perspectives. Each paper concentrates upon a critical period in American prehistory, protohistory, and history from an anthropological perspective and provides a basis for future archaeological and ethnohistorical research. Studies of this time period in the Mid-South are beginning to take advantage of the rich corpus of interdisciplinary data through new syntheses and interpretations that document Native American life-styles with greater accuracy and precision. This perspective should bring about a greater understanding and awareness of the cultural events and processes that have taken place in the Mid-South prior to European and African settlement and will outline the complex cultural developments achieved by the indigenous inhabitants of the towns and temples along the Mississippi.

1

Comments on the Late Prehistoric Societies in the Southeast

James B. Griffin

This paper is a prelude to the more detailed and authoritative presentations in this volume. The information or interpretations in this paper are based on a number of publications as well as on my own participation in archaeological research in the area. This contribution is, of course, a compression of prehistoric and ethnohistoric research and as such is bound to be unsatisfactory in quite a number of ways to various colleagues whose views cannot be precisely followed.

Pre-Mississippian Culture Growth

For at least 10,000 years before the Late Prehistoric and Early Historic periods, the ancestors of the Mississippian societies of about A.D. 700 to 1000 were in the Memphis area and indeed in most of the eastern United States. The earliest inhabitants are identified by archaeologists as Paleoindians who were hunters and gatherers living in small bands of 20 to 40 individuals, interacting with other similar bands at seasonally abundant food resource areas. In this way communication of ideas moved over large areas, producing and maintaining similar lifeways.

From about 8000 to around 1000 B.C., the changes in the recovered material culture, settlement patterns, and other data are referred to as the Archaic period. During this long span of time many behavioral changes took place as the climate moderated, affecting the vegetation and animal life. One of the more important developments in the latter half of the Archaic period was the gradual domestication of a number of plants, the first of which appears to be cucurbits. These may have been introduced from northern Mexico or perhaps from cucurbits native to the eastern United States. In addition, plants such as marsh elder (*Iva*), erect knotweed (*Polygonum erectum*), maygrass (*Phalaris caroliniana*), goosefoot (*Chenopodium sp.*), and sunflower (*Heleanthus*) were domesticated.

Settlements were occupied seasonally and at strategic locations for longer periods of time by larger societies. What may loosely be called "tribal" areas are identified. Transport and exchange of nonlocal raw materials bear witness to the development of trails and expanding knowledge by local societies of the resources available over much of the East. For example, marine shells from the Gulf of Mexico and Atlantic Coast and copper implements began to appear in areas far distant from their source. Ground and polished stone were produced as utilitarian and ornamental artifacts. The Archaic developments provide the background accomplishments for the succeeding period.

From about 1000 B.C. to A.D. 700 the Woodland societies of much of the eastern United States had ceramics and used burial mounds for some or many of the members of local groups. From the study of burial procedures it is evident that some members, particularly adult males, had leadership roles in the society and that some of the strategically located, larger social groups were "superior" to others in local areas. Trade and exchange over a wide area between about A.D. 1 and A.D. 300 represent increased knowledge of contemporary groups in the eastern United States. Large regions, such as southern Ohio, the Illinois Valley, the Lower Mississippi Valley, the Tennessee River in northern Alabama, and many other areas, can be seen as having a cultural integrity that differentiates them. This differentiation does not mean that they were political units, but the regions clearly represent a remarkable cultural complexity for societies that were still primarily hunting and gathering groups with minor dependence on the early domesticates; they did not effectively grow maize.

Mississippian Maize and Mexico

From about A.D. 350 to the early part of the Mississippian period, evidence of broad regional interaction is much less visible. Also, evidence of the emphasis on burial ceremonialism of the preceding period decreases. Some archaeologists have viewed this time period as one of decline in cultural complexity over much of the greater Mississippi Valley, but this interpretation is not unanimous. The appearance of maize in the Mississippi Valley between A.D. 700 and 900 is interpreted here as the primary addition that nurtured the growth of Mississippian societies from the eastern prairies to the Appalachians and from the Gulf of Mexico to the Illinois area and to the Ohio Valley. Maize was domesticated in Mexico by around 5000 B.C. and was a major factor in the gradual development of the later Mexican civilizations. It was introduced into the southwestern United States by 1000 to 500 B.C. and stimulated the growth of Puebloan and other prehistoric cultures in that region. The route by which maize reached the East is probably from the Southwest and not by way of northeastern Mexico. At no time during the development or life span of the Mississippian societies is there any substantive evidence of stimulus from Mexico having any input to these southeastern societies. Again, this view is not universally held because of the longtime view that earlier and more advanced Mexican societies must somehow have stimulated the Mississippian societies, which achieved the most complex levels of any of the Indian groups in the United States. This opinion is held by eastern archaeologists but not, however, by the "unenlightened" southwestern archaeologists.

Some Features of Mississippian Cultures

There are several definitions and descriptions of Mississippian (Griffin 1967, 1985a; Hudson 1976; Smith 1987b), and my characterization is a short introduction to the successive and disparate societies that occupied the Southeast for about 1,000 years in an area about the size of western Europe.

Many of the southeastern prehistoric societies called Mississippian occupied areal units that included a major town and ceremonial center; a number of subsidiary villages; a larger number of farmsteads; and fishing, hunting, and other locations where raw materials were obtained. The larger towns had more or less permanent residents in rectangular wattle and daub houses of varying sizes to accommodate a single family, with food preparation, storage, and other household necessities recognizable from excavation. Larger buildings, which were placed on platform mounds around an open courtyard or plaza, were council houses for the top warriors, shamans, medicine men, and other heads of activities or social units in the community. The top man may certainly be called a "chief." The de Soto narratives state that when major decisions were to be made that affected the whole community, the action to be taken was decided at a meeting of the council, indicating that the "chief" was not an autocratic head of state. Smaller Mississippian societies would differ significantly from the largest ones such as Moundville, near Tuscaloosa, Alabama, or the Cahokia site in the Mississippi floodplain opposite St. Louis. None of the Mississippian societies in the Memphis area were as large and complex as those two outstanding examples of American

Indian cultural achievements. Some of the mounds had buildings that were charnel houses and contained the remains of the tribal leaders and ancestors who commanded the respect of the living and were accorded much veneration. The sacred fire was maintained with four logs arranged so that the space between the logs formed a cross representing the four world quarters.

The majority of Mississippian sites are located in major river floodplains, on river levees, or on bluffs overlooking such an environment. The Mississippian Indians were then in an advantageous position to select favorable soils for their crops and to obtain fish, migratory waterfowl, many animal species, the fruits of the bottomland forest and swamps and adjacent upland trees, and tubers and the seeds of plants, including those that were domesticated during the Archaic and Woodland periods. Maize was the primary crop and constituted about 50 percent of the diet. They did not abandon the foods that had been their mainstay for thousands of years. They did not develop an alcoholic beverage.

They produced such innovations as shell tempering for their pottery vessels and began making a great variety of vessel forms for everyday use and for burial furniture. A marked increase in population occurred; hamlets grew into villages and villages into towns and ceremonial centers. Many of these sedentary larger aggregates of people were organized into hierarchical, social, political entities with a number of leaders for political and other functions. Many of their villages and towns were palisaded for defense with posts 12 to 14 feet (3.6 to 4.2 m) tall, bastions at regular intervals, and a walkway attached to the inner wall. They built substructures of earth on which their council houses, ancestor shrines, and leaders' buildings were erected. When such buildings deteriorated or were destroyed, the area was leveled and more earth was added on the top and sides. In this way the mound would both rise in height and expand. The larger mounds were built over a period of 300 years. Long-occupied sites would have additional substructure mounds and a plaza or courtyard between the largest mound and smaller ones. The plaza or courtyard was used for games and ceremonies and contained little or no occupational debris. The houses of the inhabitants were arranged in regular rows outside the plaza area, and in some excavated sites there are indications of social groups separated by unoccupied space or by some type of screening. Towns vary in size, but a population of 300–500 would probably be the norm. A population of over 1,000 would have indicated a major town, while sites like Cahokia, Moundville, or Angel in southwestern Indiana are unusual with populations of 2,000–5,000, or perhaps even 10,000 for the central Cahokia area at its peak. Villages had populations of ca. 100–300, while farmsteads, fishing camps, and other such food-harvesting locations would have had few inhabitants for part of the year.

The Powers Fort complex in the Little Black River area of southeastern Missouri had one of the smaller civic ceremonial centers, which was ca. 15 acres (6 ha) in size with a fortification wall and four mounds. Intensive archaeological suveying has identified some 80 villages, farmsteads, and other smaller societal procurement areas in the old braided channel area of the Mississippi River. The Snodgrass village site of the Powers phase had a carefully planned house alignment within the 1-acre (0.4-ha) area enclosed by a palisade. When it was excavated in 1966–74 by the University of Michigan, it was the first Mississippian village to be almost completely excavated (Price and Griffin 1979). It had 90 structures, a population of ca. 350, and an occupation dating to the first half of the fourteenth century. While a strong degree of similarity occurs in the material culture remains, village plans, and site locations, the degree of political control is not known and continues to be a question. The same uncertainty is true of almost all of the cultural units constructed by archaeologists.

The social-political structure of these societies would vary in complexity depending on the size of the society, and many of them are characterized by archaeologists as chiefdoms. The agricultural development, with maize as the main crop, helped to fuel the increase of population and the need for administrative direction and controls. Male members of the major lineages, or hereditary extended families, obtained the primary positions of political leaders, religious leaders who interceded with the forces of nature, magical practitioners, and other such posts. Control of valued exotic goods of religious importance was vested in the "elite." Ceremonial activities would have taken place at fairly regular intervals during the year, almost certainly at new fire observances about New Year's, at the reappearance of vegetation during the spring or Easter, at the first maturing of corn about July 4, and at the main harvest season, which can be likened to Halloween or Thanksgiving. Supervised rituals would occur at such times as planting or harvesting, under the belief that these would ensure bountiful returns. Supernatural forces were thought to affect every phase of their activities. Such beliefs are best known to modern Americans from the written records of Mediterranean and Near Eastern societies, and they still linger in today's world.

Trade and Exchange

Trade of raw materials and of manufactured goods was one of the main threads that helped to produce and integrate the southeastern Mississippian societies. While little archaeological evidence exists for trade in salt, the concentration of ceramics used in salt evaporation found around salt springs and the known Early Historic records certify to the probability of this condiment being traded. Large marine gastropods, especially the *Busycon* whelks found along the south Atlantic and Gulf coasts, furnished cups for the Black Drink purification purgative ceremony before meetings of the "tribal" council. These shells were also fashioned into beads, pendants, and circular cutout sections that were used as gorgets. These latter objects were engraved with designs representing various concepts of their complex ritual and religious belief. The Mill Creek flint quarries in Union County, Illinois, produced the raw material for hundreds of flint hoes in a variety of shapes. The Crescent quarries on the southwestern outskirts of St. Louis furnished an excellent white flint, out of which was made not only projectile points but also flint maces 12 to 14 inches (30 to 36 cm) long that were used as symbols of status and authority. Pottery vessels with distinctive forms made for different functions and embellished with engraved or painted decoration can be recognized as regional products. Engraved shell gorgets of distinctive areal or individual style appear far outside their usual territory. Embossed copper plates with a variety of symbolic meanings represent trade in copper from sources in the Lake Superior Basin and the southern Appalachians, and some trade activity of these finely decorated plates also took place. The large reddish sculptured human figures and human effigy pipes, once thought to have been made of Arkansas bauxite, are now believed to have been made in the Cahokia area perhaps out of Missouri fire clay, which is found a short distance north of Rolla, Missouri. Other pipe forms in Late Mississippian times, such as the catlinite disk pipe or various distinctive styles of Protohistoric Cherokee forms, were extensively traded. It has been suggested that food was traded, but this is difficult to document. For most of the Southeast, trade in food would not have been necessary except in time of local droughts, floods, or other weather vagaries such as tornadoes or hurricanes. We can be sure that the weather patterns were very close to those of today and that agricultural failure and success varied from year to year and decade to decade.

We know from the Early Historic accounts of Europeans who penetrated into the eastern United States that individuals of the several native societies actively engaged in trade and were multilingual. They followed well-established land and water routes, many of which had been followed by similar Indian tradesmen thousands of years earlier. Some of the trails were followed by European and American traders, missionaries, settlers, and our own highways and railroads. It is unlikely that trade and exchange were conducted on any regular weekly, monthly, or yearly basis or that there was a special class of individuals organized as traders. It is also not likely that the trade was controlled or directed by the leaders of a particular society, either to distribute or acquire raw materials or manufactured products. However, trade was one of the activities that served both to reflect and to produce the distinctive "world" view that identifies most of the southeastern Mississippian societies as a large interacting culture area.

The Importance of Games

From archaeological finds as well as Early Historic accounts, we know that Mississippian societies had a wide variety of games of chance and games of skill requiring considerable athletic ability. Their ball game, played by teams from different towns, developed into the game of lacrosse in Canada. It was accompanied by elaborate ceremonies, dances, rigorous training, appeal to the supernatural to affect the outcome, and prizes. Another game was played with a disk-shaped chunkey stone, which was rolled along a prepared field; the two contestants threw a notched pole, trying to land it as close as possible to the place where the stone stopped. There were many games of chance, and Indians gambled on the outcome of all of their games. The supernatural forces, which they attempted to manipulate, were responsible for the outcome.

Warfare

Just as Mississippian societies in many phases of their culture reached a high point of Indian achievement, they had a large, well-organized warfare program. This program is attested to by the fortified towns and villages, projectile points embedded in human bone, blows to the head and left forearm bones, evidence of cut neck vertebrae, engraved shell gorgets, and large marine shells with scenes of conflict between warriors or a warrior with a flint knife in one hand, carrying a human head in the other. The latter was also portrayed on embossed copper plates. Mississippian armed forces had several grades of war chiefs, were made up of units from different associated villages, had a variety of weapons, and employed warfare as a means of defeating enemies but rarely as a means for acquiring territory. Mississippians also acquired slaves and controlled the products and people of adjoining societies. It is this latter factor that indicates they were on the road to civilization.

Mississippian Art: A Reflection of Religious Beliefs

The Late Prehistoric societies in the Southeast produced outstanding craftsmen in many areas. Because of the environmental effects on wood, fabric, leather products, basketry, and matting, those products have largely disappeared. A few examples exist, however, in European and American museums, and Early Historic accounts attest to their capability. The objects that have survived, such as ground and polished stone, chipped flint, marine shell, copper, and pottery, are outstanding artistic examples that do not suffer in comparison with those of any other group at a comparable level of development. The pottery of the Mississippian societies has long been recognized as outstanding in its variations in form, skill of the application of painted, incised, and engraved designs, and realistic modeling of human and animal effigy figures. Distinctive subareas of greater ceramic similarity can be recognized, such as the Nashville Basin, Moundville, Cahokia, southeastern Missouri and northeastern Arkansas, the Lower Arkansas Valley, and several others.

Much scholarly attention has been given to the embossed copper plates and engraved and excised shell gorgets with imagery representing humans, eagles, falcons, pumas, snakes, and portrayals of combat or chunkey players. There are, even on some of the shell art, combinations of real and imagined creatures comparable to those in any other areas of the world (Phillips and Brown 1978). These latter are represented in one of their last expressions in Europe in medieval art and sculpture, appropriately on cathedrals, where such mythical creatures as griffins and gargoyles do not carry the same meaning to observers in this century as they did when they were made. It was the southeastern iconographic art on copper plates and on shell that, along with the platform mounds, were thought to represent a derivation of Mississippian culture from Mexico. However, more careful comparative analysis of the designs from both areas, and more accurate chronologies, do not support this interpretation.

Some of the human representations on engraved shell and embossed copper appear to be dancing, as a part of one of the ceremonial activities that accompanied major or even minor socioreligious observances. Recorded ethnographic belief systems of the Cherokee, Creeks, and other linguistic groups indicate concepts that seem to be represented on the iconographic art. One thing is certain, and that is that many of the items of dress and other objects portrayed in the art have been found with the upper class burials of the Mississippian societies. One type of art is noticeably absent. Representations of sexual activity, or items associated with sex, are not found, which is in contrast to portrayals and models in many other cultures.

Sixteenth-Century European Imperialism and Southeastern Indians

The Historic period in the Southeast may be said to begin with Spanish explorations and some attempts at settlement and the establishment of forts along the Gulf and South Atlantic coasts from A.D. 1513 to 1528. One of the main objectives of the Spanish was to prevent colonization by the English and French and also to safeguard their trade route from Mexico. These Spanish contacts enabled some European trade goods to reach the interior and also probably introduced European diseases. The expedition led by Hernando de Soto from A.D. 1539 to 1542 had the most profound effect on southern Mississippian societies and marked the beginning of a disastrous decline from which they did not recover. De Soto had served with Pizzaro in Peru and was familiar with the subjugation of central Mexico by Cortez. His major aim was to reap similar wealth in the Southeast.

With the discovery that the New World was inhabited by higher primates, a considerable controversy arose in Europe as to whether they were true human beings. They were not mentioned in the Bible, which was then regarded as authoritative. After some years, Pope John III stated in a bull *Sublimis Deus*, on June 9, 1537, that indeed the Indians were human and, in effect, had souls and could become Christians. This declaration placed them under the jurisdiction of the Catholic church (Hanke 1937:71–72). The de Soto expedition, which landed near present Tampa Bay, included some 700 men from Spain, Portugal, and Spanish Cuba, also Spanish women, 3 Negroes, and Indian servants. There were some 500 soldiers, craftsmen including shipbuilders, and priests. They had 225 horses when they disembarked but only about 50 when the remnants of the army prepared to abandon their hope for wealth and glory. They transported from Cuba 300 hogs, which were under de Soto's charge. "These were Spanish range pigs, long snouted, long legged, agile and accustomed to foraging for mast and acorns in woodlands and to being driven in herds. . . . De Soto's men knew how to manage them and drove them along with the army, . . . They were probably the ancestors of the feral pigs reported by European settlers in Georgia many years later, and of the characteristic rural breeds in the South today" (Parry 1985:91–92). In spite of consumption by the army, escapes, and gifts to or thefts by Indians, 500 pigs remained when de Soto died in the spring of 1542. Only about 300 of the original 700 men reached Mexico in the summer of 1542.

The de Soto expedition was one of the most devastating military invasions in southeastern history. Estimates of the number of Indians killed vary, but it is safe to say at least 3,500 were slain; others were taken captive. Towns and villages were destroyed; native food supplies were consumed or destroyed. In addition, it is believed that epidemic diseases were introduced during the 30 years of European contact before de Soto in the Southeast and by the de Soto and later seventeenth-century expeditions (Milner 1980). Southeastern native societies never really recovered from these "benefits" of European penetration into their world (Peebles 1986). Remnants of the societies in the Memphis area moved south and perhaps in other directions as well during the seventeenth century. Some, such as the Quapaw in the Lower Arkansas Valley, were left.

Late Mississippian Societies and Possible "Tribal" Affiliations in the Memphis Area

During the period just before de Soto and for about 100 years thereafter, Mississippian sites from the southeastern tip of Missouri to the mouth of the Arkansas share quite a few similarities of site plans, pottery, and other artifacts. Stephen Williams calls this area of the Central Mississippi Valley the Armorel phase for this time period. This construct perhaps should be broken up into several approximately contemporary local phases. In any event, both from archaeological and from ethnohistoric evidence, this time period is probably the high point of Mississippian development in the greater Memphis area. There is some disagreement about this, for other archaeologists refer to an earlier Mississippian period as the high point from about A.D. 1100 to 1400, which is particularly true if the archaeologist is primarily concerned with developments within 100 miles (160 km) of the mouth of the Ohio River.

Other archaeologists recognize a number of contemporary, rather large societies, composed of major towns and subsidiary hamlets and other food-procurement locations. From about Lepanto to Blytheville, Arkansas, along the Mississippi, and in the immediate area to the Little River, were populations identified as the Nodena phase. This phase is probably the province of Pacaha of de Soto's chroniclers; the Pecan Point site near Wilson, Arkansas, was perhaps the central town of Pacaha. The territory, just to the north and south of Memphis along the Mississippi on both the east and west sides, is called the Walls phase, with the later occupations at Chucalissa on the southern edge of Memphis being part of this archaeological construct. The Walls phase may be the province of Aquixo. The Walls and the Nodena phases have many similarities. South of the Walls phase and on the west side of the river and up the St. Francis about 20 miles (32 km) is the Kent phase, which could be the Quiguate province. The Parkin phase, along the St. Francis and the Tyronza rivers for about 20 miles (32 km) north of the Kent phase, is perhaps the province of the chief of Casqui. In western Mississippi, opposite the mouth of the Arkansas and along the lower reaches of that river, were sites of the Quapaw phase, with the Menard and Wallace sites between the confluence of the White and Arkansas rivers as some of the archaeologically known sites. As one of the early supporters of these sites as the culture of the Siouan Quapaw, I am still impressed with that association.

There is also the possibility, but somewhat less certain, that the Walls-Nodena populations were also Siouan speakers, and perhaps this is also the case for the Old Town phase societies north of the Arkansas River and primarily on the west side of the Mississippi. At the de Soto time line most of the authorities agree that the Chickcasaw were located south of Tupelo in Pontotoc and Lee counties and perhaps in northwestern Monroe County, Mississippi. However, none of the Chickasaw sites has been securely identified as having been visited by the Spaniards. The Choctaw were in the areas south of the Chickasaw, other Muskogean speakers were east of the Chickasaw-Choctaw, and Caddoan speakers were west of the Quapaw.

2

An Evaluation of the Biocultural Consequences of The Mississippian Transformation

George J. Armelagos
M. Cassandra Hill

Paleopathology is a powerful tool for understanding the adaptation of prehistoric populations. The analysis of health and disease in archaeological populations provides a measure of the group's ability to adjust and to respond to changes in the environment. The pattern of disease found in a group is not a matter of chance. As Calvin Wells (1964:17) so aptly states:

> The pattern of disease of injuries . . . is invariably the expression of stresses and strains to which they were exposed, a response to everything in their environment and behaviour. It reflects their genetic inheritance (which is their internal environment), the climate in which they live, the soil that gave them sustenance and the animals or plants that shared their homeland. It is influenced by their daily occupation, their habits of diet, their choice of dwelling and clothes, their social structure, even their folklore and mythology.

Because the pattern of disease reflects a group's lifeways, the lesions found on bones can be used to interpret the adaptive capacity of their owners. With this perspective, we have three objectives for this paper. The first is to demonstrate the methods used to analyze disease in prehistoric skeletal populations. The second is to discuss the ways in which the disease profile can be used to analyze the adaptation of prehistoric populations. Third, we will apply these methods to assess the adaptation of Mississippian populations living along the Mississippi River and adjacent areas.

The Ecological Model

To evaluate the adaptation of these groups we will use an ecological model (Figure 2-1). Of the five components of the model, the first considers the environment, which is the source of all resources necessary for maintaining and reproducing the population. If the environmental resources are adequate and the population is able to exploit them effectively, survival should be ensured. However, constraints in the environment often may affect survival. These constraints may include limited resources, such as shortages of food and fuel, and/or other stressors. For example, a group living in an area without an adequate source of proteins would have limits imposed on its biological development.

These constraints correspond to the elements of risk, diversity, and productivity discussed by Sanders and Webster (1978). Risk is defined as any environmental parameter essential to the production of energy (such as moisture and temperature) with wide, relatively frequent, and unpredictable variation (Sanders and Webster 1978:253). Diversity is characterized by the closeness and pattern of spacing of the contrasting environmental conditions that are significant in terms of human exploitation. This diversity may involve subsistence or nonsubsistence resources. Productivity is the potential of the landscape to produce energy in the form of subsistence products for the support of human populations (Sanders and Webster 1978:253). Although productivity may be absolutely constrained by environmental factors, it is intimately linked to such cultural factors as levels of technology and so-

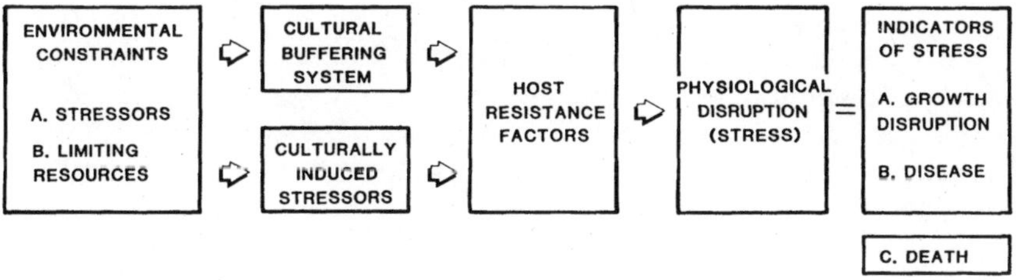

Figure 2-1. Model for the Interpretation of Stress Indicators in Paleoepidemiological Research (from Goodman et al. 1984b:278)

cial organization, energy sources available, and variety of information processed by the group (Sanders and Webster 1978:261).

In addition to these constraints, the environment also contains stressors. Stressors are any physical, chemical, biological, social, or psychological insults that threaten an individual's or population's equilibrium within the environment. These stressors can be physical factors such as extremes in temperature or physical forces that cause trauma, chemical insults such as toxins that may be found in the plants, or biological factors such as bacteria that cause disease.

Second, the adaptation of human populations is enhanced by a cultural system that buffers the population from environmental constraints (limited resources or stressors). The technology (the means by which energy is extracted from the environment), the social organization (patterns of kinship and social interaction that allow the group to function), and the ideology (attitudes, ideas, and beliefs) act to buffer the environmental stressors. In most instances, the culture alleviates the stressors, lessening their impact on the individual or population. However, in some cases, the cultural system fails to inhibit the stressors, and they will then have their full impact on the biology of the individual or population.

There is also a possibility that the cultural system may not only fail to buffer the stressors but can produce additional stressors. Changes in subsistence, for example, may have a dramatic impact by creating new stressors. The development of agriculture led to an increase in population size, population density, and sedentism, all of which increase the potential for infectious disease transmission.

The impact of the stressor on the individual or population is related to its strength and duration. A severe nutritional shortage of a day or two may be tolerated by most individuals. On the other hand, a less severe shortage that lasts months or years could have quite a severe effect on the health of the group.

The third component of the model considers the impact of stressors on the host. The host represents the individual's or population's resistance to stressors. Susceptibility to stressors can change. For example, susceptibility can increase as a consequence of changes within the system. Repeated response to stressors may decrease an individual's resistance to insults, and not all groups are equally at risk. Subgroups of the population who would be more likely to be at risk for nutritional deficiency would be infants and adult females (particularly those of reproductive age), while older individuals would be susceptible to infectious disease because of a reduction in their T-cell function.

If the host is not able to resist a stressor, a physiological disruption (stress) will occur. In living populations, the physiological disruption can be measured by changes in the vital signs. Changes in serum and other hematological values can be used to evaluate susceptibility to disease. In archaeological populations, we have only the skeletal remains from which to determine the pattern of disease. Because many of the other systems of the body can act to buffer the skeletal physiology, the disruptions that affect bone must be quite severe or of chronic duration.

The final component of the model is the individual's response to stress. The physiological disruptions in skeletal populations can be measured by biological indicators such as (a) growth disruption; (b) disease, such as nutritional disorders, infection, trauma (either accidental or nonaccidental, which may be indicative of strife), and degenerative pathologies; and ultimately (c) death, exemplified by increased number in subgroups at risk, decreased life expectancy, and increased probability of dying (see Goodman et al. 1984a; Huss-Ashmore et al. 1982; and Magennis and Martin 1985, for general discussion and examples of application of this model).

Disease in Prehistoric Populations

The recent success of paleopathology is the result of a research strategy that moved beyond the analysis of single indicators of stress. The objective of much of the earlier work in paleopathology was the diagnosis of specific diseases. In this method of analysis, the chronology and geographic distribution of a disease were the major objectives of the research. Attention was focused on finding the earliest occurrence of a specific disease in an area. However, there are inherent limits to this perspective because it fails to go beyond historical and geographic factors and does not consider the role of sociocultural behavior in the disease process. The shift to multiple stress indicators coincided with a shift to a concern for understanding the adaptation of the population. The pattern of disease (multiple stress indicators) is the means for determining the adaptive success of a population.

The diagnosis of pathological conditions in living and deceased populations is fraught with difficulties. For living populations, the disease causing insult is isolated and analyzed by studying the vital signs and living tissues for evidence of physiological disruption. Because the paleopathologist usually has only the skeletal remains, the diagnosis of disease is even more difficult. The paleopathologist can only determine diseases that leave their marks on bones. Fortunately, many diseases do leave their mark on bone, thus providing indisputable evidence of their presence. For example, syphilis and tuberculosis each leave a distinct pattern of lesions. Because this pattern is unique, the lesions are a signature of the existence of the disease in the population.

Even in instances in which the specific infection cannot be isolated, valuable information can be obtained from the bone. Generalized changes in the periosteum (the outer covering of bone) may be caused by a number of different irritants. Periostitis (the changes in the bone that result from inflammation of the periosteum) can be an important clue to the health of the population because it indicates the presence of an infectious pathogen, such as a staphylococcus.

In addition to the analysis of perisosteal reactions as a measure of infection, other stress indicators can be used to evaluate the severity of physiological disruption. The occurrence of porotic hyperostosis, which results from anemia, is one such example. The response to anemia produces a distinctive pattern of osteological changes in the bones of the skeleton that are involved in the production of red blood cells at the time of the onset of anemia (thin bones of the cranium, sternum, ribs, and long bones). The initial physiological response to the anemia is an increase in red blood cell production. This increase in the production of red blood cells will result in an expansion of the diploe (the inner portion of thin bones of the skull) or expansion of the medullary cavities of other bones, which will subsequently result in a thinning of the outer table of bone, exposing the diploe. Exposure of the porous diploe gives the lesion its name (porotic hyperostosis).

Many anemias cause porotic hyperostosis. Severe hereditary anemias such as thalassemia, sickle cell anemia, and hereditary nonspherocytic anemia will result in porotic hyperostotic changes in the thin bones of the skeleton. These hereditary anemias are usually so severe that they will also result in significant changes in the postcranial skeleton. Thalassemia causes major changes in the growth patterns of facial bones that can be easily identified radiograpahically.

While the nutritional anemias are usually less severe and are not as likely to cause postcranial changes, they do result in porotic hyperostosis of the roof of the orbits, the parietal, frontal, or occipital bones. In addition, the nutritional anemias are more likely to occur in infants over six months of age, children, and young adult females. The differential diagnosis relies on the analysis of the pattern of the lesions, their severity, their differential occurrence in males and females, and distribution in various age groups (Mensforth et al. 1978).

If there is evidence of physiological disruption such as porotic hyperostosis or infectious disease, there should be some evidence of its impact on growth. While the analysis of growth and development is a primary health indicator in modern medicine, its application to prehistoric populations is much more difficult and is only beginning to be studied. In living populations, longitudinal studies are used to develop the standards that are used to assess normal development. In prehistoric populations, the cross-sectional nature of the data makes the analysis more difficult. The lengths of the various long bones are calculated for each of the age classes. The length for each developmental period is used to construct the cross-sectional growth velocity for the population. Because age determination of the skeleton is based on developmental features, it differs from the chronological age used in living populations.

In addition, there are few studies of archaeological populations that can be used for comparison. Johnston (1962), Merchant and Ubelaker (1977), Ubelaker (1978), and Armelagos et al. (1972) have compiled growth curves from long bones of archaeological populations. Many of these studies do show evidence of growth retardation during various stages of development. In the Nubian study, Armelagos and co-workers demonstrated a "slowing" of growth between ages two and six. Huss-Ashmore (1978), in the same population, found that even though the length of long bones was only slightly retarded, the walls of the long bones (the cortex) showed abnormal thinning. Subadults who were developmentally aged at 14 years had cortical walls of the same thickness as a two-year-old.

Because the determination of growth disruption in archaeological populations is difficult, it should be corroborated with other evidence of growth disturbance. In the case of Sudanese Nubian populations, the analysis of bone microstructure corroborated the evidence that there had been a retardation of long-bone growth. The analysis of Harris lines (Harris 1933), which are transverse lines of increased mineral density in the internal structure of the medullary canals of long bones (Garn et al. 1968), and of enamel hypoplasia (Goodman et al. 1980) does offer effective methods for determining growth disruption. Although transverse lines were initially thought to be evidence of the cessation of growth, there is evidence that they occur during the period of growth recovery. It appears that if long-bone growth is disrupted for any reason, the epiphyseal cartilage plate (growth plate at the end of the bone) will decrease in thickness and a thin layer of bone-producing cells (the osteoblastic layer) is formed. Radiographically, these appear as thin, hypermineralized transverse lines across the bone's diaphysis. In earlier studies, an increase in the frequency of transverse lines was equated with an increase of growth disruption or "growth arrest." Individuals with numerous lines were considered to be less healthy than those without lines. The recent reevaluation of Harris lines (Clark et al. 1982; Hummert and Van Gerven 1985; Murchison et al. 1983) indicates that they are evidence of growth recovery and therefore of health rather than illness. Individuals without lines are seen as those who were unable to recover from chronic stress. The interpretation of Harris line data continues to be controversial.

Chronologies have been developed that allow one to determine when the disruption and recovery occurred. By knowing the rate of growth for the bone under study (growth proceeds distally and proximally from the mid-shaft), the age at which the disruption occurred can be determined. In a sense, this method of analysis provides a window into the past and allows us to investigate the physiological stress that an individual experienced during his/her lifetime. It also provides a method for assessing ages at risk for segments of the population. There is, however, one problem with transverse lines: they can be resorbed during the process of skeletal remodeling (the bones of the skeleton are always in the process of resorbing and depositing bone).

The dental enamel provides an indelible record of growth disruption because it will record the stress by failing to mineralize the enamel properly (enamel hypoplasia is the result). Because the enamel cannot be resorbed and its development chronology is well established, it provides a permanent metabolic history of the individual's growth process.

Trauma is another insult that provides information about the stress that a group faces. Even the location and the type of trauma can provide information relevant to the adaptation of a group. As an example, fractures of the forearm (the radius and ulna) that occur near the wrist area are usually the result of a person's falling and extending the arm in order to "break" the fall. A fracture of the radius or ulna usually occurs in the distal portion of these bones in the area of the wrist. A fracture of the forearm at midshaft, however, is the result of a person's raising the forearm to ward off or "parry" a blow delivered by another person. Just as the fracture at the wrist may provide an index of "klutziness," the "parry fracture" may be used as an index of strife for a population.

The wear and tear that a population experiences will be reflected in the degenerative changes that occur in the skeletal system. The most frequent pathological condition found in archaeological populations is degenerative joint disease (osteoarthritis). The severity and location are often the result of physical activity that can cause degeneration of the cartilage and subsequently result in osteological changes in the joint (Merbs 1983). Similarly, the degenerative changes in the body of the vertebrae (osteophytosis) can give clues to stress and strains that result from the aging process and physical activity.

The ultimate measure of adaptation is mortality. While all of the individuals in an archaeological sample are dead, the pattern of mortality can be very distinct. The comparison of these patterns allows one to assess differences in adaptation. Once the determination of the age at death has been made, a number of methods can be used to analyze patterns of death. The construction of a life table (see Swedlund and Armelagos 1976) provides a number of parameters for mortality comparison. The life table calculates measures such as expectation of life, probability of dying, and survivorship for each specific age category. Although the life table has been criticized because of inaccuracy of age determination (Bocquet-Appel and Masset 1982; Howell 1982), recent studies by Lovejoy and co-workers (1985) have shown that serially aging the population can minimize this source of error.

The application of the life table to study the adaptation of prehistoric populations does make a number of basic assumptions. It is assumed that the mortality pattern of the skeletal sample represents that of the living population (that no age classes are underenumerated) and that the society is not experiencing population growth. Paleodemographic studies are often criticized for their underrepresentation of infants. Moore et al. (1975) have shown, however, that underrepresentation of infants will have little impact on the expectation of life for the older age categories. They also point out that a growth in population will affect estimations of life expectancy (increase in infants will lower life expectancy). Techniques can be used to estimate the impact of growth on life expectancy.

Health and the Adaptation of Mississippian Populations

We have selected an analysis of the shift to an agricultural subsistence economy as the focus of our study. The shift from gathering-hunting to primary food production represents one of the most profound adaptive changes in human history. The increase in population size following the development of agriculture has always been assumed to be the result of improved nutrition and health. There is, however, increasing evidence that nutrition and health may have deteriorated as a result of agricultural intensification (Cohen and Armelagos 1984). The transition of prehistoric Amer-

Figure 2-2. Mississippian Culture Area ca. A.D. 1400, Including Oneota and Fort Ancient (from Dickens 1982:11)

ican Indian groups to a Middle Mississippian economy provides an excellent test of this hypothesis. Because this shift to agricultural intensification occurred in many areas of the eastern United States, it provides many case studies with which to evaluate the consequences of this adaptive shift.

In examining the cultural remnants of populations who participated in the life-style known as "Mississippian," two questions have been of particular interest: Why did this life-style develop? And what were the long-term consequences to the individuals involved? While the causes of Mississippian development continue to elude us and can only remain in the realm of speculation, the consequences have been transformed from theoretical issues to documented facts. The scientific technology available to the biological anthropologist has aided in this interpretation of the impact of the Mississippian development on the biology of the populations.

Earlier theoretical discussions maintained the notion of "progress," in which each stage of cultural development was seen as an improvement upon the latter and therefore beneficial to the descendants of the previous stage. However, more recent approaches have been critical of the notion of progress and have said that while increasing cultural complexity did have some

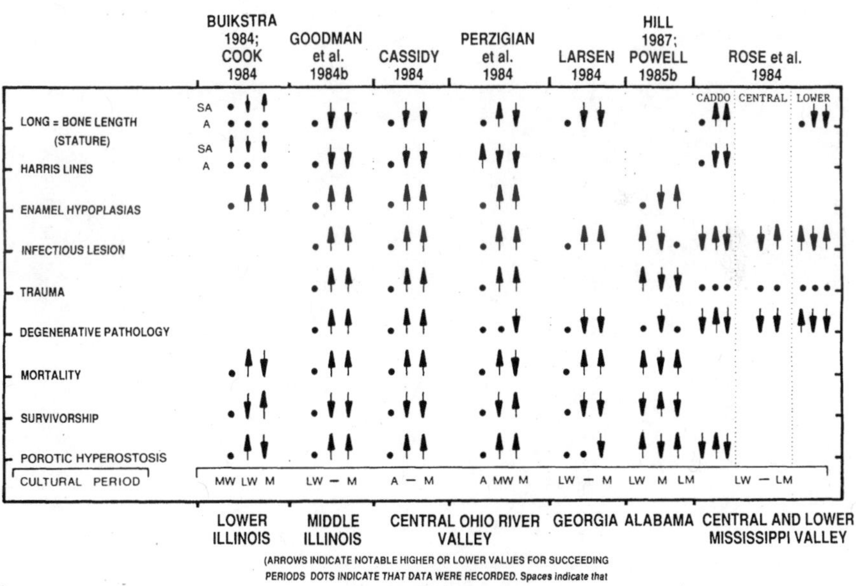

Table 2-1. Cross-cultural Secular Trends of Selected Stress Indicators

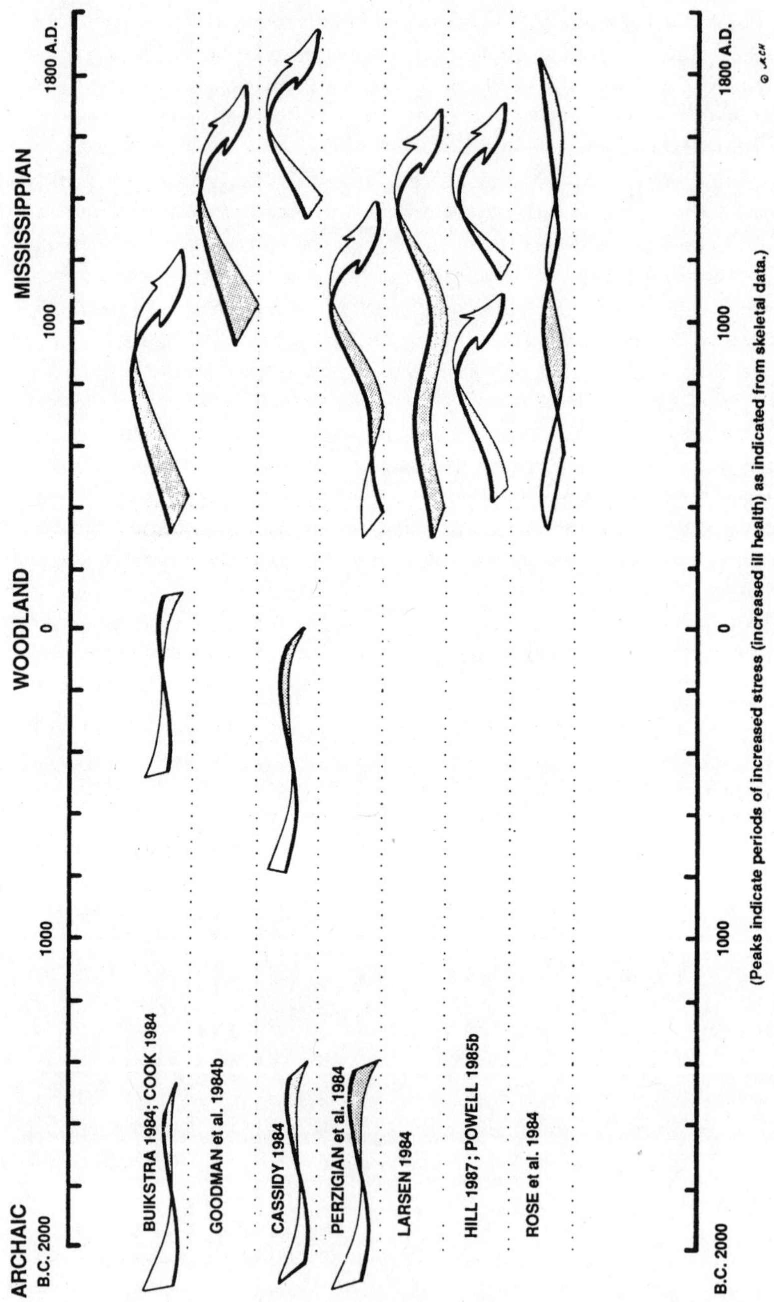

Figure 2-3. Time Spans of Analyses of Populations from the Eastern United States

positive effects, such as a secular trend toward increased longevity, many negative effects occurred as well. For the studies of prehistoric populations, these criticisms have centered on three basic components of cultural systems: (1) subsistence patterns, (2) settlement patterns, and (3) social status, all of which are intimately linked.

In a conference focused on examining the causes and consequences of agriculture in archaeological populations, several authors (Buikstra 1984; Cassidy 1984; Cook 1984; Goodman et al. 1984a, 1984c; Larsen 1984; Perzigian et al. 1984; and Rose et al. 1984) presented data on this transition (Cohen and Armelagos 1984). For our purpose here, we have selected those studies (and their respective bibliographies) with information on the Mississippian transformation in the eastern United States. In addition, we have included information from sites in Alabama reported by Hill (1979, 1981a, 1984), Cole et al. (1982), Powell (1983, 1985b), and Turner (1985) (see Figure 2-2, Table 2-1, and Figure 2-3).

The transition to a Mississippian life-style from that of Late Woodland took 250 years to accomplish at Dickson Mounds in Fulton County, Illinois, and had a significant impact on the health of the population. The Late Woodland occupation at Dickson Mounds is dated at A.D. 950–1100. It has been described as a gathering-hunting economy with a relatively small average population size of 75 to 125 persons living in seasonal campsites. During the later phases of this period, Dickson Mounds apparently came under the influence of Middle Mississippian groups at Cahokia (Fowler 1978), which is 180 km to the south. By the end of the Late Woodland period at Dickson Mounds, the influence of the Middle Mississippian culture had increased to the extent that there was a transformation of the adaptation, which has been referred to as the Mississippian Acculturated Late Woodland period (A.D. 1100–1200). The culmination of the Mississippian influence at Dickson Mounds occurred from A.D. 1200 to 1300. There were hamlets that were surrounded by support camps, work stations associated with local ceremonial centers, and sites where resources were extracted (Harn 1978). There were probably 600 to 1,170 individuals occupying the 234 habitation structures at the site at that time.

Three major trends occurred at Dickson that reached their pinnacles during the Middle Mississippian period. First, there was an increase in population size, population density, and sedentism. Second, there was an intensification of maize agriculture; and finally, trade was extended and intensified.

Lallo (1973) and Goodman and co-workers (Goodman et al. 1984a; Goodman et al. 1984b; Goodman and Armelagos 1985b) have analyzed the evidence for physiological disruption at Dickson Mounds as subsistence shifted from the Late Woodland to the Middle Mississippian period. They were able to show that there was (1) a reduction in long-bone growth and velocity, (2) an increase in porotic hyperostosis (iron-deficiency anemia) from 13 to 51 percent in the subadults, (3) an increase in infectious diseases involving the tibia (periosteal reaction) from 26 to 84 percent, (4) an increase in the frequency of degenerative lesions of the spine (osteophytosis) from 39.7 to 65.8 percent, (5) an increase in traumatic lesions from 13.5 to 19.5 percent, (6) an increase in enamel hypoplasia, and (7) a decrease in life expectancy at every age interval.

The evidence for the increase in disease stress at Dickson Mounds offers many interesting paradoxes. There is no question that a serious deterioration in the health of the population occurred. Harn (1980) has shown that the environment in this area of Fulton County, Illinois, was well balanced and should have sustained the Dickson population throughout the Middle Mississippian period. Harn has also demonstrated that gathering and hunting were still major activities during the Middle Mississippian period. If the Dickson Mounds population was left on its own in a closed cultural-ecological system, they should have done well nutritionally (Goodman et al. 1984a:300). While the increase in infectious disease may reflect the increase in population density, the decline in dietary resources requires a consideration of other factors. It may be that the Dickson Mounds population became involved in a regional system of ideological and economic exchange that may have affected their diet. It is possible that the Dickson population may not have benefited biologically from their exploitation of the local resources and the intensification of agriculture during the Middle Mississippian period.

Goodman and Armelagos (1985a) argue that the intensification of agriculture may explain only part of the decline in health of the Dickson Mounds population. The increase in arthritis and trauma during the Middle Mississippian period may have been due to increase in strenuous activity related to agricultural production and the rise of social tensions that results from the increase in population size and density. Because there is archaeological evidence that the environment could support a balanced diet for the Dickson population, it is necessary to understand the disparity between the foods available (in a sense, the menu) and what was eaten (the diet).

In search for an explanation of this paradox, Goodman and Armelagos (1985b) look further to the south and the relationships that existed between Dickson Mounds and Cahokia. They suggest that exotic items that have been uncovered at Dickson (they have found copper-covered ear spools and marine-shell necklaces) were traded into the area. If the Dickson population had sent foodstuffs such as meat to Cahokia for luxury items, then the pattern of nutritional deficiency could be explained. While this scenario is suggestive, a confirmation of the scenario would require an analysis of the flow of perishable goods. This exchange could be traced by the analysis of the faunal remains at Dickson Mounds and Cahokia. Bogan (1983) has shown that analysis of the fauna can provide a great deal of information about the use of perishable resources. He showed that at Toqua, a late Mississippian site in eastern Tennessee, there appeared to be differential access to certain animal species and preferred cuts of meat in certain parts of the site that may reflect status differences.

While the shift in subsistence resulted in significant changes in the nutrition and health of the Dickson Mounds population, it remains to be determined whether this change reflects the general case for the shift to Middle Mississippian adaptation or if it was a unique situation. In the following case studies we can assess the hypothesis that the intensification of agriculture resulted in a deterioration of nutrition and health. We should also be able to determine if there was a differential effect of agricultural intensification among the larger ceremonial centers and the smaller sites that were within their interaction spheres. It may also be possible to uncover instances in which the larger centers were exploiting other groups for subsistence goods.

In the studies of the Lower Illinois Valley populations, Buikstra (1984) and Cook (1984) define the region in terms of three distinctive ecological zones: the upland prairie, the forest, and the river floodplains, all of which are enumerated in increasing order of productivity according to their carrying capacity (Buikstra 1984:217). The majority of their data covers the time period of A.D. 400–1200. The skeletal material is associated with Middle Woodland, early Late Woodland, late Late Woodland, and Early Mississippian sites.

The skeletal data presented by Cook indicate several trends that can be summarized as follows: (1) a significant decrease in juvenile stature in the Late Woodland period, (2) no significant changes in adult stature (although notable status differences occurred in the Middle Woodland sample), (3) an increase in sexual dimorphism in the late Late Woodland period, followed by a decrease in sexual dimorphism in the Mississippian samples, (4) no changes in cortical bone maintenance (it remained low throughout Late Woodland and Early Mississippian periods), (5) a significantly higher average number of Harris lines in the Middle Woodland subadults, and a slight increase in the number of Harris lines in adults through time, and (6) an increase in enamel hypoplasia. The evidence of iron-deficiency anemia (porotic hyperostosis) was highest among the Late Woodland subadults, and there was an increase in age-specific mortality in the Late Woodland population. The data presented by Cook suggest that the populations in the Lower Illinois River Valley experienced a deterioration of health during the Late Woodland period. This deterioration may represent, as Cook suggests, the health problems associated with the adaptation to a sedentary life-style. Cook also suggests that the data would support the hypothesis that population pressure occurred prior to the development of agriculture.

Skeletal data from population samples from the Ohio Valley were presented by Cassidy (1984) and Perzigian et al. (1984). The sites cover a broad geographic zone encompassing three states: Kentucky, western Tennessee, and southeastern Ohio (Cassidy 1984) and southwestern Ohio and the general Ohio region (Perzigian et al. 1984). Similarly, their studies evaluate data representing a very large time expanse, Paleoindian through Mississippian, with the greatest amount of information coming from populations representing the Woodland/Mississippian transition. Their conclusions are drawn from cross-cultural comparisons among sites, based on the generally accepted assumptions that the geographic region is environmentally homogeneous and that the populations were genetically homogenous as well (Cassidy 1984:307).

The Central Ohio Valley is well known for its rich prehistoric cultural development known as the Adena/Hopewell–Fort Ancient complex, which represents the Woodland/Mississippian transition. In evaluating the questions of the overall effects of the long-term consequences of dependence on an agricultural subsistence base, Cassidy (1984:334–38) points out that in this region the Woodland/Mississippian transition is actually a transition from one cultivated diet to another. The relationship of health to actual diet, *not* agriculture, is then the critical issue.

Cassidy summarized the population data as (1) a decrease in life expectancy, (2) a slight decrease in infant death rate, (3) a pronounced increase in weanling-age death rate, (4) an overall steady rate for arthritis, but occurring at earlier ages in the later time periods, (5) an overall steady rate of infectious lesions, but occurring at earlier ages in later time periods, and (6) a decrease in Harris lines through time. Nutritional indicators included an increase in porotic hyperostosis and a decrease in cortical bone thickness through time. Dental pathology indicated (1) a pronounced increase in frequency of caries, (2) a slight decrease in abscess frequency, occurring at younger ages in the later groups, (3) a steady rate of enamel hypoplasia in adult dentition, with the frequency of severe cases increasing in the later groups, and (4) enamel hypoplasias in the deciduous dentition being absent or rare in the early groups, followed by a moderate rate in the later groups. Cassidy's general conclusions were that the overall quality of diet had decreased through time, with the most profound effects being evidenced in children of weanling age and females in the reproduction age range.

The skeletal samples examined by Perzigian et al. (1984) in southwestern Ohio and the general Ohio region represent three cultural phases: Archaic, Middle Woodland, and Mississippian. In summarizing their analyses, they noted (1) a fluctuation of sexual dimorphism through time (low to high to low), (2) a similar fluctuation of stature through time, (3) a decrease in frequency of Harris lines, (4) an increase in generalized periostitis and specific "crowd infections," (5) an increase in porotic hyperostosis, (6) an increase in nonaccidental trauma, (7) a slight decrease in degenerative pathology, and (8) a decrease in mortality. (The demography is taken from Lovejoy et al. [1977]. Age at death does increase [decrease in mortality], but Lovejoy et al. believe that this occurrence may be due to sampling error.) Dental pathology evidenced increases in enamel hypoplasia and caries frequency. The authors concluded that the skeletal data were indicative of an increasing synergism among diet, population density, and disease (Perzigian et al. 1984:361).

Larsen's (1984) analysis was based on populations from the coastal area of Georgia. The skeletal stress indicators were similarly subdivided into those reflective of disease-nutritional stress and those indicative of mechanical stress. The analysis showed several trends through time: (1) an increase in frequency of periosteal reactions, (2) an increase in caries frequency, (3) a decrease in degenerative pathology, (4) a decrease in bone size, (5) a decrease in skeletal robusticity, and (6) a decrease in stature. The rates in caries, degenerative pathology, size reduction, robusticity reduction, and stature reduction were more notable in females in each instance. The increase in periosteal reactions was attributed to increased sedentism. The decrease in degenerative pathology was attributed to a combination of a decrease of age at death and a decrease in mechanical stress load related to the transition from gathering-hunting to agriculture. It was suggested that these effects were more notable in females than those of males because the males essentially continued the hunting activities to which they were adapted, while females changed behavioral activity from that associated with gathering to increasingly strenuous cultivation activities. The decreases in size, robusticity, and stature were attributed to a combination of decreased nutritive value and decreased mechanical stress, being more pronounced in females because of differential access to high nutritive foodstuffs. The overall decrease in age at death was also attributed to a decrease in quality and variety of diet as a consequence of increased dependency on maize horticulture.

The analyses of populations from Alabama conducted by Hill (1979, 1981a, 1981b, 1983, 1984, 1985a, 1985b), Cole et al. (1982), and Powell (1983, 1985b) indicate a pattern of peak stress very similar to that of the populations from Illinois analyzed by Cook (see Figure 2–3). The geographic area encompasses the entire central portion of the state and includes three of the four major waterways: the Tombigbee, Black Warrior, and Alabama rivers. The selected population samples are all either contemporaneous or ancestral to each other.

Individuals obtained from excavations of the Gainesville section of the Tennessee-Tombigbee Waterway represent a chronological and biocultural continuum from Late Woodland to Protohistoric. Data from the region indicate a dramatic population increase and increased sedentism during the Late Woodland period. The subsistence strategy was one of central-based wandering. Faunal assemblages and catchment analyses from the area indicate increased dependence on second-line resources, with pronounced overlap of catchment zones. The increased stress that would have occurred as increased population density competed for available resources is reflected in the individuals themselves.

Comparison of early Late Woodland, late Late Woodland, and Early Mississippian samples show that the late Late Woodland individuals were under pronounced environmental stress (Cole et al. 1982; Hill 1985, 1987). In the mortality profile, infants and children made up the largest portion of the late Late Woodland group. Although the highest incidence of nonaccidental trauma occurred during the early Late Woodland period, there was an overall 3 percent increase in trauma during the late Late Woodland period followed by a decrease of 20.6 percent in the Early Mississippian. Nutritional stress increased 11.7 percent from early Late Woodland to late Late Woodland and decreased 6.4 percent in the Early Mississippian. (Nutritional stress indicators analyzed by Hill [1981a] included porotic hyperostosis and severe dental pathology that would have inhibited normal masticatory function. Because of the high incidence of dental pathology in the Early Mississippian sample, the percentages of nutritional stress between late Late Woodland and Early Mississippian appear to be similar. However, the incidence of porotic hyperostosis was much higher in the late Late Woodland group.) The pathologies indicative of degeneration of the skeletal tissues were those common to all geriatric individuals. Other than the normal processes of aging, however, degeneration can be affected by such factors as habitual behavior, diet, or prolonged illness. Degenerative pathology increased 3.6 percent from early Late Woodland to late Late Woodland and decreased 9.4 percent in the Early Mississippian.

Analysis conducted by Powell (1983) of individuals representative of Early Mississippian to Protohistoric time periods from the same location showed trends of improved health throughout the Mississippian period. Mortality rate essentially remained unchanged from Early to Late Mississippian. Traumatic injury decreased dramatically (5.3 percent of the entire population). Porotic hyperostosis remained low (6.9 percent); and the rate of infection remained essentially unchanged (36.8 percent) and was generally mild in terms of severity.

Overall incidence of caries increased from Late Woodland to Late Mississippian; however, the caries rate was much higher for upper status individuals in the Mississippian samples than for lower status individuals whose rates more closely approximated those of the Late Woodland samples (Powell 1983). No enamel hypoplasias were observed on the Mississippian subadult dentitions. Sixty percent of the Mississippian adults exhibited at least one hypoplastic episode, with the three-to-five age period being the most likely for time of occurrence. Powell (1983) reported that all hypoplasias observed were of a very mild form, appearing most typically as amorphously distributed pits rather than as discrete lines.

The relatively high level of quality of life as reflected in skeletal stress indicators seen in these Mississippian individuals were also observed by Powell (1985b) in her analysis of the Moundville population. In fact, frequencies of each stress indicator in Tombigbee Mississippians and Moundville closely approximated each other. Most notable in this study was the absence of significant differences in frequencies of occurrence of stress indicators across status boundaries. With the exception of sexual dimorphism, significant differences were also not noted across boundaries of sex and age, although age was noted as the principal contributing factor in such categories as trauma and degenerative pathology, as would be expected.

Examination of Mature Mississippian/Protohistoric population samples from sites along the Tombigbee, Black Warrior, and Alabama rivers by Hill (1979, 1981b, 1984) and Powell (1983, 1985b) indicates that the quality of life decreased during this time, particularly with reference to nutritional stress. Populations from these sites are considered to be descendants of the last vestiges of the large Mississippian urban-center populations (most notably Moundville). During the Late or Terminal Mississippian period, subadult mortality had begun to increase. This trend is seen in periostitic reactions and enamel hypoplasias (Powell 1983). Powell also noted that the number of severe cases of periostitis occurred more frequently in the Protohistoric individuals. The most prominent indicator of increased stress during this time period is the dramatic increase in number of individuals affected by porotic hyperostosis, not only in absolute frequencies but in severity and duration as well. All age categories, including senile adults, had evidence of the pathology (Hill 1981b, 1985a). This notable increase is further substantiated by Turner's analysis (1985), which examined Protohistoric and Historic samples and found that the Protohistoric frequencies of porotic hyperostosis and enamel hypoplasias were also significantly higher than those of the Historic samples, while frequencies of other stress indicators remained essentially unchanged across chronological/cultural parameters.

The western expanse of the Mississippian culture area was examined by Rose and co-workers (1984). Because of the large geographic area and consequent environmental variability, the authors subdivided their study into three ecological/cultural zones: Caddoan, Central Mississippi Valley, and Lower Mississippi Valley. The bioarchaeolgical data were considered within the context of four ecological variables: the environmental carrying capacity, climatic stability, the absence of environmental circumstances (that is, geographic extent of the most productive ecological niche), and the distribution of resources within the environment.

In attempting to synthesize the vast amount of data that had been collected by many different researchers over a long period of time, Rose and co-workers were compelled to make generalizations in trend analysis because of vagaries and inconsistencies in data computation. One such generalization was the use of caries rate as an indication of the development of an agricultural subsistence base. In summarizing their finds they noted that the data were highly variable. Differing rates occurred within and between the geographic zones, the populations, and even the individuals constituting the populations (males and females, adults and subadults). Their summaries were presented according to their tripartite subdivision.

Within the Caddoan area were (1) a low rate of infection, followed by a pronounced increase in the Early Mississippian, with a subsequent decrease in the Late Mississippian; (2) increased sexual dimorphism; (3) increased overall size, particularly in females; (4) fluctuating osteoarthritis and osteophytosis rates (low, high, then low); (5) fluctuating porotic hyperostosis (low, high, low); (6) a decrease in Harris lines; (7) no change in the rate of trauma; and (8) an increase in caries.

Although no preagricultural data samples were available for the Central Mississippi Valley, several trends were noted for the agricultural samples. These included (1) an increase in caries, (2) a slight increase in sexual dimorphism, (3) a decrease in osteophytosis and osteoarthritis, (4) a slight decrease followed by a dramatic increase in infection, (5) and no general trend in trauma.

Trends for preagricultural and agricultural samples in the Lower Mississippi

Valley included (1) an increase in caries, (2) a low rate followed by a slight decrease and then dramatic increase in osteophytosis, (3) a high rate followed by a continual decrease in osteoarthritis, (4) a fluctuating rate of infection (high, low, high), (5) an increase in sexual dimorphism, (6) a decrease in physical size, and (7) no general trend in trauma.

Rose and co-workers noted the great variability of pathology data associated with the advent and continuation of an agricultural subsistence base. For example, the increases in size and sexual dimorphism were attributed to mechanical stress and distribution of resources (which had a positive effect on the populations in some areas, while in other areas the mechanical stress and distribution of resources were related to a decrease in physical stress). Similarly, access to corn was suggested for different segments of populations and, therefore, related to pathology. For example, in the Red River (Caddo) series, caries rates at the ceremonial center of Belcher Mound were lower than those of the dispersed communities of Kaufman-Williams and Roden, implying to the authors that the lower status individuals ate more maize. Conversely, in the Central Mississippi Valley the caries rates led the authors to speculate that the farmstead residents ate less maize and had greater access to other food resources (and, therefore, lower caries rates).

In their conclusions, the authors suggested that, at least in certain areas, the fluctuations in pathology rates may be indicative of greater stress during the transition from one subsistence cultural system to another (Rose et al. 1984:406). A similar argument has been presented by Cook and Buikstra (1979) and Hill (1983, 1985b).

Summary

Because of the physiological nature of bone as a hard tissue not easily perturbed by mild external insult, any perturbation that is severe enough or of sufficient duration to result in an alteration of the skeleton of an individual was certainly a stressful event during that individual's lifetime. Thus, the skeleton may be viewed as representing the biological "memory" of most, if not all, such events that occurred while that individual was alive and that may have ultimately led to the death of the individual. Skeletal populations are, therefore, extremely valuable for inference about past behavior and for evaluation of the success or failure of the complexes of behavior to which we loosely refer as "life-style." Unlike living populations in public health studies that tend to be clusters of "case studies," prehistoric populations essentially practiced a single life-style at a given point in time, with variability in behavior occurring primarily along age and sex lines. In analyses, then, subsistence and individuality or idiosyncrasy of behavioral patterns are more easily controlled, allowing the researcher to identify critical variables more easily. Ultimately, through the accumulation of information on these critical variables, we can evaluate the impact of particular stressors and determine trends in the process of human adaptation.

In this chapter, we have chosen to examine the Mississippian life-style in the eastern United States and whether it had any positive or negative effects on those populations that were active participants. A central question is whether the intensification of agriculture resulted in an overall increase or decrease in health for the populations. Peripheral to this central issue are the questions that pertain to any differential effects either between groups or among different segments of single populations. In other words, if intensification did have a specific effect on overall health, was there a differential effect between cultural centers and peripheral populations, and/or was there a differential effect along status, gender, or age delineations? If differences do exist, are these differences any indication of exploitation or differential access to subsistence resources?

Obviously, the selection of samples is critical. Comparisons of groups over large expanses of time or large geographic areas are difficult to do with any degree of significance in results. Rose and co-workers (1984) noted that they were unable to establish a secular trend for any of their selected stress indicators because of the large geographic expanse (environmental variability) and variability in data available. A comparison of the secular trends (Table 2–1 and Figure 2–3) appears to reflect the variability of these trends and thus the plasticity of adaptive strategies. However, great variability also exists in samples selected by the researchers. Consequently, the variability in peak periods of stress (Figure 2–3) may be an artifact of sample selection.

In summarizing the secular trend analyses, there is evidence that tends to support the hypothesis concerning differential access. There is no indication that intensification of agriculture resulted in decreased health across the board. There is evidence, however, that the cultural centers were to a greater or lesser extent exploiting the peripheral populations (Goodman and Armelagos 1985a; Hill 1981a; Larsen 1984; Powell 1985b; Rose et al. 1984). Information about differential access at the individual level, across status, gender, and age delineations, is site specific. In some areas there is evidence of differential access along these lines (e.g., Larsen 1984), and in other areas there is no evidence for this level of differential access (e.g., Powell 1985b).

In concluding this discussion, we reiterate that paleopathology is indeed a powerful tool for analyzing prehistoric groups. Through an understanding of the causes and consequences of certain pathological conditions, we can make inferences about behavioral strategies and their success or failure in the adaptive process. Our knowledge of the most culturally significant prehistoric subsistence/settlement pattern in the eastern United States has been greatly enhanced by the information obtained from these skeletal populations, and it will continue as the sophistication and integration of the levels of analysis improve.

3

The Late Prehistory of the Ohio–Mississippi Rivers Confluence Region, Kentucky and Missouri

R. Barry Lewis

The Ohio–Mississippi rivers confluence region (Figure 3-1) stretches from the city of Cairo, Illinois, down the Mississippi River to the Kentucky-Tennessee state line. It takes in the Cane Hills and Big Bottoms (Davis 1923) on the Kentucky side of the river and the Cairo Lowland in Missouri. The Kentucky portion is predominately upland, and the floodplain is narrow except at its northern and southern ends where river bends create its maximum width. The steep valley bluffs average about 50 m high, and gently rolling uplands stretch to the east from the bluff crests. The native cover was hardwood forest with dense stands of cane as the major undergrowth along the bluffs and the most well-drained parts of the floodplain. Today, most of the uplands are in pasture or woodlots, and the bottoms are cultivated. Across the river in Missouri, the Cairo Lowland presents a stark geographical contrast. It is a flat floodplain that lies mostly within the present meander belt of the river. It was once covered by a mixture of hardwood forest, cypress swamps, and lakes, but it is now almost entirely in row crop cultivation (Lewis 1974).

This chapter summarizes current understanding of Mississippi period archaeology in the study region with particular emphasis on recent research results of the University of Illinois's "Western Kentucky Research Project" (UIUC-WKY), which began there in 1982. My focus is specifically on the last half of the period, A.D. 1300–1700, referred to here simply for descriptive ease as the "late" Mississippi period. I also treat the Mississippi period strictly as a unit of archaeological time and keep it separate from the concept of a Mississippian Cultural Tradition. The latter is a unit of cultural similarity and is more than simply a synonym for the period of similar name. This distinction is necessary to avoid several problems, foremost among them being

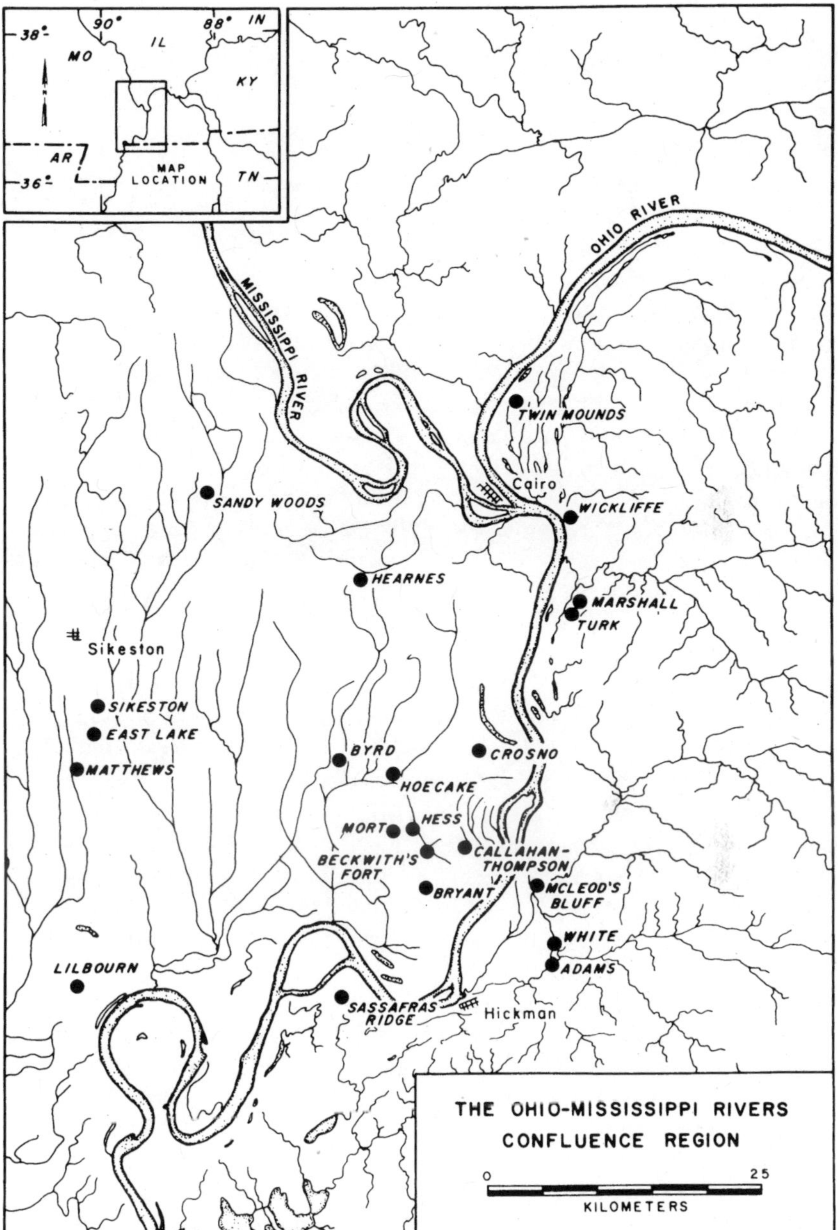

Figure 3-1. Mississippi Period Sites in the Ohio–Mississippi Rivers Confluence Region

the implication that the temporal cutpoints for the Mississippi period are somehow discoverable with great precision rather than being essentially arbitrary.

Late Mississippi Period Research

The history of archaeological research in the Ohio–Mississippi rivers confluence region, as well as the folklore of local relic hunters, is dominated by an interest in Mississippi period research problems, sites, and artifacts. This focus on Mississippi period research is not surprising, because many investigators have favorably compared this archaeological manifestation of the Mississippi period with that of the American Bottom in Illinois (e.g., Morse 1977; Phillips 1970:926). In spite of its apparent archaeological richness, the scientific understanding of the late prehistory of this region has developed rather slowly. Although ostensibly part of the vast study area that Phillips et al. (1951) carved out for themselves in the 1940s, the Ohio-Mississippi rivers confluence region was not actually investigated by that project. Their site survey coverage ended at about the latitude of Osceola, Arkansas. The first regional sequence for southeastern Missouri did not emerge until Williams's (1954) important doctoral dissertation filled in the gap left by the Phillips, Ford, and Griffin (1951) survey report. Williams laid out the basic dimensions of one Mississippi period phase, called the Cairo Lowland phase, that would be the basic temporal unit for Mississippi period archaeological research in this region for the following 30 years.

The present sequence (Lewis 1983, 1987), a portion of which is described below, was developed to replace the Cairo Lowland phase, which had become a regional synonym for the Mississippi period. It also supersedes Phillips's (1970:912–13, 925–26) Beckwith and Cairo Lowland phases, which were proposed to make local Mississippi period ceramic assemblages fit notions that still prevailed in the early 1970s concerning the temporal priority of "clay" over "shell" tempering.

The sequence divides the Mississippi period into four phases (Figure 3-2). Each phase spans 200 years, and the temporal cutpoints are drawn at the beginning of every other century encompassed by the period. This treatment of the temporal dimension of phases differs enough from current practice in the Lower Mississippi Valley to warrant explanation.

At the risk of oversimplifying something that is actually a very complex process, phase definition in the Lower Mississippi Valley tends to be based on the delineation of a bundle of material culture attributes that co-occur within the archaeological record of a region. Some subsets of those attri-

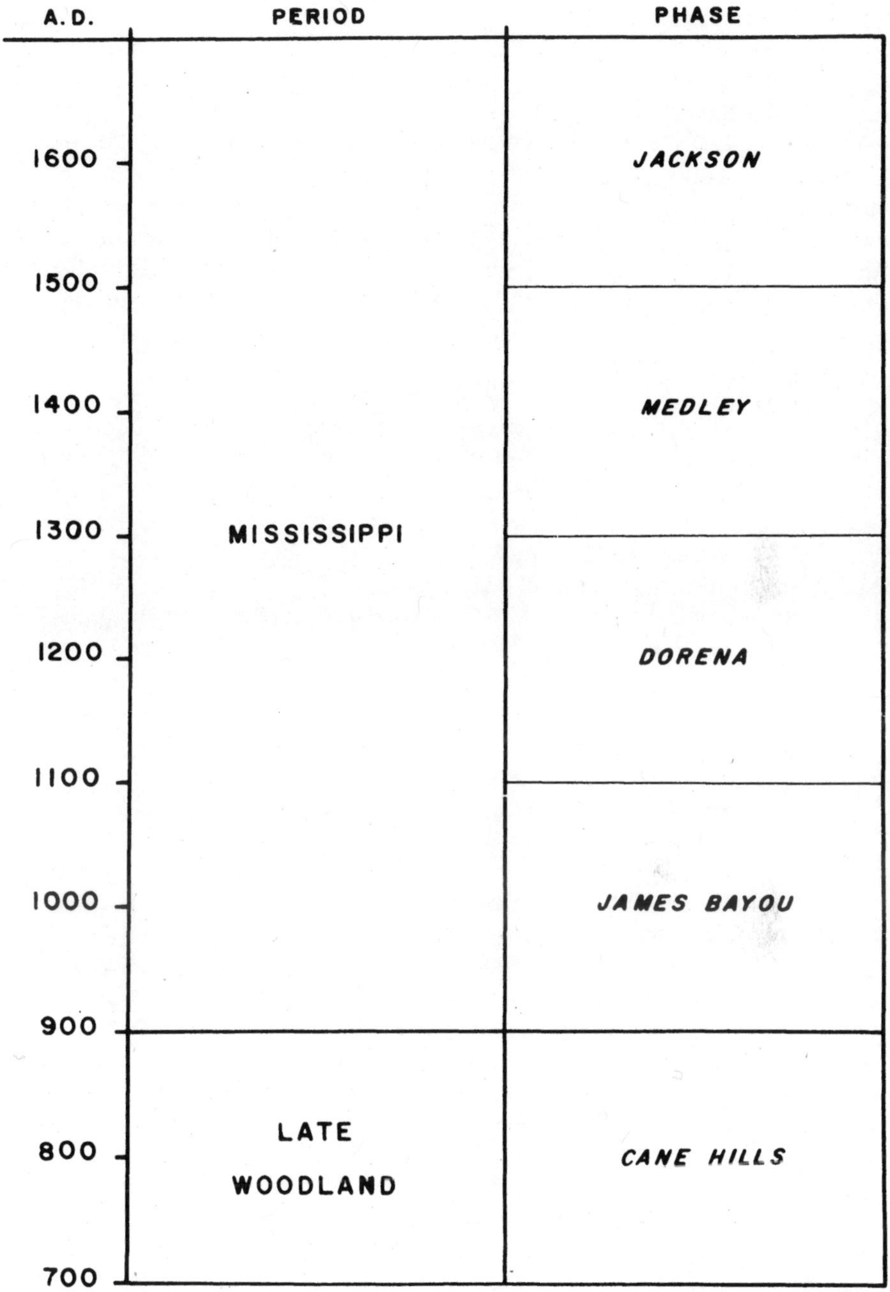

Figure 3-2. The Late Prehistoric Chronological Sequence in the Ohio–Mississippi Rivers Confluence Region

butes are taken to be diagnostic criteria for the identification of new components. The temporal dimensions of a phase, so defined, are continually reassessed through cross-dating the diagnostic criteria, interpreting the absolute dating of contexts within which those criteria occur, and the analysis of stratigraphy. The approach followed here differs in that the temporal dimension of a phase is arbitrarily set. Its cultural content is then continually reassessed by cross-dating the material culture attributes that appear to be useful diagnostic features of that phase, interpreting the absolute dating of contexts within which those diagnostic features occur, and analyzing stratigraphy.

I adopted this approach for the following reasons: first, I believe that it is useful analytically to emphasize the temporal nature of units that are defined at the end of the prehistoric portion of the archaeological record in order to facilitate linking to the historical record. For example, the "Vacant Quarter Hypothesis" (Williams 1980, 1985), an argument that a large portion of the Mid-continent was essentially abandoned by aboriginal populations during late prehistory, is primarily a chronological problem. No one disputes the fact that eighteenth-century historical data show a much smaller population in this large region than one would expect from the late prehistory. The quibble, if one can call it that, is when did the population density become smaller? The correct answer has fundamental implications for the interpretation of late prehistory across portions of several states.

Second, my decision to emphasize time over traits (but not to exclude either from consideration) also reflects my concern that the invaluable data offered by radiocarbon and other inexpensive forms of absolute dating in the Lower Mississippi Valley are often poorly integrated into cultural chronologies. A full discussion of this concern is too long to introduce here and must be treated separately. The approach that I have taken in the study region forces the archaeologist to integrate absolute dating into the contextual interpretation of a component much more than would otherwise be necessary, and it removes the possibility that those data are dismissed without investigation when they conflict with the chronological expectations generated from the analysis of other data (for example, cross-dated artifacts, seriations).

The temporal approach to phase construction simply requires one to adopt a more explicit perspective on the problems of component dating and the measurement of cultural changes over time. It is not asserted, as one reviewer mistakenly inferred, that archaeologists who define phases principally on the basis of diagnostic material culture do not continue to revise their understanding of the content of those phases on the basis of new data. The two approaches differ primarily in the relative classificatory weight that is assigned to the major categories of archaeological data. The temporal approach is compatible in principle with Willey and Phillips's (1958:21) definition of the phase concept and with the operational definition of phase as it has traditionally been applied in the Southwest (Olson 1962). Operationally, the approach does not inhibit or constrain interregional comparisons with phases based primarily on diagnostic material culture attributes. Qualitative differences between phase types are also insignificant at the level of interassemblage comparisons.

Third, a fixed phase interval of 200 years was chosen because it is the minimum unit of prehistoric time that can currently be examined with reasonable chronometric accuracy in this region (that is, it is roughly the resolution limit of available absolute dating techniques). No assumptions about the rate or uniformity of cultural change are implied by the choice of a fixed interval.

Fourth, the phase cutpoints are anchored at the beginning of every other century in the Gregorian calendar for the sake of convenience and to serve as a reminder that the sequence depends fundamentally on the accurate measurement of the age of interpretable archaeological contexts.

Certain aspects of the temporal approach could be described as yielding subperiods rather than phases. For example, one can create phases that are multiethnic. A given phase might subsume under one phase name the contemporaneous archaeological remains of European colonists and Native Americans in a region. This approach also readily permits the construction (but, of course, not the full definition) of phases for which few archaeological data are available from a region. The Jackson phase, as described below, exemplifies aspects of both points.

The Late Phases

By A.D. 900, the major elements of a settlement hierarchy had emerged in the study region, and, by implication, many of the cultural changes that are widely characterized as "Mississippian" (Griffin 1985a) were already present. Nevertheless, as B. Smith (1984) has pointed out, the region was not a "heartland" for Mississippian cultural developments any more than other parts of the East. It also does not appear to have been the source for, nor the recipient of, waves of migrants to or from other regions at that time. In James Bayou (A.D. 900–1100), the oldest Mississippi period phase, large villages such as the Marshall site in Carlisle County, Kentucky (Sussenbach and Lewis 1987), were thriving communities built on an economic base of maize horticulture, hunting, and gathering. Site surveys have revealed few such villages, but their scarcity is more apparent than real inasmuch as most of them appear to have been continuously occupied throughout the period. Their presence is therefore usually masked by more recent archaeological deposits.

Cultural developments over the next 800 years continued the developmental trends that had been set in motion during the Late Woodland period. The economic base of the regional cultural adaptation also appears to have remained relatively stable during those centuries. Carbonized maize (*Zea mays*) grains, kernels, and cupules tend to be ubiquitous in Mississippi period middens, and this crop was clearly of great economic importance. Beans (*Phaseolus vulgaris*), squash (*Cucurbita* sp.), and gourds (*Lagenaria siceraria*) were also cultivated. Important gathered plant foods included hickory (*Carya* sp.) nuts, persimmons (*Diospyros virginiana*), and the seeds of goosefoot (*Chenopodium* sp.) and erect knotweed (*Polygonum erectum*). Staple animal food species were the white-tailed deer (*Odocoileus virginianus*), raccoon (*Procyon lotor*), wild turkey (*Meleagris gallopavo*), turtles, and fish.

The top of the settlement hierarchy was dominated by large villages and towns, which were designed around the mound-and-plaza arrangement of public space that is so fundamentally a part of the Mississippian concept. At the bottom of the hierarchy, small hamlets and isolated households occur throughout the region but were most common along the old natural levees or "cane ridges" of the floodplain. Intermediate settlement hierarchy levels appear to have existed, but little is known about them. A recently initiated doctoral dissertation research project by Paul Kreisa, a University of Illinois graduate student, focuses specifically on improving our understanding of those "second order" settlement levels in this region.

Medley Phase (A.D. *1300–1500)*

The situation during A.D. 1300–1500 does not appear to have differed significantly from that of the preceding phases. Fundamental continuities along several dimensions, especially the economic and settlement systems, have been demonstrated archaeologically by recent UIUC-WKY investigations at numerous Kentucky sites (Edging 1985; Lewis 1986; Stout 1985; Sussenbach and Lewis 1987).

Major excavated components include the Adams town site (Lewis 1986; Lewis and Mackin 1984; Stout 1984a, 1984b, 1985, 1987) along the Mississippi Valley edge in Kentucky, and the Callahan-Thompson and Hess hamlets in the Cairo Lowland (Lewis 1974, 1982). Other components for which there are important recent test excavation or surface-collected data include the

Wickliffe (Lewis 1986; Wesler 1985), Turk (Edging 1985), McLeod's Bluff (Webb and Funkhouser 1933), Sassafras Ridge (Lewis 1986), and possibly the Beckwith's Fort (Southard and Cottier 1973) towns, and the White (Sussenbach and Lewis 1987) and Bryant (J. Williams 1967) second-order habitation sites. Calibrated radiocarbon dates for those locations are listed in Table 3-1.

Sample No.	Context	Dated Material	DCA	ADC	Laboratory Estimate	Corrected Date or Date Range*	Reference
KENTUCKY							
Turk (15Ca6)							
ISGS-1288	Midden	CW	C	C	710±90	1280±90	Edging 1985:15
ISGS-1289	Midden	CW	C	C	700±70	1280±70	Edging 1985:15
ISGS-1323	Midden	CW	C	C	910±70	1040±72 to 1160±72	Edging 1985:15
ISGS-1324	Midden	CW	C	C	710±70	1280±70	Edging 1985:18
White (15Fu24)							
ISGS-1543	Structure	CW	C	C	640±70	1300±70 or 1380±70	Sussenbach 1987
ISGS-1544	Structure	CW	A	C	560±70	1330±70 or 1400±70	Sussenbach 1987
Adams (15Fu4)							
ISGS-1141	Midden	CW	C	C	610±70	1320±70 or1390±70	Lewis 1986a:156
ISGS-1149	Midden	CW	D	C	700±70	1290±70	Lewis 1986a:156
ISGS-1150	Structure	CW	B	C	820±70	1220±73	Lewis 1986a:156
ISGS-1151	Structure	CW	A	C	610±70	1320±70 to 1390±70	Lewis 1986a:156
ISGS-1161	Structure	CW	C	C	900±70	1060±73 to 1160±70	Lewis 1986a:156
ISGS-1172	Refuse Pit	CW	C	C	810±80	1220±81 to 1260±81	Lewis 1986a:156
Sassafras Ridge (15Fu3)00							
ISGS-1142	Structure	CW	B	C	660±80	1290±80 or 1380±80	Lewis 1986a:156
Wickliffe (15Ba4)							
ISGS-1143	Md. A Structure	CW	B	C	830±70	1210±70	Lewis 1986a:156
ISGS-1152	Md. A Structure	CW	B	C	760±70	1260±71	Lewis 1986a:156
ISGS-1171	Midden	CW	C	C	720±70	1280±70	Lewis 1986a:156
Beta-12529	Md. A, ref. pit	CW	C	C	520±70	1420±71	Lewis 1985:12
MISSOURI							
Hess (23Mi55)							
UGA-147	Refuse pit	CW	C	C	355±75	1500-1520±76 or 1610±76	Lewis 1982:152
GAK-1309	Structure	CW	A	C	350±90	1500-1520±91 or 1610±91	Williams 1968:167
Callahan-Thompson (23Mi71)							
UGA-147	Structure	CW	A	C	480±65	1420±66	Lewis 1982:10
UGA-148	Structure	CW	C	C	570±90	1350±91 to 1390±91	Lewis 1982:10
Bryant (23Mi59)							
GAK - 1310	Mound fill	CW	C	C	610±80	1320±81 to 1390±81	Williams 1968:189
Beckwith's Fort (23Mi2)							
UGA-244	Stockade A	CW	A	C	675±70	1290±71	Southard & Cottier 1973:11
N-1250	Stockade A	CW	A	C	835±85	1210±86	Southard & Cottier 1973:11
N-1251	Stockade A	CW	A	C	1090±85	900±86 to 980±86	Southard & Cottier 1973:11
N-1253	Stockade A	CW	A	C	1240±145	720±45 to 780±145	Southard & Cottier 1973:11
N-1252	Stockade B	CW	A	C	995±100	1000±101	Southard & Cottier 1973:11

Context
CW - Carbonized Wood
CN - Carbonized Wood & Nutshell
MC - Maize, cobs, & Grain
/B - Bone

DCA (Degree of Certainity of Association)
A - Full Certainity
B - High Probability
C - Probable
D - Reasonable Possibility
E - Unknown, Not Reported

ADC (Possible Age Differential Class)
a - Less than 20 years
B - About 20 years to 100 years
C - More than 100 years
D - Unknown; Not Reported

*Based on Sturvier 1982; Calibrated range given for those samples with radiocarbon contents that have multiple dendroyear calibrations.

Table 3-1. Ohio–Mississippi Rivers Confluence Region Late Mississippi Period Radiocarbon Dates

The Adams site is a large (7.25 ha) Mississipian town and ceremonial center located on an isolated terrace of the Lower Bayou de Chien Valley in Fulton County (Figures 3-3, 3-4). It consists of a central mound group, plaza, and two distinct village segments, one located to the east of the plaza and the other situated south of Mound A. The site was initially reported in 1888 by Loughridge, but it remained uninvestigated until 1983 when the University of Illinois began research there. Test excavations and other archaeological investigations have shown that the location contains major Late Woodland and Mississippi period components, including thick Dorena and Medley phase deposits.

In 1984, Stout (1984a, 1984b, 1985, 1987) completed a controlled surface collection of the entire site (2,257 five-m^2 collection units) exclusive of the

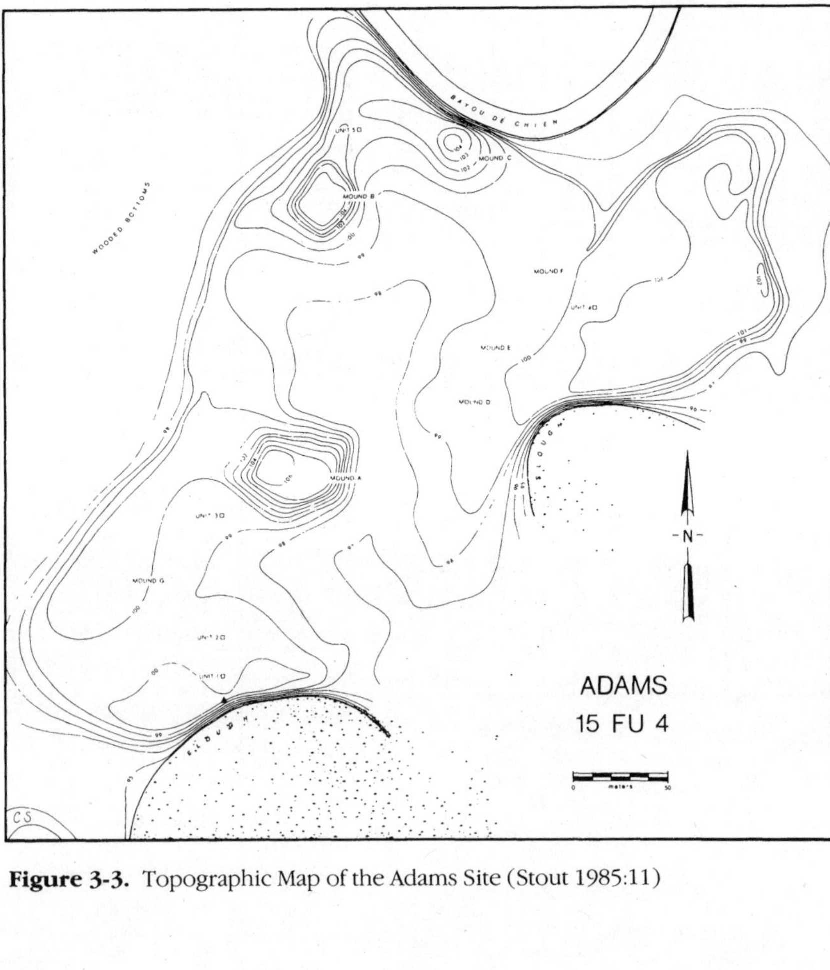

Figure 3-3. Topographic Map of the Adams Site (Stout 1985:11)

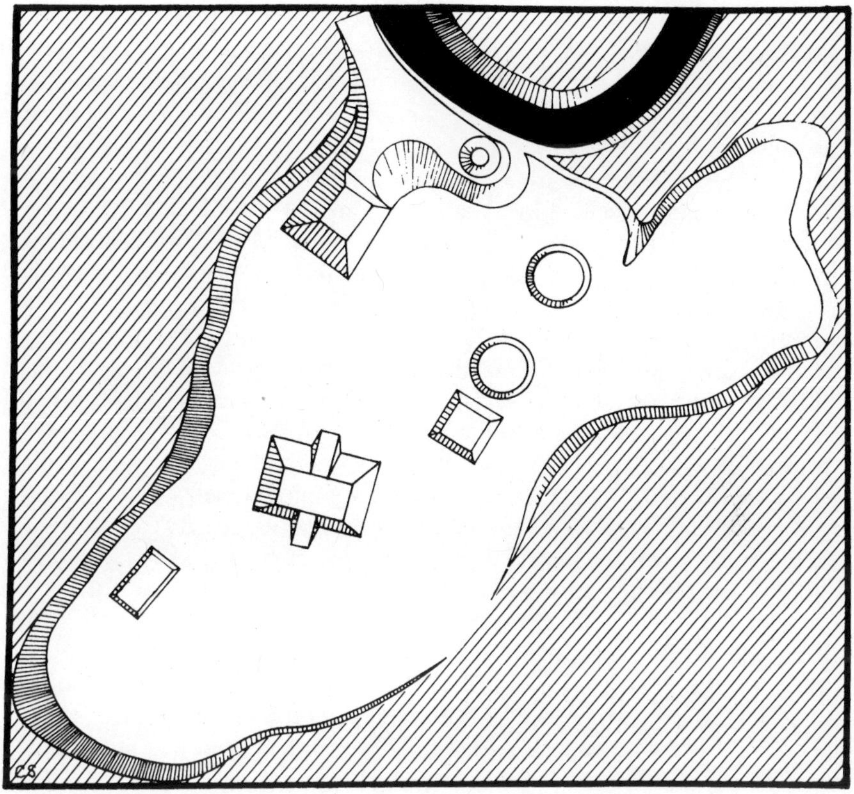

Figure 3-4. Reconstruction of the Adams Site Mound Group (courtesy of Charles B. Stout)

mounds. This project yielded data on the spatial patterning of approximately 100,000 artifacts, including about 35,000 potsherds. The analysis of those data, which began only recently after more than two and a half years of washing, cataloging, and sorting the massive artifact collection, promises to yield significant new insights into the organization of one late Mississippi period town, viewed as a case study for the Mississippian towns of the region.

As expected on the basis of our other investigations, preliminary tests of the surface-collection data show that most artifact class concentrations are strongly patterned in their distribution. High artifact densities were found on the east and west villages, along the low ridge that connects Mounds A and G, and along the southern edges of the three mounds (D–F) that flank the east side of the plaza (Figure 3-3). The large plaza, which is defined by the placement of Mounds A–F, and the small plaza, which lies next to the south side of Mound A, were kept relatively clear of village debris.

Midden development in the village segments is extensive and ranges from 1 to 1.5 m thick. In the Medley component, as in the rest of the archaeological record of this community, excavations have revealed evidence of numerous superimposed wall-trench houses, pits, fire basins, and other features (Figure 3-5). A great deal of information is available about the economy and food preferences of the site inhabitants. Cultivated plants included maize, beans, squash, gourds, and possibly several starchy or oily seeds, including goosefoot, erect knotweed, and maygrass (*Phalaris caroliniana*) (Edging and Dunavan 1986:98). Most of the maize cobs are 8- or 10-row varieties (Edging and Dunavan 1986:99). Gathered plant foods included persimmons and hickory nuts. White-tailed deer, raccoons, waterfowl, wild turkeys, turtles, and fish were the most commonly hunted animals.

The Medley phase pottery from the Adams component is typical of the phase as a whole. In this, as in older Mississippi components, most vessels are Mississippi Plain or Bell Plain utility wares. The incised types, Matthews Incised, *vars. Beckwith*, *Matthews*, and *Manly*, Barton Incised, *vars. Barton* and *Kent*, and O'Byam Incised, *var. O'Byam*, are common relative to their representation in older components (Figures 3-6, 3-7). Nevertheless, decorated sherds seldom account for more than 3 to 5 percent of ceramic assemblages during the Mississippi period. The diversity of vessel shape modes also continues to increase slowly, but differences represent the elaboration of existing shape classes rather than the development of new vessel forms. Plates, mostly of a fine Bell paste, become common relative to preceding phases.

Nonpottery artifacts that are common on Adams and other late sites include recycled hoe fragments and resharpening flakes, projectile points, sandstone abraders, metates, adz fragments, bone awls, pottery ear spools and pins, and drills.

Across the river in the Cairo Lowland, investigations at two small villages or hamlets, the Callahan-Thompson and Hess sites (Lewis 1974, 1982), revealed the remains of late Medley and early Jackson phase wall-trench-type structures that had been constructed in old house basins that had been dug out and reused (Figure 3-8). The occupation at the Callahan-Thompson site was intensive enough to lead to the formation of a shallow midden. The ceramic assemblages of both hamlets are basically the same as that of the Adams component. The fundamental continuity of the settlement pattern from older phases can be seen in the strong association of Mississippi period habitation sites with the sandy loams of natural levee remnants (Lewis 1974). An important material culture difference appears to be the relative scarcity of

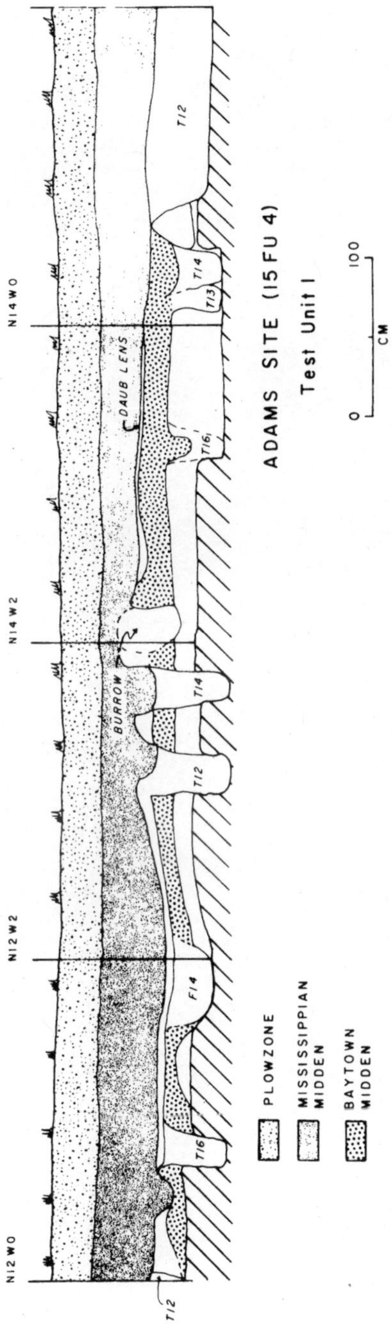

Figure 3-5. Vertical Profile of Test Unit 1 at the Adams Site (Lewis 1986:18)

Figure 3-6. Late Mississippi Period Ceramics: a-c. Matthews Incised, *var. Beckwith*; d-e. Matthews Incised, *var. Matthews*; f-g. Matthews Incised, *var. Manby*; h, j. Barton Incised, *var. Barton*; i, k. Barton Incised, *var. Kent*; 1-m. O'Byam Incised, *var. O'Byam*

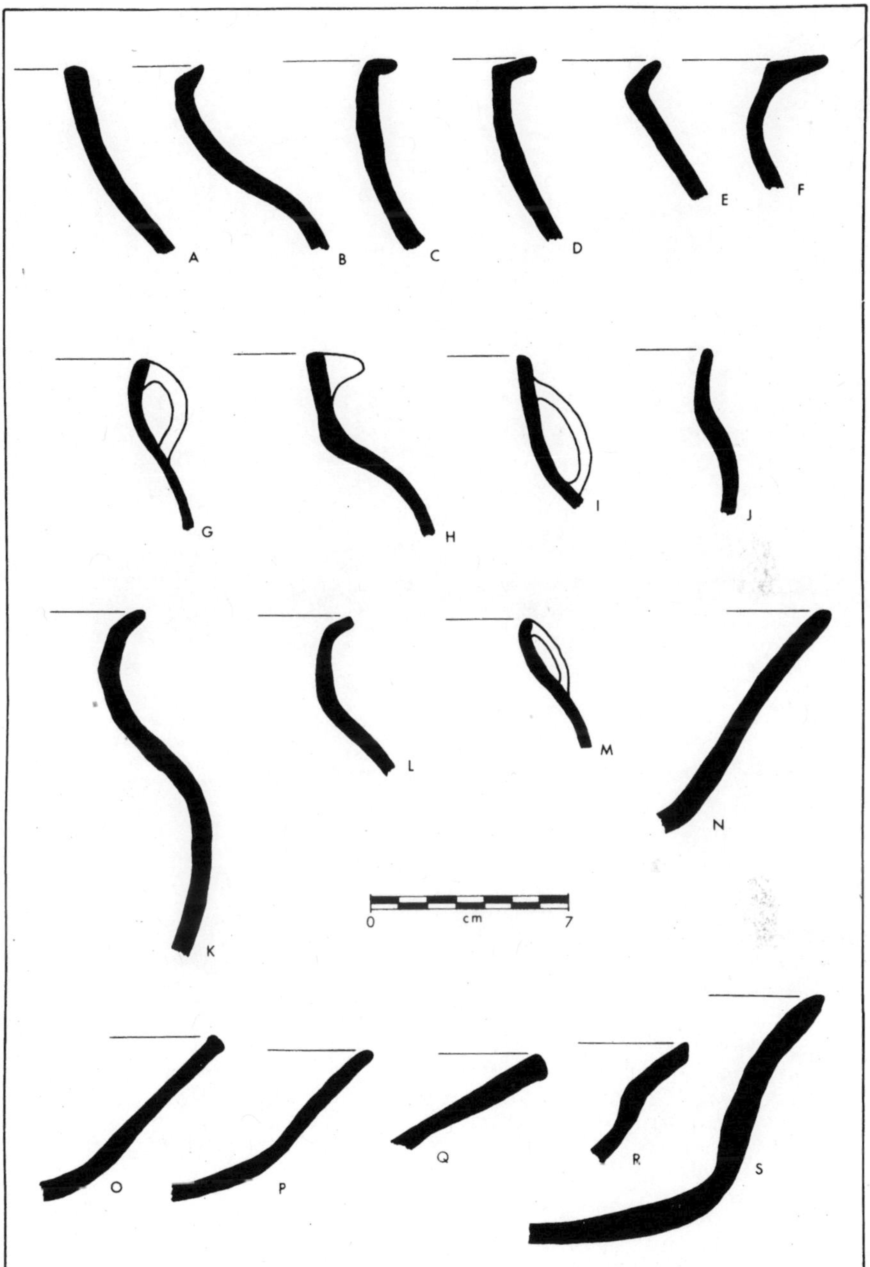

Figure 3-7. Rim Profiles: a-f. Matthews Incised, *var. Beckwith*; g-h. Matthews Incised, *var. Matthews*; i-j. Matthews Incised, *var. Manly*; k-l. Barton Incised, *var. Barton*; m. Barton Incised, *var. Kent*; n-s. O'Byam Incised, *var. O'Byam*

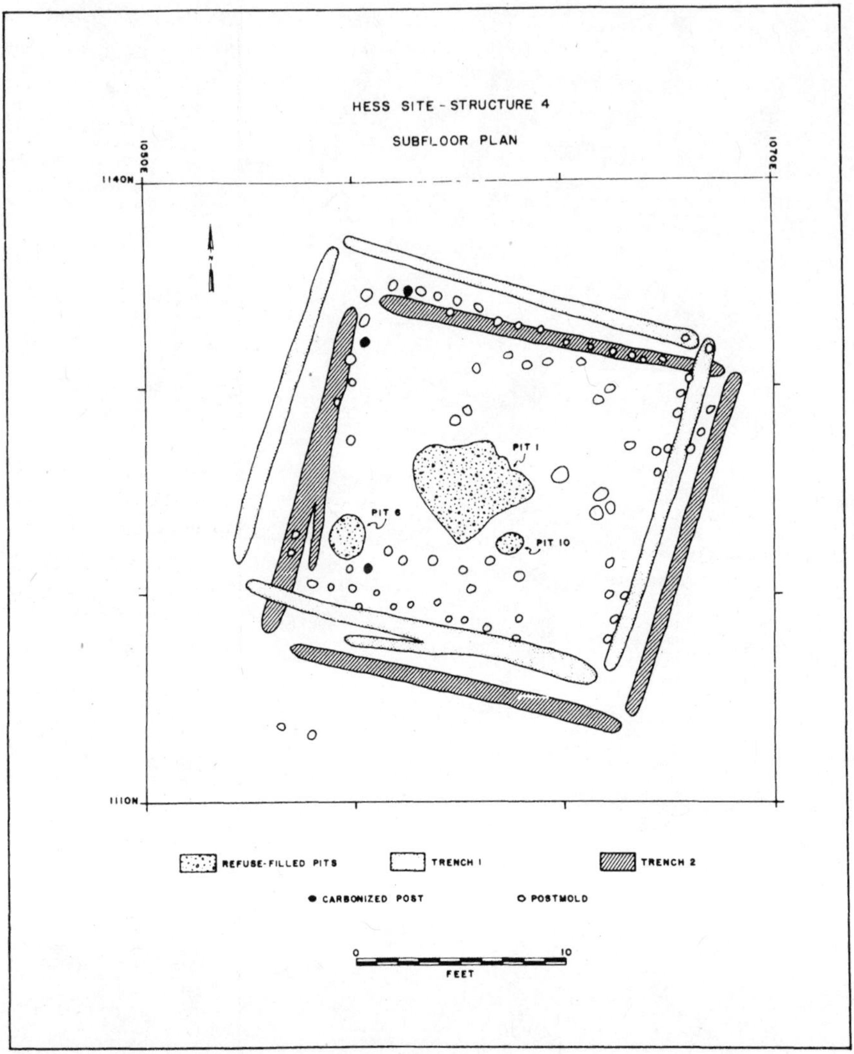

Figure 3-8. Plan of Subfloor Features, Structure 4, Hess Site (Lewis 1982:56)

stone artifacts and debitage on the Cairo Lowland sites and the extent to which those artifacts show greater evidence of recycling than material from comparable sites on or near the Mississippi Valley bluffs in Kentucky. This difference is easily explained by the absence of local stone sources in the Cairo Lowland and the presence of at least some sources along the Kentucky bluffs.

The subsistence economy of the hamlets also appears to have been essentially the same as that documented at Adams and comparable sites in this region. Apparent differences such as the absence of the starchy/oily seed complex and the relative scarcity of fish remains in the hamlets (Lewis 1974) are simply functions of sample recovery techniques. All of the soil samples from Callahan-Thompson and Hess were washed through window screen rather than being sent through a flotation process. The collections that resulted are therefore biased against small remains.

Jackson Phase (A.D. *1500–1700)*

This is the final aboriginal archaeological phase in the region. Its definition is consistent with my bias, described earlier, toward the construction of phases in which the temporal dimension is given greater priority than are bundles of diagnostic material traits of presumed contemporaneity. This point is crucial to understanding why the Jackson phase was defined: the temporal priority requires that the regional sequence span the entire record, including those intervals of time that are little known or poorly understood archaeologically.

The Jackson phase brackets the time interval during which many aboriginal populations across eastern North America became extinct, were decimated, or were otherwise affected culturally and biologically by introduced diseases from the Old World (cf. Ramenofsky 1982). It is also reasonable to assume that the local effects of those contacts occurred primarily during the 1500s. Ethnohistorical data from the late 1600s suggest that the aboriginal population density of this region was low and that it had been low for a long time (Lewis 1986).

In the Ohio–Mississippi confluence region, as in many other parts of the Southeast, the lifeways of native populations were profoundly affected by the events of those centuries. Towns, villages, and hamlets were abandoned as the inhabitants died from the combined effects of diseases, starvation, and exposure. The distinctive ceramic technology of the study region essentially ceased and left no stylistic successors.

There are few Jackson phase archaeological data to discuss, and given the nature of events that occurred during this phase and the prevailing dogma of our own discipline, significant new data may be slow to emerge. Among archaeologists working in the East, for example, there appear to be no clear-cut expectations concerning the archaeological characteristics of native communities that became extinct following contacts with Europeans or their diseases. How would those sites differ archaeologically from older sites of the same type? Among groups that survived the effects of this contact, what were the economic, sociopolitical, and material culture changes? With a few notable exceptions, such as the recent dissertation by M. Smith (1984), these and other questions concerning the effects of contact have not yet been addressed in the East. They were not problems when most archaeologists agreed with the American Historical School's interpretation that the effects of contact were initially minimal. Now that this interpretation has been seriously questioned, if not largely discredited (Ramenofsky 1982), they loom large.

In this and adjacent regions, there are few data from town sites, a fact that the proponents of the "Vacant Quarter Hypothesis" interpret as support for the view that much of the Lower Ohio, Central Mississippi, and Northern Lower Mississippi valleys were abandoned by native populations prior to the beginning of the Jackson phase (Morse and Morse 1983; Williams 1980, 1982, 1985). On the basis of the analysis of test excavation and controlled surface-collection data, Lewis (1986) has argued that the occupations of the Adams and Sassafras Ridge towns extended into the Jackson phase but probably ended sometime during the first half of that phase. There is also a large Jackson phase component at Wolf Island, Missouri (ca. 15 km northwest of the Adams site), that has been mostly destroyed by looting and by the construction of farm buildings, a store, and houses. In their recent book, Morse and Morse (1983:299) describe this location as the "Hickman site" and attribute its existence to "reuse by hunters" of an older Mississippi period site. In spite of the fact that Morse and Morse attribute this interpretation to me, I have no reason to believe that the location was anything other than a large Mississippian village with a Jackson phase component, among others.

At the hamlet end of the settlement hierarchy, the Hess and Callahan-Thompson sites both have Jackson phase components. The radiocarbon dates for the Hess site structures (Table 3-1) are the most recent age estimates for Native American dwellings in the region. I have always thought that they were particularly "strong" dates because they had been collected from different structures by different excavation teams and dated by different laboratories. I was also uneasy about them because they worked out to about A.D. 1600 in uncalibrated calendar years, and that seemed too recent as age estimates for cutting dates for wood from those contexts. The publication of a high precision calibration for the recent end of the radiocarbon time scale (Stuiver and Pearson 1986) clarifies matters greatly by demonstrating that the cutting dates could actually have been about A.D. 1500 with a 1 sigma range of A.D. 1440–1640.

The domestic architecture, economic organization, material culture, and other archaeologically delineable aspects of the most recent, and presumably terminal, occupations at the Hess and Callahan-Thompson sites show no changes from the stable pattern of older Mississippi period occupations. There is no evidence of cultural stresses, no hint of impending disaster. The occupations appear merely to end. It is entirely possible that this abrupt, but almost orderly, truncation of the archaeological record may eventually prove to be among the few markers by which we are able to delineate regional extinctions of native communities and populations in the century or two that followed the first landfall by Europeans in the New World.

In the study region, UIUC-WKY researchers are currently working with a model of the "Contact Era" in which it is assumed that the effects of initial contacts, probably all of an indirect nature, tended to be manifested in the archaeological record as abrupt changes. It is anticipated that those cultural changes will appear archaeologically to have occurred instantaneously even though the actual sequence of events may have been drawn out over decades.

If subsequent tests demonstrate that this model does not fit the culture contact effects in the study region, Jackson phase components will nevertheless continue to be difficult to find simply because this phase spans the terminal episode of prehistoric aboriginal occupations. Those components may bracket a brief interval of archaeological time. The material culture is also probably quite similar to Medley phase components except for a few artifacts such as astragalus dice, "Nodena" points, and disk pipes (Lewis 1987, 1988) and the possible, but frankly unlikely, occurrence of European artifacts. Many of Williams's (1980) diagnostic nonpottery artifacts for the Armorel phase in northeastern Arkansas should also be applicable in this region.

Discussion

The Mississippi period in the Ohio–Mississippi confluence region is difficult to interpret in the Mid-South context. Part of the problem stems from the region's location in the Lower Mississippi Valley, which provided a central axis for the movement of people, goods, and ideas throughout prehistory. Given the nature of the geography of this massive valley, the logistics of communication were far easier up it and down it than between its walls. For example, assemblages from sites in southwestern Mississippi can be quite similar to contemporaneous assemblages in western Tennessee some 600 km up the valley, and yet they may be noticeably different from eastern Mississippi material only 200 km away. This loose pattern of intravalley cohesiveness breaks down, however, in the northern reaches of the Lower Valley (including the study region) and on the deltaic plain of coastal Louisiana.

Both of those peripheral regions were less Lower Valley–centered than the core, presumably because trade relations and other ties were maintained with a wider range of people whose lives were not inexorably linked to the Mississippi Valley.

The cohesiveness of the core may help to explain why some researchers who work in the Lower Mississippi Valley core may perceive the Ohio–Mississippi rivers confluence region as an aberrant place. Throughout the Mississippi period, communities situated around the mouth of the Ohio River shared far stronger material culture similarities (and, by implication, other aspects of their lifeways) with the inhabitants of the Tennessee and Cumberland valleys of western Kentucky and central Tennessee (Griffin 1952a; Lewis 1987; Thomas 1894; Thruston 1897) than with groups in the rest of the Lower Mississippi Valley. Viewed from the Memphis subarea, the study region lacks many of the major late Mississippi period ceramic types that are typical of Armorel, Parkin, Walls, and other local phases. The absence of these ceramic types reflects differences in the cultural networks in which the communities of those regions participated and not, as some investigators have inferred, that the Ohio–Mississippi rivers confluence region lacks a significant late prehistory.

The latter view is reflected in the Morse and Morse (1983:271, 280) book on the Central Mississippi Valley and is compounded by errors in cross-dating Mississippi period assemblages from the study region with those of northeastern Arkansas. For example, Morse and Morse (1983:266–67) inferred that O'Byam Incised, *var. O'Byam* (Phillips 1970:144), and the plate vessel form occur early during the Mississippi period and drop out of the Northern Lower Valley sequence by about A.D. 1350. This interpretation appears to stem ultimately from the misclassification of similar but apparently older material from the Hazel and Cherry Valley sites in northeastern Arkansas. Regardless of the source of the errors, once those chronological links between the two regions were forged, it was reasonable to view the Ohio-Mississippi rivers confluence region sequence as terminating around A.D 1400. It also explained why few late Mississippi period ceramic types of the Memphis subarea show up there.

Research completed by the UIUC-WKY project subsequent to the publication of the Morses' book has not supported their interpretations of the late prehistory of the study region. Numerous stratigraphic cuts and other data demonstrate unequivocally that O'Byam Incised, *var. O'Byam*, did not become common in assemblages until after about A.D. 1300 (Lewis 1986; Lewis and Mackin 1984). Plates occur in small numbers in Dorena (A.D. 1100–1300) assemblages but are most common during the Medley and Jackson phases. The important points that emerge from this example are these: first, the relative scarcity of "late" types typical of the Lower Valley core does not necessarily imply the absence of a late prehistory in the study region; second, two regions can lie within reasonable geographical proximity to one another, but considerable cultural distance may occur between them; and third, cross-dating sequences between peripheral and core regions in the Lower Valley requires a greater level of stratigraphic control than similar comparisons within those regions.

If the study region's closest cultural ties are to the east rather than the south, what do those relationships look like during the late Mississippi period? I have addressed this question in detail elsewhere (Lewis 1986, 1987) and will offer only a few comments here. In the Tennessee-Cumberland region of the Jackson Purchase, the Tinsley Hill phase covers essentially the same centuries as the Medley and Jackson phases. The temporal cutpoints of this phase are not well defined inasmuch as it is primarily a cultural similarities–based unit rather than a temporal one, but a range of A.D. 1300–1600 is a reasonable estimate. There are several excavated components, including Tinsley Hill and Rodgers in the Cumberland Valley and Birmingham, Goheen, Roach, and possibly Jonathan Creek in the Tennessee Valley (Clay 1979). The settlement system and economy of Tinsley Hill communities appear to differ little from those of the older Jonathan Creek phase settlements in the region. Most of the pottery from those sites is Mississippi Plain or Bell Plain. The common decorated types are the several varieties of Matthews Incised and O'Byam Incised, *var. Stewart* (Clay 1979:123), although they account for only a minor proportion of any assemblage. Tinsley Hill, the phase's type site, was a village that covered ca. 7 ha of the Cumberland Valley floodplain and three low bluffs near the mouth of Eddy Creek. It may have been fortified. Several seasons of excavations in the early 1960s yielded data on the village, cemetery, and a substructure mound (Clay 1961, 1963a, 1963b; Schwartz 1961).

The radiocarbon dates associated with Tinsley Hill phase contexts are comparable to those for Medley and Jackson phase components (Lewis 1987:Appendix 1). Much of the discussion of those dates (e.g., Butler 1983; Clay 1979, 1984) has centered on the fact that they are simply too recent (in uncorrected calendar years) for the associated archaeological contexts. Calibration of the dates by the Stuiver and Pearson (1986) method removes the basis for those objections. The effective end of the aboriginal portion of the archaeological record appears to have occurred at roughly the same time in this region as around the mouth of the Ohio. There are a few artifacts, such as ground astragalus dice from the Tinsley Hill (Clay 1961:63) and the Stone sites in the Tennessee portion of the Barkley Reservoir (Coe and Fischer 1959), that occur in Jackson phase components and in sixteenth-century contexts elsewhere in the Mid-South (Lewis 1988). Both regions, unfortunately, lie outside that part of the South in which one can realistically harbor hopes of recovering early European artifacts in association with interpretable aboriginal contexts.

Acknowledgments

Funding for the Western Kentucky Research Project has been provided by the University of Illinois and matching grants from the Kentucky Heritage Council, the State Historic Preservation Office, Frankfort, Kentucky. The continued support of both organizations is gratefully acknowledged. Thanks also to my graduate student advisees, Paul Kreisa, Chuck Stout, and Tom Sussenbach, for enduring my frequent nagging and questions about their dissertation research projects that are currently underway in western Kentucky. Portions of this chapter are from my recently completed overview of the Mississippi period in Kentucky (Lewis 1987).

4

Protohistoric/Early Historic Manifestations in Southeastern Missouri

James E. Price
Cynthia R. Price

Most of the southeastern Missouri lowlands appear to have been abandoned during the Late Prehistoric period, ca. A.D. 1350–1450. Evidence of Protohistoric and Early Historic occupation, ca. A.D. 1550–1650, has been found only at a tightly clustered complex of relatively small sites situated along Pemiscot Bayou in the extreme southeastern part of the Missouri Bootheel. This pattern is in marked contrast to site frequency and distribution in southeastern Missouri ca. A.D. 1250–1350 when the Eastern Lowlands, Sikeston Ridge, the Malden Plain, and the Western Lowlands were blanketed with large, fortified civic-ceremonial centers and their outlying villages, hamlets, farmsteads, and extractive sites, evidencing a high population density and apparent exploitation of almost all arable sandy loam soil associations through the region (Price 1974). For some yet unexplained reason this seemingly stable settlement system and agrarian lifeway came to a rather abrupt end ca. A.D. 1350 (Price 1976:50). By ca. A.D. 1450 the great civic-ceremonial centers were completely abandoned and left to ruin. This abandonment phenomenon has been termed the "Vacant Quarter" (Williams 1985), and while it can be documented by the lack of Protohistoric sites around the mouth of the Ohio River and adjacent areas, the prime mover that brought it about can only be conjecture given the current data base. The Protohistoric manifestation that prevailed in the southeastern tip of southeastern Missouri after ca. A.D. 1550 exhibits a settlement pattern and material culture quite unlike that of the Mississippian populations of the region some 200 years earlier. These late sites exhibit affinities not to southeastern Missouri antecedents but to late complexes to the south. They are included in the Armorel phase (Williams 1980) based on an assemblage of artifacts indicative of the Late Prehistoric/Protohistoric period centered in the Lower St. Francis River and Chickasaw Bluffs areas and are the northernmost representatives of this phase.

The Sites

There are five reported villages in the cluster of very late southeastern Missouri sites: Campbell or Cooter, McCoy or Chute, Berry, Maclin or Holland, and Kersey II or Cagle Lake. There is also a poorly documented small component of similar cultural materials at the Murphy Mound near Caruthersville. The village sites are located on natural levees of Pemiscot Bayou that meander across the southeastern tip of the Missouri Bootheel. The exact extent of these sites has never been fully determined, but it is known that they are generally smaller than the earlier fortified civic-ceremonial centers in the region. All but the Berry site are reported to have contained at least one flat-topped pyramidal mound. The dimensions of the mound at the Campbell site were reported as 75 feet (22.5 m) by 150 feet (45 m) by 15 feet high (4.5 m) (Houck 1908:1:56). Campbell and McCoy had plazas immediately east of their respective mounds. The Campbell site covered ca. 40 acres (16 ha). Chapman and Anderson (1955:13) noted the absence of evidence for fortifications at the Campbell site; no data exist on the presence or absence of fortifications at the other sites. To our knowledge no actual excavations have been conducted to test for the presence of fortifications at any of these sites.

Detailed data are lacking on the nature and arrangement of structures, pits, and other features on these sites although they are known to be present in great numbers as evidenced by a map of the McCoy site made by J. Raymond Williams (1968:101) after the site had been chisel plowed. Structures at Cagle Lake, Campbell, and McCoy were square to rectangular. Those at Cagle Lake had wall trenches, and those at Campbell had puddled clay floors and were reported to be of wattle and daub construction (Williams 1968:100, 102, 108–10: Chapman and Anderson 1955:100, 103, 106).

Burials at the Campbell site were found in a cemetery and scattered across the village (Chapman and Anderson 1955:49–50); those at McCoy were scattered and did not appear to be in a formal burial area (Williams 1968:101).

No data exist on the subsistence patterns of these Protohistoric peoples other than a recent study of human collagen from Campbell site burials that revealed a high level of maize consumption in that population's diet (Boutton et al. 1984).

Unfortunately, little is known about the settlement pattern of these peoples; only the large villages have been reported. It is presently unknown if there are smaller sites in the form of hamlets or farmsteads associated with the villages. The tight cluster of village sites in the southeastern tip of the Missouri Bootheel represents the most extensive and intensive occupation in the southeastern part of the state during Protohistoric times. However, a few artifacts that probably date to this time period also have been reported from the eastern Ozark Highland as well. Recent examination of an amateur archaeologist's collection in Shannon County, deep in the Ozarks, provided evidence of at least sporadic activity of Protohistoric peoples in this area (Price and Price 1983). The collection contains a number of Nodena points as well as triangular points from a number of open sites in the Current and Jacks Fork drainages. He also has what appears to be a fin from a late shell-tempered fish effigy bowl. A Nodena point, not unlike those from the Campbell site, was also found in Shannon County, Missouri, embedded in the skeleton of a bear from Bat Cave, which overlooks the Current River. The bear skeleton with the projectile point embedded in a vertebra is on display in the foyer of the Missouri Department of Natural Resources building in Rolla.

To our knowledge, no other artifacts dating from this late time period have been found in the southeastern Ozarks of Missouri. The arrow points in the Shannon County collection are from a few open terrace sites and occur in small numbers at each site. This evidence suggests that the area was visited during hunting forays into the highland interior. There appears never to have been any permanent or long-term occupation by these late peoples of this rugged Ozark landscape.

Archaeological Investigations

Although one of the sites, Campbell, was reported in the literature in the early twentieth century (Houck 1908:1:56), no professional fieldwork was conducted on any of these sites until the late 1940s and early 1950s. Three of the sites, Campbell, McCoy, and Holland, were visited by Stephen Williams during his survey in southeastern Missouri for his doctoral dissertation research at Yale University (Williams 1954). The Cagle Lake site was reported and limited test excavations were carried out there by J. Raymond Williams during land-leveling salvage work funded by the National Park Service through the University of Missouri in the 1960s (Williams 1968). At this time the McCoy site, which had been disturbed by chisel plowing, was mapped (Williams 1968:100–116).

During the 1950s and 1960s an amateur archaeologist and his wife, Leo and Mary Ellen Anderson, of Van Buren, Missouri, excavated over 200 burials at the Campbell site and obtained surface collections from the Berry and McCoy sites as well. Between 1951 and 1969 the Andersons drove from Van Buren in the Ozarks each weekend and worked on the sites in the Bootheel. During this time they excavated 215 burials at the Campbell site. Each burial with its accompanying grave goods was carefully drawn and described in a set of field notebooks laboriously compiled by Leo Anderson, copies of which are in the possession of the authors and the University of Missouri American Archaeology Division. For a short time in 1954 and 1955, the Andersons were joined by Carl and Eleanor Chapman at the Campbell site. Information on the first 35 burials excavated was published in the *Missouri Archaeologist* (Chapman and Anderson 1955). The remaining work is unpublished. The human skeletal materials resulting from these excavations are curated at the University of Missouri.

Leo Anderson also worked briefly at the Berry site while it was being looted by pothunters; but because looting was proceeding at such a rapid pace and was so extensive, he found it impossible to maintain any records.

Leo and Mary Ellen Anderson built a museum to house and display the collection they had obtained. Because the Andersons did not participate in the buying and selling of artifacts, the collection was maintained intact until 1978. The Andersons offered to donate the collection to the National Park Service but the offer was declined. Selected human burials exhibited in museum cases were donated to Southeast Missouri State University.

In 1978, Big Springs Village, Incorporated, purchased the museum and collection from the Andersons. In 1979, working under contract with the new owners, the Center for Archaeological Research at Southwest Missouri State University made a detailed catalog, with photographs, of the collection. The cataloging was completed by James E. Price and Cynthia R. Price with the photography conducted by Jeep Helm. A report was prepared (Price and Price 1979) that described and indexed the field notes, inventoried the cultural materials, and provided recommendations for museum interpretation. In 1980, the corporation suffered financial difficulty and the collection was divided and sold at auction. The "better" specimens were sold at auction through an art gallery in New York City and the remaining materials were auctioned in St. Louis, Missouri.

The collection made by Leo and Mary Ellen Anderson was the largest single collection of fully provenienced ceramic vessels and other artifacts from this group of sites in southeastern Missouri. The collection contained the best data on the material culture associated with this late complex. In conjunction with the limited work by Stephen Williams and the University of Missouri salvage operations, the Anderson collection and its accompanying data constituted the principal source of information on the Protohistoric period in southeastern Missouri. Now, the field notes, a catalog of the collection, and the human skeletal materials at the University of Missouri are the only remaining sources of information. Because of commercial looting and destructive agricultural practices, we may never have the opportunity to add significantly to the data base on this important cultural complex of southeastern Missouri.

Material Culture

Data on the material culture have been principally derived from the Anderson collection—from the burial excavations at the Campbell site and surface collections from the Campbell, McCoy, and Berry sites. These data are supplemented by those from the University of Missouri salvage work at McCoy and Cagle Lake. The material culture assemblage appears to be remarkably similar for the four sites from which there are reliable data.

Ceramics

Data on ceramics from these prehistoric sites are probably not representative of the entire ceramic assemblage because most of the known specimens are complete vessels from mortuary contexts. However, surface collections from Campbell, McCoy, and Cagle Lake and limited excavations at Campbell and McCoy have provided data on the material culture present on these village sites. Table 4-1 presents sherd percentages from surface collections from the three sites from which data are available. Descriptions of these pottery types can be found in Phillips (1970:37–178) and in Phillips, Ford, and Griffin (1951:61–153). Almost all are ceramic types associated with and diagnostic of a very Late Prehistoric, Protohistoric, or Early Historic occupation in the Lower Mississippi Alluvial Valley (Brain et al. 1974:236–37). Based on ceramic similarities, the southeastern Missouri sites are assumed to have been occupied during the A.D. 1500–1650 period.

The ceramic assemblage, based on surface collections that may be biased toward finer wares, are presumably representative of utilitarian vessels employed primarily in a domestic context. This assemblage is quite unlike that found on sites in the region occupied ca. A.D. 1250–1350. On these earlier sites the overwhelming majority of domestic ceramics are of a Neeley's Ferry Plain or Mississippi Plain type with only a minority of the more finely made Bell Plain ware being present. If the sherd collection data are not overly biased, quite the reverse is the prevailing pattern on the Protohistoric sites

	Sites		
Ceramic Types	Campbell	Cagle Lake	McCoy
Bell Plain	38	44	32
Neely's Ferry Plain	19	27	23
Parkin Punctated	7	9	24
Old Town Red	4	4	6
Barton Incised	-	3	1
Ranch Incised	3	-	.4
Rhodes Incised	-	.2	.4
Kent Incised	1	-	.4
Vernon Paul Applique	T	.2	-
Campbell Applique	16	.7	1
Campbell Incised	2	-	-
Campbell Punctated	1	.2	1
Walls Engraved	3	.4	.7
Hollywood White Filmed	2	.2	-
Carson Red on Buff	T	-	-
Mound Place Incised	-	.4	-
Fortune Noded	-	.2	.4
Nodena Red and White	-	-	.4
Varney Red Filmed*	-	5	5
Manly Punctated	-	.2	.4
Matthews Incised	-	.2	-

*An earlier ceramic type, the identification of this type may be an error in sherd classification. Earlier types often associated with Varney Red Filmed are not reported.

Table 4-1. Sherd Percentages (to the Nearest Whole Percent) in Surface Collections from Three Protohistoric-Historic Sites in Southeastern Missouri. Sherd percentages are not included for miscellaneous or untyped sherds (Chapman and Anderson 1955:29-48; Williams 1968:103, 197)

of the Armorel phase where Bell Plain wares are more common. Generally, Protohistoric ceramics exhibit finer workmanship and better paste than other regional antecedents.

Whereas punctating, incising, and noding occur earlier, they become much more common in a variety of expressions on the later wares. In Protohistoric times these coexist in an assemblage replete with various forms of applique, engraving, and both monochrome and polychrome painting, attributes almost totally lacking in earlier assemblages.

Examination of whole ceramic vessels in the Anderson collection from mortuary contexts reveal that while the basic vessels (that is, jars, bowls, bottles, and various effigies) continue to be made in the Protohistoric period, they took on new forms and were executed and artistically enhanced in new and innovative ways. Jars in the earlier period, ca. A.D. 1250–1350, commonly had two strap handles on opposing sides of the vessel rim. During later Protohistoric times such handles were expanded in number to encircle jar rims until they formed an arcade of low relief handles that were probably no longer functional. These in turn were expressed on some late vessels as mere applique arcades on jar rims. Vertical applique fillets were often added to vessel rims (Campbell Applique). Certain surface treatments such as noding (Fortune Noded) and punctating (Parkin Punctated) were employed as a motif over most of the exterior surface of jars.

Shallow bowls continued to be made in this late time period, taking on several forms of applique rim treatment unknown earlier. Bottles differ in several morphological attributes from earlier forms. The narrow carafe-necked bottles so ubiquitous ca. A.D. 1200–1400 gave way to bottles with necks of a much larger diameter. Gadrooning became popular on the exterior of these vessels. Several forms of motifs were applied to bottles, including painting (Nodena Red and White, and Carson Red on Buff) as well as engraving in the form of spiraled pinwheels or swastikas (Walls Engraved). Another unique feature associated with this vessel form in the late time period is iconographic figures scratched on bottle necks. These take the form of birds, arrows with fletching, and zigzag lightning motifs. Such scratchings are strikingly reminiscent of similar engraved motifs on Oneota catlinite tablets (Bray 1963).

Various forms of effigy vessels are relatively common in mortuary contexts during the Protohistoric period. Perhaps the most famous are the human effigy head pots. The Anderson collection contained a single complete specimen of this type as well as a basal section of another specimen recovered from the surface. Other effigy forms occur on bowls. The most striking of these are "upstream" and "downstream" fish effigies often painted and exhibiting skilled workmanship. Bird effigies occur in some numbers on deeper bowls, sometimes with a small rimrider on the tabtail opposite the dominant effigy appendage. Bat and "cat monster" effigies also occur on this form. The Campbell site produced a unique leaping deer effigy bowl. Conch effigy bowls also occur.

Effigy bottles are present in the form of turtles, fish, and dogs, and one bottle from this complex consists of three human head effigies fused at the back to create a tripod vessel.

One striking characteristic of these late vessel assemblages is that some specimens were made as compound vessels representing two vessel forms. Both horizontal and vertical compound vessels occur in the assemblages, usually representing a bottle fused on top of a jar, or a jar within a jar, or a bottle or jar within a bowl. "Jar-neck" bottles also occur. Horizontal compound vessels usually represent small short-necked bottles fused at the fullest lateral extent of their bodies.

Lithics

Lithic assemblages from this complex of Late Prehistoric sites reveal continuity from earlier Mississippian tool kits as well as new forms. Projectile points characteristic of the Late Prehistoric time period occur in two forms, triangular and willow leaf. These forms reflect a dramatic departure from earlier Mississippian types in southeastern Missouri. Earlier arrow points are typically small corner-notched specimens, often with serrated blade margins. Typical Late Prehistoric points on the other hand are somewhat larger and exhibit superior workmanship. They are typically quite thin with nearly perfect symmetry and well-executed pressure flaking. This holds true for both triangular and willow-leaf specimens that were represented in the Anderson collection in approximately equal numbers. Functional or cultural reasons that these points occur in two morphological types are unknown.

Another flaked-stone tool type prevalent on these Late Mississippian sites are unifacial snub-nosed end scrapers. These were common in the region during Dalton times, approximately 7,000–9,000 years ago (Price and Krakker 1975) and only again witnessed wide usage ca. A.D. 1500–1650. These occur in large numbers on late sites and probably represent tools employed in scraping animal hides. Other flaked-stone tools on these sites take the form of long, nearly cylindrical drills or perforators commonly referred to as "pipe drills."

Ground-stone tools occur primarily in the forms of adzes, chisels, and celts. Large adzes were typically manufactured of dark basalt, which was probably procured from outcrops of this material in the Upper St. Francis River Valley in present-day St. Francis County, Missouri. These pecked and ground specimens are plano-convex in cross-section and have well-honed cutting edges. Celts of the same material were also frequently made. Other, commonly smaller adzes were flaked and ground of Illinois novaculite, as were small slender adzlike tools that probably functioned as chisels. Other ground-stone artifacts include discoidals in a variety of forms and at least one perforated spatulate celt. Saws fashioned from thin sandstone pebbles and pumice shaft and awl abraders also occur. Raw materials for these latter two kinds of artifacts appear to have been gathered from Mississippi River sandbars.

One of the most spectacular lithic artifacts recovered by Leo Anderson came from the McCoy site. It is a large catlinite disk pipe that was fragmented. Leo Anderson collected the pieces of this specimen from the surface of the site over a period of several years. He even constructed a sifter and screened a large area of plowzone on the site in search of fragments and recovered all of the specimens with the exception of a few fragments of the disk. This artifact is a good hallmark for the Late Prehistoric/Protohistoric time frame. Similar but less spectacular specimens have been found associated with the Caborn-Welborn complex at the mouth of the Wabash River (Munson and Green 1970) and in various Oneota contexts (Hamilton 1967).

Shell

A variety of shell artifacts were recovered from the Campbell and Berry sites by Leo Anderson. Common river mussel shells were fashioned into spoons. Several such specimens were discovered in bowls in a mortuary context. Marine-shell ornaments were also recovered in considerable quantity from the Campbell site. These take the form of disk beads ground from conch shells. Mask gorgets with holes for eyes also occur, as do long columella beads. These and long shell ear pins with flat naillike heads are diagnostic of late occupation on the sites.

Bone

Bone tools recovered from the sites include fishhooks, antler wrenches or shaft straighteners, needles, and awls. The bone tool assemblage is not unlike those of earlier Mississippian assemblages in the region.

European Trade Goods

Proof positive that this Late Prehistoric occupation continued into the Protohistoric era takes the form of a limited number of European trade goods recovered from the Campbell site by Leo Anderson; other specimens are reported to have been found by recent looters active on this site.

Those found by Anderson include blue glass beads, sheet-iron tubular beads, and fragments of rusted iron that appear to be pieces of a knife blade. These artifacts were recovered from two burials at the Campbell site. Historic trade goods more recently reported to have been recovered are additional glass beads, among which there is a chevron bead and a Clarksdale bell.

Historic artifacts are extremely rare on these sites, but the presence of the limited number of specimens proves that this late cultural complex was either in limited contact with Europeans or traded with other aboriginal groups who had encountered European travelers.

Summary and Conclusions

The Protohistoric period, ca. A.D. 1550–1650, is primarily represented in southeastern Missouri by a group of archaeological village sites along Pemiscot Bayou in the extreme southeastern corner of the Missouri Bootheel. These sites have witnessed relatively few professional archaeological investigations, and the major data base on this aboriginal culture in the region is embodied in the field notes, photographs, and other records associated with the Anderson collection. Although the collection was sold, the data base remains as a substantial volume of information on these late archaeological materials, which exhibit only general affinities with regional Mississippian antecedents. The ceramic and lithic artifacts from these sites indicate that they belong to the Armorel phase, whose principal territory lay to the south in northeastern Arkansas. The occupation of these sites continued into the Historic era; at least two burials from the Campbell site had Historic European artifacts associated with them.

5

The Nodena Phase

Dan F. Morse

The extinct societies of northeastern Arkansas are classified by archaeologists into broad cultural periods. The cultural period for the Nodena site is the Mississippian period of ca. A.D. 800–1650 ± 100 years. During Mississippian times a special environmental adaptation took place in much of the eastern United States. Maize floodplain agriculture, the bow and arrow, shell-tempered pottery, and a complex tribal organization called chiefdom (Morse and Morse 1983:201–13) are the major characteristics. By virtue of particular styles of artifacts, the people who inhabited Nodena phase sites can be more securely classified as "late period Mississippian" or as dating somewhere within the A.D. 1400–1650 period.

Because of Dr. James K. Hampson's work at the Upper Nodena site and his generosity in sharing his work with professional archaeologists, the site was chosen in 1954 by Stephen Williams as the type site for a geographical expression named the Nodena phase (Phillips 1970:933–34). "Nodena phase" is not what those people called themselves; the name is an archaeological classification. However, we think the people who lived near Nodena in 1541 called themselves the inhabitants of the province of "Pacaha" (Hudson 1985).

"Pacaha" is thought generally to encompass our archaeological "Nodena phase" to the extent that we are confident we have been able to isolate a functioning extinct society. Pacaha was the major chiefdom of the Central Mississippi Valley, and if Nodena was indeed Pacaha, then we are translating a most unusual story. Unfortunately, many pages are missing from the book, due to vandalism and to land leveling. The funds spent to investigate the Nodena phase professionally would barely purchase a single exotic pot of the thousands dug by treasure hunters over the past century. There is a war between science and greed, and so far greed is winning. Dr. Hampson was one of the very few early avocational archaeologists of means who opted for science, and it is fitting that Nodena lives on in his memory.

Stephen Williams named the Nodena phase in a paper read at the Southeastern Archaeological Conference in 1954 (Phillips 1970:933–34). Previously, Griffin (1952a:233) had included it as part of the Walls–Pecan Point complex on the basis of a tight ceramic similarity that extended from the Pecan Point site in Mississippi County, Arkansas, to the Walls site in De Soto County, Mississippi. Williams, however, made an extensive study of the pottery in Dr. Hampson's collection from Nodena and concluded that a further subdivision was necessary.

The Nodena phase has never been adequately defined. As Phillips states: "Nodena is no more in need of exact definition than are three-fourths of the other phases dealt with in this section. . . . This will continue until that far-off time when the formulation of cultural units on the firm basis of excavation data and stratigraphy becomes the rule rather than the exception" (1970:934–35).

The Upper Nodena Site

The site of Upper Nodena, type site for the Nodena phase, was first scientifically investigated by James K. Hampson from 1897 until about 1941 (Morse 1973a; Williams 1957). The University of Arkansas Museum and the Alabama State Museum of Natural History excavated at the site during the late winter and early spring of 1932 as a joint expedition. Dan F. Morse directed a test excavation at Upper Nodena in the summer of 1973 as part of the University of Arkansas and Arkansas State University combined field schools (Morse 1973b).

On the basis of Dr. Hampson's notes, some older aerial photographs, and a general surface collection made in 1970 at the site, we formulated a tentative village plan (Figure 5-1). These data were verified to a large degree by a photograph of a scale model constructed by Dr. Hampson (Ferguson and Atkinson 1966; Morse 1973b:Figure 39). The data indicate an organized and well-demarcated rectangular village measuring ca. 15.5 acres (6.2 ha) in extent and situated next to a relict meander channel (Morse 1973a:Figure 38). This village plan should be verified by a controlled surface collection and through test excavation of specific expected village features.

We began the research with Dr. Hampson's lone remaining map and his unpublished manuscripts. His large table model of Upper Nodena and his field maps disappeared several years ago. Farming has altered the landscape

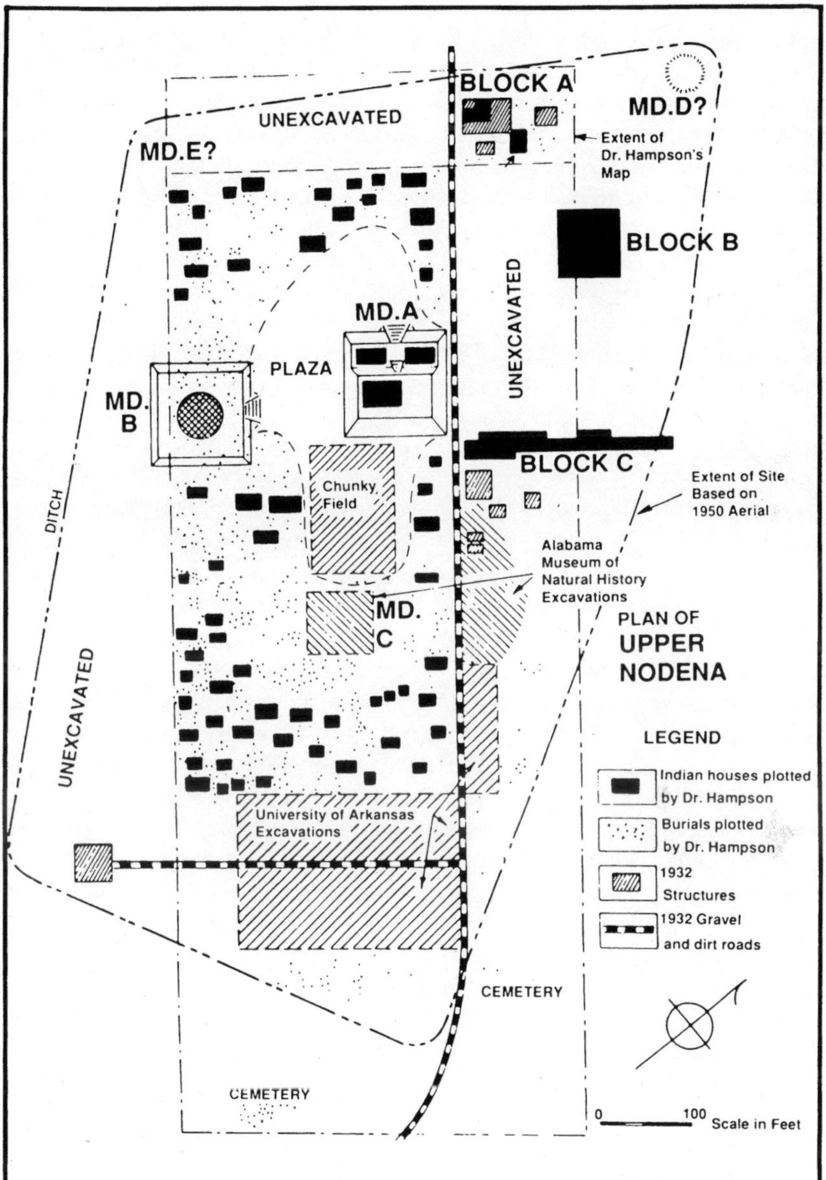

Figure 5-1. Upper Nodena Site

so much that Mrs. Hampson was not even sure just where the large mound was located. As a matter of fact, we first missed its location by almost 800 feet (240 m). We discovered that we were working with a reduced blueprint of the original Hampson map and had to compute a scale on the basis of an average of five measurements of features found in the manuscripts. Harold F. Ohlendorf, of Osceola, lent us an aerial photograph dating from 1950, and Donald Miller, soil conservationist with the Soil Conservation Service, Osceola, got us prints of an aerial photograph dating from 1954. The 1950 photograph (Morse 1973b:Figure 37) is a very dirty print or negative, but because it was taken soon after the spring cultivation, it clearly shows the site. The 1954 photograph verifies the 1950 photograph, although it was taken at a less favorable time of year. Two roads and one modern drainage ditch were the features used to correlate Dr. Hampson's map and the aerial photographs.

We lettered by size the mounds described by Dr. Hampson at the begining of his paper, beginning with the largest mound and ending with the smallest or at least smaller mounds (Figure 5-1). Mound A was identified and found to match the location on Dr. Hampson's map. We then attempted to locate Mound B, described by Dr. Hampson as being west of Mound A, on the premise that it was the low mound with the large circular structure indicated on his map. The aerial photographs suggested the former existence of a large, low mound where expected. Mound C, located south of Mound A by Dr. Hampson, is presumed to be the one excavated by the Alabama State Museum of Natural History. This excavation is shown on the Hampson map. Mound D was located by Dr. Hampson as being "to the northeast." An elevation in the extreme northeastern corner of the site is suggested as being this mound. Mound E was located by Dr. Hampson as being "to the northwest." Presumably it is the northwestern corner of the site, but this location cannot be confirmed on the aerial photographs. The group of small mounds located "to the southwest" probably were near the southwestern corner of the site. They are not visible on the aerial photographs and quite possibly had disappeared by 1927 when Dr. Hampson returned to Nodena.

Mounds A, B, and C were adjacent to an area apparently devoid of houses and burials. This area probably is a large plaza or public square. Immediately west of Mound A on Dr. Hampson's map is an unnumbered burial, but no more burials are plotted nearby. Beneath Mound B are several burials, but none are plotted east of it. Mound A must have been built early in the history of the site, inasmuch as it apparently does not overlap graves. However, because Mound B does apparently overlap several graves, the site probably had been in existence for a while before Mound B was constructed. Mound B probably was built after Mound A. Mound C involved a unique burial situation. Reportedly, 316 adult (all but 2 of them male) skeletons were found there, but this report is not correct (see chap. 6). However, such a large number of skeletons in such a small space is very unusual.

Of 18 extant skeletons identified as being found in Mound C, 2 are children and 16 are adult. Of those 16 adults, 5 are female and 8 are male. The large number of skeletons found in such a small area might indicate de Soto–period epidemics or warfare, but neither explanation can be tested with the small sample now available. Upper Nodena probably terminated as an occupied site and Middle Nodena began its existence about the time of de Soto.

The only possible definitive evidence of a traumatic encounter between Indians and Europeans at Nodena is an adult male skull from Middle Nodena (see chap. 6). In this case the evidence consists of a Caucasian skull with an unhealed arrow wound. Indian skeletons with sword wounds are found elsewhere in the Southeast, but this is the first possible incidence of a white skull with an arrow wound.

Dr. Hampson described a "chunkey field" south of Mound A but did not orient it or place it more exactly. The aerial photographs have not helped. We took the liberty of orienting the "field" on the same axis as Mound A and placed it within the southern part of the plaza, where it best fits. Its presence should be confirmed by excavation, but surface debris is indeed very light in this area.

A road bisects the site. Three modern houses, two with associated outbuildings, are known to have been present on the site, based on Dr. Hampson's notes, the aerial photographs, old quadrangle sheets, photographs provided by the Alabama State Museum of Natural History, and visual observation. One house was near the southwestern corner on a ridge and connected to the main road by a farm road that bisects the main University of Arkansas excavation.

Because the site boundary is so distinct, the site was probably enclosed within a wall. The western side of the site is very distinct. The dark midden of the site contrasts sharply with the light-colored soil outside the site, suggesting a palisade wall and ditch. Interestingly, the northern half of this side parallels a drainage ditch on the aerial photograph. The eastern border of the site has been generally verified by visual observation. However, a test in 1975 failed to reveal a palisade ditch. The southern border of the site, as indicated by the aerial photograph, has been verified by field checking on the surface. Dark soil and midden debris occur north but not south of the border indicated in the aerial photographs. A drainage ditch, considered with the clear definition seen in Figure 5-1, again suggests a palisade wall and ditch here. Some very scattered debris exists south of the line indicating the southern border, but this debris is thought to have resulted from farming and from digging the modern drainage ditch. North, inside the site, very little debris exists along the line, a situation similar to the western border of the site.

The northern border of the site is also clearly delineated by contrasting soils on the aerial photograph and is suggestive of a palisade wall and ditch; however, our surface field check revealed that the area north of the drainage ditch is a ridge with little surface indication of houses and midden. The most concentrated site midden is represented by the houses plotted by Dr. Hampson (Figure 5-1), and this concentration may be due to this soil having been dug through. A small area along the western side of the road between Mound A and the drainage ditch has very concentrated midden; the soil is very dark, in contrast to the light-colored, almost sterile plaza area. The drainage ditch could have been the northern edge of the village in an early construction phase of the site. Excavations by us in 1973 revealed only sparse debris. In support of the possibility of later habitation is the finding by Howard Pierce, of Jonesboro, of a portion of a catlinite disk pipe in this area of the site. Catlinite disk pipes are thought to have been a sixteenth-century trade item in the Central Mississippi Valley based on their presence at several Late Prehistoric and/or Protohistoric sites (cf. Gibbon 1986:328).

The houses and Mound A in Figure 5-1 are drawn as Dr. Hampson had them on his map. The chunkey field and Mound B are oriented along the same axis. Dr. Hampson assumed the main road was straight north and south. Every single house on his map is exactly perpendicular to the road, but Dr. Hampson's large comprehensive map, made to construct a museum table model of a village, may not be accurate. We wish we had his field maps to check house orientation. We would expect houses to be parallel to or perpendicular with the long axis of the site. As far as the records show, the Universities of Arkansas and Alabama did not excavate or map houses. The major areas not excavated by Dr. Hampson or the two museum expeditions are the extreme north, the northeastern and the southwestern portions of the site. In 1975 we tested the northeastern sector; we could not excavate west of the road due to the alfalfa crop.

Dr. Hampson described the typical Upper Nodena site house as having a rectangular pattern of postholes around a smooth hard-burned floor. He also noted smaller postholes in lines near the edge of one or more sides, presumably to support storage and/or sleeping platforms. This is the same kind of house pattern found by Perino (1966:Figures 6, 7, 10, 11, 84) at Nodena phase sites near Wapanocca Lake.

In 1973, a complete house pattern (rebuilt once) was uncovered at Upper Nodena. The houses measured 5.3 m by 5.2 m and 5.8 m by 5.7 m. Both patterns contained wall trenches and roof-support posts and were oriented northwest to southeast or parallel with the site. Other wall trenches were found in a second excavation unit. Considerable daub was also collected. An interesting discovery was the remains of what is thought to have been an aboveground granary. The recovered charred corn was extraordinary in its size (Blake and Cutler 1979).

Houses in the Nodena phase are difficult to reconstruct. The Quapaw reportedly built dome-shaped, bark-covered houses rather than wattle and daub types; yet the examples we found at the pre–Quapaw period Upper Nodena site were wattle and daub. Perino (1966), however, found few wall-trench houses at Banks. There, the prevalent house had a posthole pattern similar to those found at Quapaw phase sites (Ford 1961:Figure 9). Possibly, houses were dome-shaped and only partly wattle and daub (Nash 1968:Plate 2).

A total of 1,755 skeletons have been reported from the Nodena site and an unknown additional number destroyed. Approximately one-half or more of the site has been excavated. Based on Dr. Hampson's records, the skeletons usually were buried in groups of around 15 to 20 or more, suggesting family cemeteries. Normally, each skeleton was extended on its back. Many, according to the Hampson map, seem to have been oriented parallel to the nearby slough on the east, on a general north-south axis. Usually a bottle and bowl were buried near the head. A total of four disturbed adult graves, including one with 200 probe holes in its vicinity, and one child and three infant skeletons were found in 1973. None appeared to be associated with a cemetery, and no artifacts were associated, except for several large sherds placed over two of the infants perhaps to keep them from being dug up by dogs.

In conclusion, the Upper Nodena site was a 15.5-acre (6.2-ha), palisaded village (Figure 5-1). In the center, offset slightly to the north, was a ceremonial component consisting of two pyramidal mounds with associated structures. According to Dr. Hampson's paper, Mound A had two levels with structures—the upper level with a single structure and the lower level with two structures. Mound B had a circular structure in association. At the Fatherland site, the Grand Village of the Natchez, one pyramidal mound served to support the chief's house, while the other supported the temple (Neitzel 1965:62). Both of these Natchez mounds were at opposite sides of a plaza. Connecting the two pyramidal mounds at Upper Nodena is an apparent plaza or public square, which generally lacks midden. Southward, the midden-less area continues to Mound C; however, this part of the sterile area is thought to have been a chunkey field according to Dr. Hampson's notes. A chunkey field was a specially prepared playing field for rolling a stone discoidal while attempting to throw an 8-foot (2.4-m) shaft so that its distal end landed near where the stone stopped rolling (Culin 1907:485–88). Mound C seems to have been a special burial situation. The field observation seems to indicate contemporary burial of many people. There is the possibility that a large group of people were killed in a raid; for example, the Indians with de Soto, when they entered Pacaha, "slew whatever men they could lay their hands on, more than a hundred and fifty in all" (Garcilaso 1951:437). Another possibility to account for such a sizable burial situation is an epidemic. Surrounding the ceremonial component of the site were houses and small cemeteries (Figure 5-1). Larger houses tended to be nearer the plaza, but we do not have any information about associated artifacts that might indicate the probabilities of these being commercial craft structures, special houses for important kinsmen or craftsmen, or storage houses.

The Dating of the Nodena Phase

The Nodena phase and the Nodena site date within the arbitrary limits of A.D. 1400–1650. The site is thought to have existed during no more than a century, based on the shallowness of deposit and upon a relative lack of superimposed features. Houses probably lasted around 20 and no more than 30 years, based partly on archaeomagnetic samples recovered from the profile of the Hazel site (Wolfman 1982:287). The example discovered in 1973 had been rebuilt only twice. The only radiocarbon date from the Upper Nodena site yielded a radiocarbon age of 630 ± 250: A.D. 1320 (Crane 1956:666). The sample was willow (*Salix*) provided by Dr. Hampson to Dr. James B. Griffin, but the specific context of the sample is unknown. Dr. Hampson (1973:9) mentions counting 526 rings in an oak cut down in 1897 on top of Mound A. Many tree rings, however, are impossible to read accurately without special laboratory treatment (Stokes and Smiley 1968), and the count has little actual significance beyond possibly indicating a post-A.D. 1371 date for Mound A. The radiocarbon date only indicates that probably the site dated somewhere between A.D. 1070 and 1570.

The Upper Nodena site was constructed upon a soil used immediately before occupation as a field for cultivated crops. Colluvial episodes were clearly discernible in the "subsoil" at the location of the house recovered in 1973. The indication here is that the site began its existence sometime after A.D. 1400. The number of skeletons and size of site indicate a population of around 1,000–1,500 at any one time, given a time span of approximately one century. Based on figures for Quapaw villages, this general estimate would suffice for a large major site. Quapaw villages shifted through time, probably accelerated by the French presence and later by smallpox epidemics. But one Quapaw village, New Kappa, became a virtual cemetery and was abandoned soon after 1698 (Kellogg 1917:359). Perhaps the Upper Nodena site was abandoned for similar reasons.

The occupation of the Upper Nodena site may have been terminated around the de Soto time line of 1541. The reasons for thinking the Upper Nodena site was not occupied past the mid-sixteenth century are based on artifactual evidence. For instance, a general lack of end scrapers, thought to be characteristic of the Protohistoric period (Williams 1980), may indicate a pre-1541 date. However, lack of end scrapers may simply indicate that the Nodena inhabitants were not directly involved in deer procurement or that they processed deerskins elsewhere. Very little deer bone was recovered in the 1973 excavations. A complete disk pipe and a broken one have been recovered from the Upper Nodena site. Catlinite disk pipes may have been part of the "Siouan" trade in the sixteenth-century period and/or may be part of a broad horizon style in that period (Gibbon 1986:329). Their presence indicates a terminus date of at least ca. 1541. The nearby Middle Nodena site may have primarily existed immediately after Upper Nodena was abandoned. At least, there are a few more end scrapers known from this site. Middle Nodena is approximately one-half the size of Upper Nodena (ca. 7.75 acres or 3.1 ha), a fact that fits well with a concept of decreasing population after 1541. Proving this theory will require a significant increase in the level of investigation going on today.

That the Nodena phase lasted until 1650 or even later is indicated by possible French trade goods, which include blue glass beads, brass rings, iron beads, an iron knife, nails, and a knife scabbard ornament (Morse 1971). Sites in the Nodena phase that may have lasted well into the seventeenth century, based on the presence of historic artifacts, are Blytheville, Bradley, Campbell, and Wildy. Differentiating between European trade goods of the Protohistoric and the Early Colonial periods can be difficult. Upper Nodena itself almost certainly terminated before 1650. We tentatively date the Upper Nodena site to ca. A.D. 1450–1550, but on rather sparse evidence.

The Nodena Phase Settlement Pattern

No systematic survey similar to the Parkin project (Morse 1981) has ever been conducted to map archaeological sites in this area. Leads provided by landowners, collectors, and in particular Dr. Hampson were followed to the larger Nodena phase sites. Even some large significant sites were missed in the 1940–41 survey by Phillips, Ford, and Griffin (1951) and were not recorded until after the inception of the Arkansas Archeological Survey of 1967.

Northeastern Arkansas has been cultivated for a long period of time and activities associated with increased production have been detrimental to site preservation (Medford 1972). On the other hand, because the area is so intensively farmed, we have an excellent chance to locate a large number of the sites before they disappear. A number of intensive surveys have been conducted in certain portions of the geographical area of the Nodena phase.

Three distinct geographical clusters of Nodena phase sites west of the Mississippi River are shown in Figure 5-2. An attempt to designate these clusters as separate named phases is premature at best (see chap. 8). Our investigations indicate a single phase (Morse 1973b). Southward is the Wapanocca Lake cluster in Crittenden County, Arkansas. The best-known site here is Bradley. Earlier Nodena phase remains at Banks have been found, indicating the presence of some kind of developmental sequence (Perino 1967:76–77). Upper Nodena belongs to the Wilson-Joiner cluster in southeastern Mississippi County, Arkansas. No developmental sequence has been recognized for this region. Pecan Point (Moore 1911:447–74) and Upper and Middle Nodena are probably the most widely known sites in this subdistrict. A third cluster of Nodena phase sites is the Pemiscot Bayou cluster in Mississippi County, Arkansas, and Pemiscot County, Missouri. The best-known sites are Chickasawba or Gosnell, located within the city limits of Blytheville; Campbell, located in Missouri north of Blytheville; Gant, and Wildy, located along Little River in Arkansas.

Wapanocca Lake Cluster. Six sites are recorded in the area immediately around Wapanocca Lake. The largest site, Bradley, contained four mounds. Only a single mound and a mere acre remnant of the site remain today. The site was used as levee fill in the late nineteenth century. The preserved portion is the location of a cemetery of the early town of Greenock, Arkansas.

The Bradley site actually continues with sites 3CT9 and 3CT43 over a 2-km length of Bradley Ridge, and if these constituted a single site, it must have been a spectacular one. Moore (1911:431, 324) found "abundant" European trade goods at Bradley, so this portion of the complex of sites seems to have been the latest of the six known.

At the opposite (western) end of Bradley Ridge are the Banks site and site 3CT17, which consists of 10 to 15 widely spaced house patterns (Perino 1966:158) and probably was roughly contemporaneous with Bradley. Banks appears clearly to predate Bradley while still being Nodena phase in nature. Banks was ca. 10 acres (4 ha) in extent, with a plaza but no extant mounds. A mound nearby (3CT14) is thought to belong to a much earlier period (Perino 1966:4) but may be contemporaneous with Banks. A sixth definite Nodena phase site (3CT139) located on the opposite side of Wapanocca Lake measures 2 to 4 acres (0.4 to 0.8 ha) in extent and seems to date to about the time period of the Banks site.

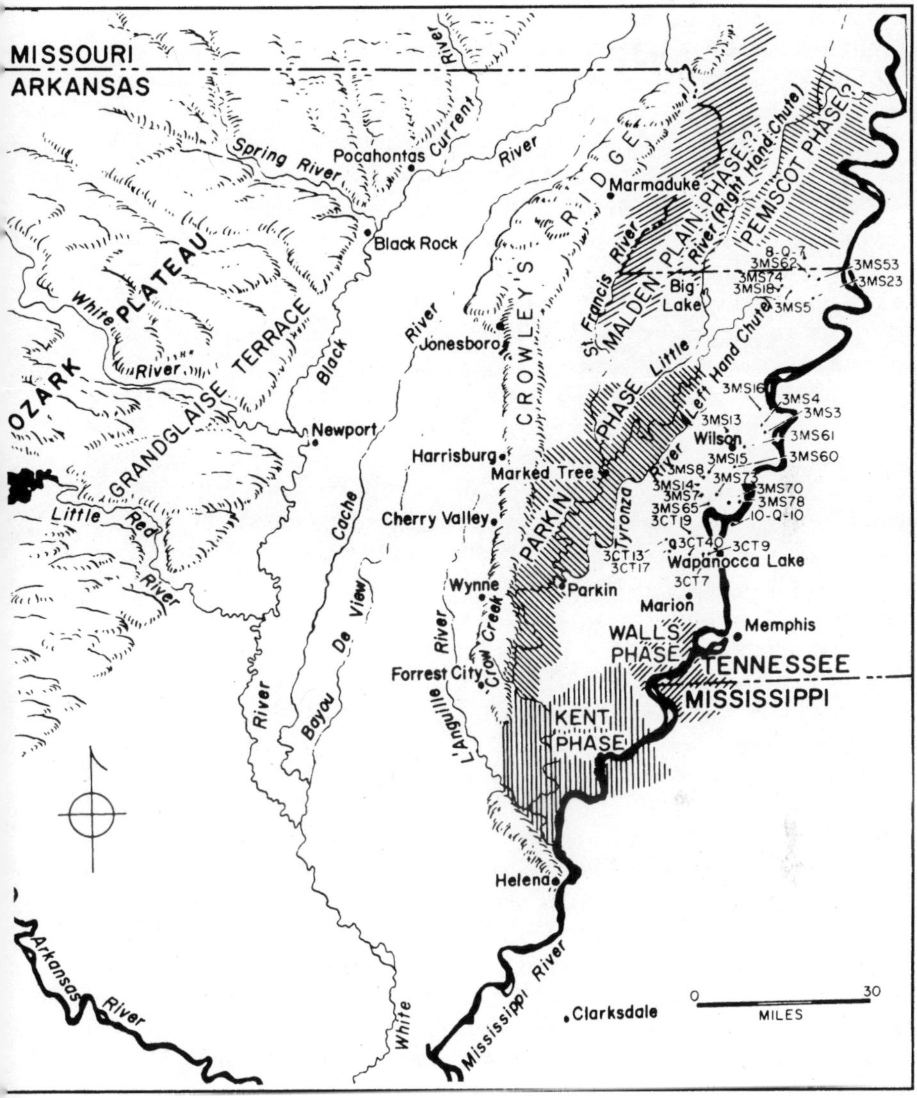

Figure 5-2. Nodena Phase Sites

It would appear that all of the developmental stages of the Nodena phase are represented in the known Wapanocca Lake cluster of six sites and that a population increase occurred in the latter part of this period. Given that our reconstruction of the 1541 de Soto route is correct (Hudson 1985), Bradley was Pacaha, the capital of the province of Pacaha ("Capaha" according to the Garcilaso account) and of what we know as the Nodena phase. Aquixo and Quizquiz (Walls and Belle Meade phases) were vassal to Pacaha. The capital was strategically located at the southern end of the province, where its vassals could be more easily controlled. Even the Parkin phase (Casqui in 1541) was threatened by the expanding powerful Pacaha chiefdom or incipient state. The Quapaw principal site of *Kappa* (or "Capaha" according to La Metairie) was located ca. 100 miles (160 km) to the south in 1673. C. B. Moore collected brass finger rings at Bradley (David H. Dye, personal communication), indicating the possibility of a seventeenth-century survival, and it is not unreasonable to hypothesize that Bradley shifted south to become Kappa in the seventeenth century.

Wilson-Joiner Cluster. Twenty-one known sites that are clearly Nodena phase exist in the vicinity of Joiner and Wilson, Arkansas. Pecan Point is the most spectacular despite major destruction. Just the fact that nine "head pots" were known in 1911 to have been found there indicates the importance of Pecan Point. Its size is not known (Thomas 1894:220), and only one prominent mound was described by Moore (1911:447–75). Yet we believe the site was on a par with Bradley, also not measured by early explorers. The coexistence of such major sites in the same phase may reflect changing political conditions.

Besides Pecan Point, the two largest sites measured to date are Upper Nodena (6.2 ha or 15.5 acres) and site 3MS7 (5.1 ha or 12.75 acres). Five sites measure between 7 and 10 acres (2.8 and 4.0 ha): 3MS14, Notgrass, 3MS17, 3MS73, and Middle Nodena. Nine sites measure between 2 and 4 acres (0.8 and 1.6 ha): 3MS8, Carson Lake, 3MS60(?), 3MS61, 3MS65, 3MS69, 3MS71, 3MS80, and 3MS91. One site, not yet mapped, but which probably falls between 2 and 10 acres (0.8 and 4.0 ha) in size is site 3MS57. Site 3MS340 is located 6 miles (9.6 km) west of the cluster, and the Nodena component is of unknown size. Site 3MS94, originally thought to measure only 0.25 acre (0.01 ha) in size, has been revealed (via a backhoe and increased treasure hunting by the landowner during February 1987) to be mostly buried and to be larger than 2.5 acres (1 ha) in actual size. With the exception of Upper Nodena, none of these sites has been accurately measured similar to what Morse (1981) has done for the Parkin phase.

In the entire Parkin phase, 21 known sites cluster as follows: one site of 17 acres (6.8 ha), four sites of 7–9 acres (2.8–3.6 ha), 10 sites of 4–6 acres (1.6–2.4 ha), and six sites of 1–2 acres (0.4–0.8 ha) (Morse 1981). None was 0.25 acre (0.1 ha) in size. Thus, the entire Parkin phase (Pecan Point was not included), as measured by total site size, is roughly equivalent to or slightly smaller than the Nodena phase Wilson-Joiner cluster of sites. A few more

Parkin phase sites may be identified to the south but probably not enough to change this comparison significantly (Phyllis Morse, personal communication).

Evidently this cluster of Nodena phase sites near Joiner and Wilson was not directly visited by the de Soto expedition (Hudson 1985). European contact with its population was probably made by Spanish scouting parties out from Pacaha as well as by visitation to Pacaha by the people of the Wilson-Joiner region. These kinds of activities are abundantly referenced in the de Soto accounts and the expedition spent a month in Pacaha in 1541. None of the 21 known sites appears to have survived into the seventeenth century, based on current archaeological data, which do not include seventeenth-century artifacts. The indication is that this population became extinct by about the seventeenth century.

In the latter part of the seventeenth century, isolated Indian structures were observed near this region by the French. Direct evidence of Indian presence at the end of the seventeenth century are in the form of trap wire bracelets, which were collected near Osceola and deposited in the Smithsonian Institution and Putnam Museum. By about 1673 the indigenous Indian population of northeastern Arkansas had become extinct or had shifted downstream to or near the Arkansas River. The only known Indian tribe for the region to the north of the Arkansas in 1673 were the Michigamea.

Pemiscot Bayou Cluster. Thirty-four sites are identified along Pemiscot Bayou and its other name, Little River. Twenty are located in Arkansas—3MS2, Chickasawba/Gosnell, Wildy, Gant, 18, 22, 23, 24, 53, 55, 62, 93, 100, 101, 104, 105, 106, 108, 109, and 110—and 14 are known for Missouri (Richard Marshall, personal communication). An additional two sites, Stott and Potter, are located between the Parkin and Nodena phase districts halfway between the sites of Hazel and Wildy. Both are Late Mississippian in artifact content and may be Nodena and/or Parkin phase sites. At least one other unmapped site, located at Dell, Arkansas, has not been counted as Nodena phase because of lack of verification. A few other mapped sites without representative artifact samples may also be Nodena phase. The Campbell site in southeastern Missouri has produced considerable European trade goods, according to local treasure hunters. Leo Anderson, the only known reliable informant (see chap. 4), found bluc glass beads, iron beads, and iron knife fragments at Campbell. A central burial crypt contained an extraordinary amount of very early-sixteenth-century trade goods (unfortunately some commercial salting occurred after the initial discovery), which included a large catlinite disk pipe, reportedly brass hawk bells, 24 chevron glass beads, and other items (Gregory Perino, personal communication). A number of chevron beads, an amber bead, and a pendant made from a cloisonné book clasp are reported from the site. The Campbell site probably was the frontier trading site visited by Garcilaso de la Vega's informants in 1541 when they traded beans and European items for salt and copper (Garcilaso 1951:449–50). The Spanish period artifacts evidently were mostly found in one deposit. Blue glass beads and iron, including nails, two knives, and beads, found with individual burials may indicate that the site terminus was well within the seventeenth century when French goods had initially become available via the Great Lakes and St. Lawrence River (Mason 1986; Quimby 1960). Recently it has been discovered that the 1673 Michigamea summer village on the Marquette map was located near Pocahontas, Arkansas (3RA262). Similar kinds of European artifacts have been recovered at this site. Because of the emphasis on indirect European trading, Campbell may well have moved south where it became the site of Arkansas post, Osotouy, in 1686. Two other sites known to have produced seventeenth-century French and/or English trade goods, Blytheville and Wildy, also indicate seventeenth-century survival of certain Nodena phase sites, some of which may have become Quapaw villages by 1673. However, the blue glass bead found at Blytheville and the brass pendant found at Wildy might be sixteenth-century Spanish artifacts.

All of these 34 sites have been commercially dug, and most continue to be vandalized today. We know very little about site size except that it is somewhat similar to the Wilson-Joiner cluster situation. Preliminary data indicate a population size in the Pemiscot Bayou area that is more than twice that in the Wilson-Joiner cluster. The total Nodena phase population based on site acreage was ca. 3.5 times that of the Parkin phase, assuming site longevity in both areas to be comparable. There may be one single-residence site in the Arkansas sample and one other site may have been a hamlet. In contrast to earlier Mississippian phases, Nodena sites of a quarter acre or smaller are very rare in spite of a concentrated effort to find them. Morse found this same situation in her study of the Parkin phase, where she failed to discover a single identifiable Parkin phase single-residence site (1981). These populations were nucleated into fortified towns and very rarely occupied hamlets or farmsteads. The Quapaw settlement pattern as described in the seventeenth-century French accounts (Morse 1986a) is identical to the Protohistoric pattern seen archaeologically and as reflected in the de Soto accounts. This settlement pattern made these people very vulnerable to epidemic diseases.

On the east side of the Mississippi River are four sites (Morse and Morse 1983) included by other investigators in the Nodena phase. They almost certainly are contemporary with Nodena, based on artifacts, but are distinctive in having the burial urn as a prominent trait, similar to the Alabama River phase in Mississippi and Alabama (Sheldon 1974).

The three distinct clusters of Nodena phase sites west of the Mississippi River may help explain the report in Garcilaso's account of an 11-day round trip from Pacaha to the trading site that "the lands they had seen were not good, but sterile and poorly populated" (Garcilaso 1951:450). The account could pertain, as suggested earlier (Morse and Morse 1983), to the uninhabited area between the Nodena phase and the Ste. François Mountains. On the other hand, if the trading party traveled essentially along a modern route (Interstate 55) the shortest straight distance (ca. 85 miles or 136 km) between Bradley and Campbell, they would have only passed near a site near Joiner and Blytheville. Unfortunately, Arkansas did not have a highway salvage program when I-55 was constructed and this route was not surveyed. However, most of the route is on the braided surface between the Mississippi and Little rivers. This region is not noted for good farming soils, which are more concentrated near the major river. Modern surveys indicate a lack of Protohistoric habitation on this braided stream terrace.

Environmental Setting and Economic Behavior

An understanding of the environmental setting and the expected economic behavior should also help us understand the Nodena phase settlement pattern. The immediate geological topography of the Nodena phase is a river meander belt. Saucier describes a meander belt as including

> all of the area and landforms resulting from lateral migration or meandering of a stream while occupying a single course. In the case of freely meandering streams like the Mississippi . . . accretion topography consisting of parallel arcuate ridges and swales comprises most of the meander belt area. However, abandoned channels in various stages of filling from mostly open water (oxbow lakes) to completely filled (referred to as "clay plugs") are the most conspicuous meander belt features. Depending in part on the length of time a particular meander belt was active, the number of abandoned channels varies from a dozen to well over a hundred in the case of the Mississippi River meander belts. (Saucier 1974:27)

Mississippi River meander belt systems (Saucier 1970:Figure 2) are classified as either abandoned or active. Saucier distinguishes between two basic kinds of soils, clay in the abandoned channels and well-drained silty and sandy clays over much of the remaining surface deposited as levees (Saucier 1974:27). Pemiscot Bayou and Little River occupy an abandoned Mississippi River crevasse channel, which duplicates on a smaller scale typical basic Mississippi River meander belt features. Ferguson and Gray (1971:General Soil Map) indicate the general location of soil associations.

The Wilson-Joiner site cluster, for instance, is characterized by the Convent-Morganfield-Crevasse and the Tunica-Bowdre-Sharkey associations; sites in the Pemiscot Bayou cluster are in an area known primarily as the Amagon-Dundee-Crevasse soil association; and sites in the Wapanocca Lake cluster are made up basically of the Alligator-Earle association. We can speculate that these fairly distinct soil and site associations are meaningful.

Soils are important for understanding vegetation, wildlife, and the potential for agriculture. Saucier states:

> [clay plugs] typically support a dense swamp or forest vegetation.... Widespread cultivation and habitation is possible along the meander belts because of the well-developed natural levees that occur as low ridges along all abandoned channels and courses and as a thin veneer over most of the accretion topography. (Saucier 1974:27)

Ferguson and Gray state:

> Dense stands of bald cypress filled the swamp area, whereas hardwoods covered most of the loamy and sandy... and somewhat poorly drained to excessively drained soils. Many of the wetter soils... also were covered by hardwoods. The trees growing at the higher elevations were chiefly hickory, pecan, white oak, post oak, red oak, blackgum, and winged elm. In the swales and low places that were wet but not swampy, most of the trees were water tupelo, sweetgum, soft elm, green ash, hackberry, cottonwood, overcup oak, and willow oak... canebrakes... covered many of the flats between the swamps and the sloughs and bayous. (Ferguson and Gray 1971:48)

Major wildlife associated with this kind of vegetation and topography includes white-tailed deer, raccoon, muskrat, squirrel, rabbit, turkey, mallard duck, Canada goose, mussel, gar, catfish, drum, and turtle. This list can be reinforced by wildlife hunted at Wapanocca Lake (Parmalee 1966:143–45) and by ceramic animal effigies found at Nodena phase sites. There is little doubt that during the time of the Nodena phase a rich and varied environment was present. We now need to examine what advantage was gained by people from this environment and how such challenges as the lack of stone were met.

Immediately available in the environment was a variety of animal and plant products that provided wood, cane for arrow shafts, and food. A deer does not represent mere food. Lower jaws can be hafted as sickles, humeri (upper arm bones) make good scrapers, long bones can be split and made into awls and needles, astragali (heel bones) can be made into dice, antlers may become arrow points or stone-chipping tools, and even the very brain of a deer can be used to "cure" its skin. Other animals also provide turtle-shell containers, mussel-shell spoons, beaver-jaw chisels, gar-scale arrow points, bird long-bone beads, and many other goods. Local backswamp clay was used to make pottery with crushed mussel shell used as temper. Clay also was used to make clay beads and to plaster the walls of houses. Corn, beans, squash, gourds, sunflowers, and other plant foods were grown and used intensively in the diet. While agriculture provided a large percentage of the diet, considerable wild plant food was also collected.

Yet, this was a "stone age" culture, and stone for tools was unavailable locally, although necessary for survival. Stone is found to the west on Crowley's Ridge and the Ozark Highlands and to the north in the Ozark Highlands and Ste. François Mountains of southeastern Missouri. Gravel is also available across the Mississippi River, but we do not know very much about the nature and location or actual use of the outcroppings. The kinds of stone most desired were "chert" or flint, which can be easily chipped; basalt to be chipped, pecked, and ground into ax and adz blades; sandstone and ironstone to make into abraders; and fossil wood, slate, and limestone for pecking and grinding into a variety of artifacts. We know from surface collections that raw stone material was derived mainly from the north and west. Some may have been procured directly. Other trade items included Gulf Coast shells. We do not know what local surpluses were traded for these items, perhaps a foodstuff or some kinds of finished plant and/or wooden artifacts (for example, baskets or canoes). Crowley's Ridge is also known to contain an excellent red ocher, useful for paint. Salt and copper were available north of the Nodena phase and may have been directly obtained. Williams's Vacant Quarter Hypothesis indicates that no one was living there to exploit those resources.

A chiefdom may specialize in overproducing for trade a specific product characteristic of its province. A complex tribal organization can also collect specialties and redistribute them within the society without having to depend consistently on outside trade. This arrangement involves trade between kin within a single political organization. Basalt tools were produced in the Pemiscot Bayou cluster for trade. We do not know how much specialization was present in Nodena society. The main economy was still based on the household; each household produced for itself. There were warriors, but no permanent standing army. There probably were craft specialists and certainly, as today, a grain surplus was kept by the chief and his administrators.

It is difficult to assess the differential agricultural productivity between soils, particularly the three clusters of soil associations, for corn, beans, squash, sunflowers, and other aboriginal crops. Soil types may also cause differential access to different kinds of wildlife. The natural levees and point bars of recently deactivated river channels would have been the areas most easily cleared for agriculture. We suspect that the permanent sites were located along sloughs back from the river, somewhat similar to the settlement pattern followed during the past century. Upper Nodena is on a point bar characterized by Commerce silt loam, Convent fine sandy loam, and Morganfield fine sandy loam, three of the five most productive soils in the county. The others are Tiptonville and Dubbs silt loams found in the Pemiscot Bayou site cluster (Ferguson and Gray 1971:Table 2) in association with known sites. The two soil associations mentioned in connection with the Pemiscot Bayou and Wapanocca Lake clusters are not as fertile or extensive, particularly in the case of the latter. In that case, however, the sites are located on relatively higher ground and the capital, Bradley, might have been receiving tribute.

Nodena Society

Archaeological evidence of a complex political organization for Nodena society is abundant but difficult to analyze specifically. The population was mostly nucleated into fortified towns with pyramidal mounds and central plazas. The economic base was domesticated cereal grain, corn or maize, as has been proved by carbon isotope analyses (Boutton et al. 1984) and preserved specimens (Blake and Cutler 1979). This is intensive agriculture. Beans and squash were also grown. The town settlements were occupied year-round, but hunting and other exploitative camps were probably occupied seasonally away from the town sites.

Exotic artifacts do not necessarily prove a complex chiefdom system was present. They do help formulate such a conclusion when interpreted together with site complexity and obvious multiple hierarchy, population size and permanence, agricultural base, and inherent technological sophistication of the artifacts themselves. Craft specialization is indicated by the prevalence of really finely made tools, given the fact that they are made of stone, plus a monopoly on basalt inferred for the northern tier of sites. At the two investigated sites that date into the Protohistoric period, abundant primary basalt debitage together with the presence of specialized hammerstones, thought to have been used for the manufacture of heavy basalt-edged tools, plus the presence of "trough abraders" thought to have been used to finish basalt tools, all indicate direct exploitation of the basalt deposits from volcanic cones in the Ste. François Mountains and manufacture of the basalt tools. Nodena phase sites located to the south lack the primary basalt and the basalt manufacturing tools, but they do contain finished basalt products.

The lack of a Natchez-style society characterized by extreme symbolic control including the execution of retainers to accompany a deceased paramount chief is puzzling. However, such behavior in the Central Mississippi Valley may have been somewhat more characteristic of a chiefdom where authority needed to be openly symbolized. Cruelty is often characteristic of complex societies. But the Quapaw, who also reportedly lacked cruelty, have often been mistakenly interpreted as being primitive in nature. Cruelty and political complexity may not have to coexist.

Fortunately, we have some eyewitness European accounts of the Nodena phase, assuming our reconstruction of the de Soto expedition is correct (Hudson 1985). The expedition spent a month in the capital of Pacaha. One of the several smaller exploratory expeditions was sent to obtain salt and copper by barter, probably to the Campbell site. But it is the descriptions of life at Pacaha that interest us here.

Ranjel (1904:137) described an army of 7,000 men in about 200 canoes. Biedma (1922:25) counted 250 very large canoes. The warriors had bows and shields and behaved in an orderly manner by command. The canoes arrived at 3:00 P.M. every day to oppose the crossing of the Mississippi River by the expedition, so de Soto simply crossed before that time. The bureaucratic nature of this society of Pacaha even impressed the Spanish, who knew how to cope with it. It also must have impressed them with the potential of riches and the obstacles to overcome to acquire them. So de Soto marched to Casqui to raise an army to attack Pacaha. He immediately involved himself in the local power struggle to his advantage, another indication of the political complexity of northeastern Arkansas and of its ready comprehension to the Spanish.

"On Wednesday they came to the village of Pacaha, a village and lord of wide repute and highly thought of in that country" (Ranjel 1904:139). The "town" was stockaded with a ditch and towers. Pacaha (the chief of the Pacaha village and chiefdom) stated, and Casqui (the chief of the Casqui village and chiefdom) and de Soto agreed, that Pacaha outranked Casqui by virtue of kinship and courtly behavior. Pacaha presented de Soto with one of his wives, a sister, and another "woman of rank." This behavior may be interpreted as supporting a theory of partilineal descent for the Pacaha. Unfortunately, Biedma led a major side expedition to the northwest and we do not have a detailed impression of Pacaha by him. The Garcilaso and Fidalgo of Elvas accounts provide color but little more detail useful to us here.

Artifacts of the Nodena Phase

Pottery vessels buried in graves certainly constitute the best-known Nodena phase artifacts. There is little disagreement that this pottery is equal, if not superior in construction and design, to pottery found almost anywhere else in the United States; for instance, it is much sought after by collectors (Williams 1972). In addition, in northeastern Arkansas, only the Parkin phase can match the quantity of vessels interred with the dead. This is a mixed blessing, because treasure hunters have been attracted to these sites for a long period of time, and we do not have any records on probably 75 to 80 percent of the vessels found.

Two basic kinds of ware are present. Each has associated with it a variety of distinctive decorative motifs and techniques as well as a number of others that are shared. Mississippi Plain is normally not well-smoothed or polished on the exterior surface, is buff colored, and by definition the paste contains shell fragments 1 mm or greater in diameter. Bell Plain, named after the Bell Place, often has an obviously smoothed or polished exterior surface, is dark gray in color, and has small (less than 1 mm) shell and grog particles in the paste. The temper is so fine in some pottery that the vessels seem to be untempered. The grog often contains shell temper that can be seen only under a microscope. Mississippi Plain is the majority ware for both the Parkin and the Nodena phases. It is the basic domestic or kitchen pottery such as cooking jars and food bowls. Bell Plain is the primary ware in the Walls phase, while in the Nodena phase its percentage appears to be intermediate between the Walls and Parkin phase site counts. Nodena phase sites present a strange range of variability (Phillips 1970:Figure 448b). In Phillips's figure, Carson Lake and Notgrass are similar to each other while Upper Nodena is considerably different, with much more Bell Plain pottery. The reason for greater amounts of Bell Plain at some sites is that those sites have been extensively dug by treasure hunters who redistribute broken grave pottery onto the site surface. Surface collections at Upper Nodena confirmed this practice because those parts of the site dug more recently and for treasure contained much greater amounts of Bell Plain sherds.

The University of Arkansas and the Alabama State Museum of Natural History collected 910 vessels at Upper Nodena. Most of the pottery is plain; decorated and effigy vessels form a minority. Almost half the vessels are bottles and over a third are bowls. Jars constitute only about one-sixth to one-seventh of the total. Pottery in the Nodena phase can be separated into kitchen and ceremonial. We do not know much about kitchen pottery, and much of this discussion is based on ceremonial ware. Symmetry in shape and design is apparent. Some modifications are functional—for example, the ringed foot support to stabilize the vessel and internal beveling on bottle lips to facilitate the pouring of liquid from the bottles. A variety of bottle forms are present, including stirrup-necked (Figure 5-3a) and double (Figure 5-3b) forms, the "horned" bottle (Figure 5-3c), and others. Approximately one-

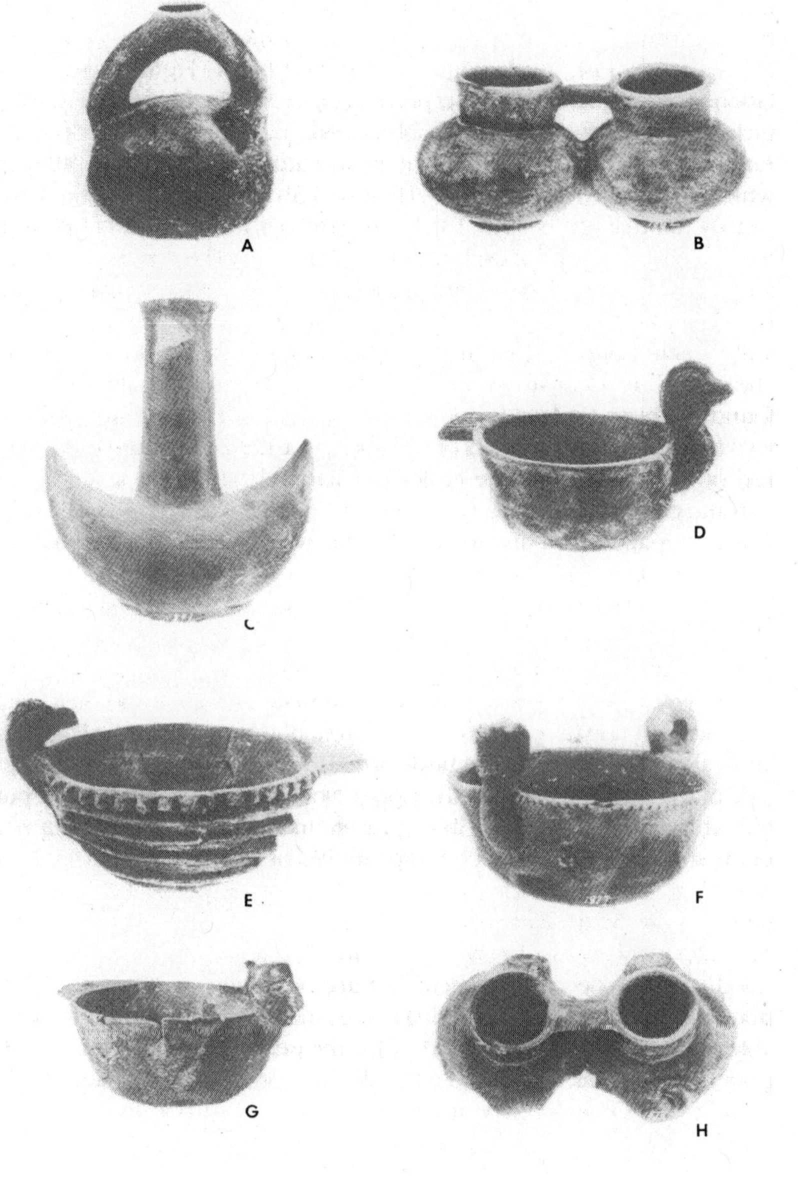

Figure 5-3. Nodena Phase Vessel Forms

tenth of the grave vessels are effigy forms. Bowls, usually with a head (sometimes made as a rattle with clay pellets inside the hollow head) and tail protruding from the lip on opposite sides, are by far the most common form of effigy vessels. Fish, duck (Figure 5-3d), and hawk (Figure 5-3e) representations are frequent while "serpents" (Figure 5-3f), humans (Figure 5-3g), turkey vultures, turkeys, dogs, rabbits, owls, bats, and passenger pigeons are rare. The effigy is usually upright, as in nature, with the exception of fish, which are molded on their sides (Figure 5-3h); occasionally effigy bowls depict the animal on its back. Fairly rare are jars in the form of opossums or frogs. Bottles are the rarest form of effigy pottery. The hooded (Figure 5-4a) and gourd (Figure 5-4b) effigy bottles and the kneeling human figure (Figure 5-4c) are present at Upper Nodena, as well as a fish with an engraved scale motif. Engraving is rare but characteristic of the Nodena phase. The "head pot" is a consistent but rare Nodena phase trait, although the type found at Upper Nodena is not a classic head pot (Figure 5-4d). Painted pottery is also rare and consists of red (ocher or hematite-based) slipped ware, red on buff, red and white (galena or kaolin-based), and a very rare polychrome pottery with black (graphite-based) as the third color. The two most common painted motifs are a swastika whorl (Figure 5-4e) and vertical stripes (Figure 5-4f).

The swastika whorl motif is also represented by incising on jars in the Nodena phase (Figure 5-4g). Another favorite incised motif consists of interrupted concentric circles (Figure 5-4h). Both a general punctated body motif (Figure 5-4h) and an incised line-filled triangle motif (Figure 5-5a) are used but nowhere are these as common as they are in the Parkin phase. Vertical lines incised on the body are also present but are more characteristic of the Parkin phase. On the Upper Nodena pottery, nodes were put over the whole vessel (Figure 5-5b) or, rarely, in a single row on the upper part of the shoulder (Figure 5-5c). Strap handles or arcaded handles are found on most jars. Notched, nicked, pinched, incised, or punctated lips are common (Figure 5-5d), particularly on Bell Plain vessels. Beaded or noded lips or nodes applied just below the lip are also a common decoration. Appliqued strips as decoration occur but are not frequent. The hand and long-bone motif in relief (Figure 5-5e) is present at Upper Nodena. Women probably made the pottery; at least 11 polishing pebbles and a mushroom-shaped pottery anvil were found with a female skeleton at Upper Nodena by the Alabama crew. Other pottery artifacts include but are not limited to toy pots (Figure 5-5f, g) and elbow smoking pipes (Figure 5-5h).

Numerous other artifacts are characteristic of this time period and of the Nodena phase (Morse and Morse 1983:271–80). Next to pottery vessels, probably the most characteristic artifact of the Nodena phase is the Nodena

Figure 5-4. Nodena Phase Vessel Forms

Figure 5-5. Nodena Phase Vessel Forms

point (Figure 5-6a). This arrow point is leaf-shaped and very characteristic of the Central Mississippi Valley. The stone chisel (Figure 5-6b) is a narrow stone tool ca. 10 cm long with a concave-convex bit. Most commonly, chisels are made of tan Crowley's Ridge chert. Chert, which is very abundant and extensively varicolored at these sites, was traded in from the Ozark High-

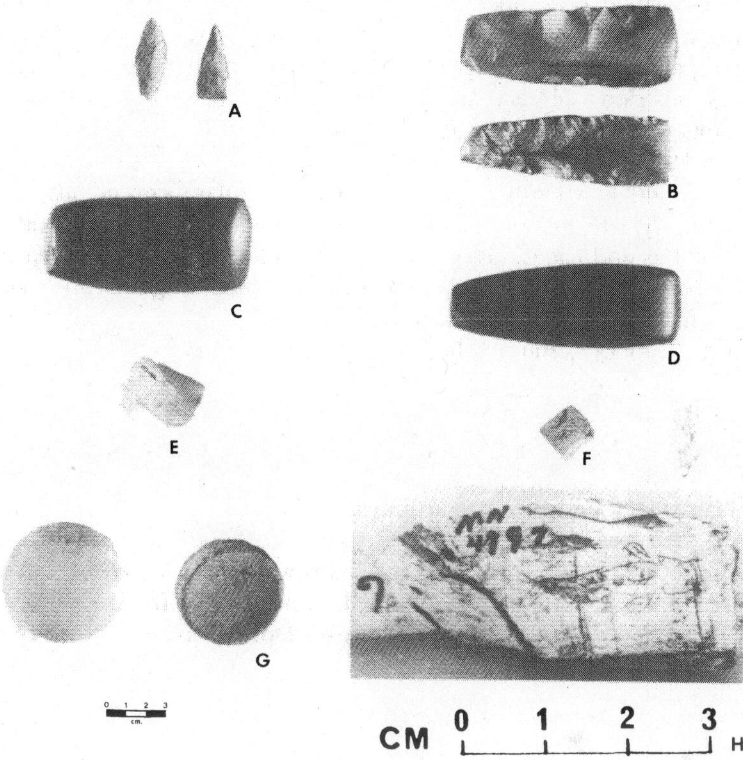

Figure 5-6. Nodena Phase Chipped- and Ground-stone Tools

lands and Crowley's Ridge to the west and northwest. The chisels were made mainly from Crowley's Ridge pebbles, which were chipped at the site. The chipping scars were ground smooth with ironstone abraders (made from the concretions that contained red ocher), often resulting in a red polished surface. At some sites, basalt adzes (Figure 5-6c) were also manufactured and the debitage is found strewn over the ground surface. These adzes and chisels, along with chert knives, are basic woodworking tools and undoubtedly indicate an industry oriented toward the manufacture and upkeep of dugout canoes. There may have been some craft specialization involved here that should be tested by making extensive collections of debitage at several sites.

Celts (Figure 5-6d) for felling and trimming trees are also present. A variety of abraders and hammerstones (used to break animal bone for marrow and to make stone tools), including some made from exhausted chert cores, are in our collections from the surface of Nodena phase sites. Knives and scrapers, particularly end scrapers (Figure 5-6e), for butchering and making bone and wooden tools, and arrow points for hunting have been found. Arrow points were made from gar scales, deer antler tines, and stone. Stone points include either the leaf-shaped Nodena point or a modified version that has a straight rather than rounded base. Triangular forms are also a frequent point style. Stone points are common on some sites and rare at others. Stone drill bits and flat, sandstone abrader slabs for shell bead manufacture are rare but present. Shell beads, ear plugs, and bird-wing fans are characteristic ornaments. Other artifacts include what may be deer-jaw sickles plus mussel-shell spoons, mussel-shell and stone hoes or spades, bone awls, deer- and elk-astragali dice (Figure 5-f), discoidal chunkey stones (Figure 5-6g), perforated and unperforated disks made from potsherds and possibly used in a guessing game or as counters or spinning tops, pottery cones to support cooking vessels, bone fishhooks, and stone palettes.

Dr. Hampson found a unique artifact at Middle Nodena in 1932 (Figure 5-6h). A design was incised on a small splintered bone fragment, part of a deer long bone broken to retrieve the marrow content. The design, which measures no more than 23 mm long and 9 mm wide, is the dorsal view of a catfish. Few lines are used, and it is an excellent example of primitive art.

Development and Aftermath of the Nodena Phase

The development of the Mississippian period in the Central Mississippi Valley is just becoming known in outline (Morse and Morse 1980, 1983). Evidently a primary evolution of Emergent Mississippian took place within and adjacent to the Cairo Lowland in the eighth century A.D. During the ninth and tenth centuries, Mississippian sites, characterized by Varney Red ceramics and Crescent Quarry cherts, were distributed throughout much of the valley to at least the mouth of the St. Francis River and into northwest Mississippi. Villages, some with mounds, of moderate sizes and small residential sites were characteristic. Corn was grown. The bow and arrow and the bow drill were adopted. Shell-tempered pottery with all its technological advantages (Morse and Morse 1983) was now universal. The Mill Creek hoe was prevalent. The discoidal and chunkey stone game had developed. Small rectangular houses were occupied within villages oriented with the solstices.

Within the eleventh century, ceramic changes included handles, long-necked bottles, plates, beakers, rim effigies on bowls, and a distinct funerary ware possibly connected with Black Drink ceremonialism. Council houses indicate the existence of a council-led or simple chiefdom system. In the twelfth century and during the next one or two hundred years, there developed distinct hierarchies in sites, ranging from large ceremonial centers to farmsteads, with exotic ceremonial artifacts (Southeastern Ceremonial Complex), larger houses, and very exotic ceramics. There was also a growing dependence upon corn, as indicated by carbon isotope analyses. During the fourteenth century, population nucleation into town sites was taking place. Whether due to strictly ecological or political (warfare) causes, villages were now well stockaded and contained structures for the entire supporting population. By this time the transition to a complex or paramount chief-led chiefdom was complete for much if not all of the Central Mississippi Valley.

An extraordinary thing happened some time before or just within the fifteenth century. Almost all of southeastern Missouri was abandoned to permanent habitation (Williams 1982). All of the Western Lowlands and much of the Braided Stream terraces of the Eastern Lowlands of northeastern Arkansas were similarly abandoned (Morse and Morse 1983) (see chap. 3 for a rare argument against this interpretation). Populations were nucleated at the mouth of the Little Red River, known as the Little Red River phase (de Soto's province of Palisema), the Greenbrier phase at the mouth of the White River (de Soto's Coligua), the Lower Kent phase on the St. Francis River (de Soto's Quiguate), the Parkin phase along the Central St. Francis River (de Soto's Casqui), the Walls and Belle Meade phases adjacent to the Mississippi River south of Memphis (de Soto's Quizquiz and Aquixo), and the Nodena phase along the Mississippi River above Memphis (de Soto's Pacaha). No other provinces, with the exception of Caluc, Calpista, and Quixila in northeastern Arkansas and southeastern Missouri, were mentioned in the de Soto accounts and no other archaeological complexes are known that date to this time period. Biedma's expedition to Caluc may have been to an abandoned province used by Protohistoric hunters based at town sites, possibly at Coligua to the south. Nodena points and end scrapers are found at apparent hunting sites dating to this period in the region indicated (Morse 1986b). Calpista may be merged by us with Palisema as the Little Red River phase. Quixila is an area not yet surveyed.

Nucleated populations meant that only certain locations with sufficient expanses of prime farming soils to support town populations could be occupied on a permanent basis. A development or expansion of prairies may have decreased possible choices, such as in the northern part of the Central Valley. Important to note is the near-perfect match between the de Soto accounts of 1541 and the archaeological reconstruction of population distribution. The unoccupied regions to the north and west of Pacaha are particularly noteworthy.

Evidently a severe population decline of the northeastern Arkansas inhabitants occurred after 1541. Phillips (Phillips et al. 1951:419) hypothesized a population decline of at least 80 percent. It probably was more on the order of 95 percent, inasmuch as probably no more than about 3,500 Quapaw survived to greet the French 132 years later (Morse 1986a). That at least some of the northeastern Arkansas population survived to become the Colonial Quapaw is indicated by several lines of evidence, but it is by no means certain (see chap. 11 for the traditional view of Quapaw migration from Ohio). The traditional view of Quapaw origin is based on (1) a realization that Siouan was spoken east and west of Ohio (Griffin 1943:310); (2) that there were no known historic tribes to place in southern Ohio (Kroeber 1953:89–91); and (3) that southern Ohio was the main region of the "Mound Builders" (Willey and Sabloff 1970). In addition, there was a need in early-nineteenth-century Arkansas to move the Quapaw west. Not only were they identified by the racist Territorial Governor Izard as wanderers but also as uncivilized and thus incapable of the feats of the Indians visited by the de Soto expedition (Bizzell 1982).

These nineteenth- and early-twentieth-century concepts are repeated even today (Baird 1980). The argument is based primarily on a folktale and partly on inaccurate translations and/or interpretations of French accounts (see chap. 11 for a different view). Linguistic evidence is that Pacaha or Capaha (Garcilaso 1951:449) was the home of Pacaha and that 132 years later Kappa or Capaha (Margry 1879:2:189) was the home of Paheka (Bizzell 1982). Site evidence is that the villages and possibly the houses were very similar. The house pattern at Menard (Ford 1961:154) is very similar to house patterns at Banks (Perino 1966:29) and to the reconstructed house at Chucalissa by Nash (1963:Figure 1). The "Marquette" journal and subsequent descriptions do not disprove the basic archaeological house pattern found at Menard as essentially Quapaw. The Quapaw built mounds (Gravier 1900:117) and fortified at least Kappa if not all four villages (Gravier 1900:116). The Quapaw only lived in four villages, which seem from the French accounts to have been similar in design to earlier Mississippian sites as reconstructed by archaeologists. Artifactual evidence is very strong and includes Nodena points, specific Mississippian ceramics, and other artifacts. Social organization appears to be similar (patrilineality) and unusual for most of the Southeast. The traditional view of social organization in northeastern Arkansas is matrilineality, but the de Soto accounts indicate the distinct possibility of partilineal descent at Pacaha. Some Nodena phase sites contain apparent seventeenth-century European trade goods, and when the Quapaw were first met by the French, they had already been acquiring trade goods from the Michigamea, an Illini group who had their summer hunting village in northeastern Arkansas. There really is no compelling reason to suppose that the Quapaw did not develop out of the Nodena phase or at least a very closely related phase or perhaps (less probably) a conglomeration of linguistically related phases. The de Soto accounts seem to indicate clearly that at least the easternmost of the northeastern Arkansas groups spoke the same language and, furthermore, that the chiefs of Casqui and Pacaha were kin. To the oft-repeated question "What happened to the Nodena phase?" we answer with considerable confidence, "They became the Quapaw." However, Swanton (1939:61) identified "Pacaha" as a Tunica word. For those people to have spoken Siouan we would have to suppose that one of de Soto's translators spoke Tunican. Furthermore, we do not yet have a definite seventeenth-century Quapaw assemblage to compare to the sixteenth–seventeenth-century Nodena phase. Our confidence may be considerable but it is by no means certain.

6

Health and Disease at Nodena: A Late Mississippian Community in Northeastern Arkansas

Mary Lucas Powell

According to Rodrigo Ranjel, one of the chroniclers of the de Soto expedition, the Spanish observed in the great chiefdoms of Casqui and Pacaha along the banks of the Mississippi River in its Central Valley "the best villages seen up to that time, better stockaded and fortified, and the people . . . of finer quality, excepting those of Cofitachequi" (Bourne 1904:2:140). The archaeological locality of Nodena, identified by Morse and Morse (1983:308) as one of Pacaha's larger settlements, is located adjacent to a relict meander channel of the Mississippi River, some 8 miles (12.8 km) northeast of Wilson, Arkansas (see chap. 5). The portion of the site known as Upper Nodena represents a 15.5-acre (6.2-ha) palisaded village, including three truncated pyramidal mounds flanking a plaza, with residential sectors and small cemeteries located around the outer periphery (see chap. 5). The area known as Middle Nodena lies 1.3 miles (2.1 km) distant on the same point bar swale. The similarity of archaeological and biological features from the two areas suggests that they represent different sectors of one extensive community. The span of occupation has been estimated as approximately one hundred years, beginning in the late fifteenth or early sixteenth century.

The inhabitants of Nodena followed the basic Mississippian subsistence regime, combining cultivated and gathered plant foods, game and aquatic animals. The warm climate, ample rainfall, and dynamic hydrologic systems of this section of the Central Mississippi Valley combine to create an environment extremely rich in both botanical and faunal resources. Morse and Morse (1983:289) note that Nodena and other Central Mississippi Valley sites tend to yield 12-row maize rather than "the more stress-resistant eastern 8-row" variety, indicating that "ideal conditions . . . must have been present for corn cultivation."

Mound C at Upper Nodena contained 316 burials, but the majority of interments were located in groups of 15 to 20 clustered around houses and probably represent family cemetery plots. This pattern has been reported at other villages elsewhere in the region (Morse and Morse 1983). Mortuary analysis of the Nodena burials has not yet been undertaken, and therefore the social dimensions of health and disease patterns in this community must await future attention.

Human Skeletal Remains from Nodena

Over the past three-quarters of a century, more than 1,700 burials have been excavated at Upper Nodena (Morse 1973a, 1973b). At the present time, the three largest collections of skeletal material from the site total 159 skeletal individuals: 32 (crania only) in the Hampson Museum, 72 in the University of Arkansas Museum at Fayetteville, and 55 in the Alabama State Museum of Natural History (curated at the Laboratory for Human Osteology), The University of Alabama, Tuscaloosa. Sixty-nine individuals (out of several hundred excavated) from Middle Nodena have been preserved in these collections: 17 (crania only) in the Hampson Museum, 13 in the University of Arkansas Museum, and 39 at The University of Alabama. The majority of these individuals were excavated in 1932 by the University of Arkansas Museum and the Alabama State Museum of Natural History.

Of the hundreds of burials excavated over a period of 50 years by Dr. James Hampson, an owner of Nodena plantation and a respected avocational archaeologist, only 49 remain in the Hampson Museum, established by the state of Arkansas to curate and exhibit his collections. This attrition resulted partly from Dr. Hampson's exceptional generosity toward research institutions that showed an interest in acquiring specimens of aboriginal populations from northeastern Arkansas. An article in the *Arkansas Gazette* (June 8, 1956) noted with pride that "A great number of these skeletons have found their way into archaeological exhibits at Harvard, Yale and various other universities."

Such fractional representation incorporates several significant biases, all of which represent logical consequences of the explicit research goals of the various excavators and the limitations of time, money, labor, and curation space under which they operated. The most obvious bias in the Nodena series is demographic. Subadults make up 4 percent of the series, 10 out of 228 individuals, instead of the 30 to 50 percent typical of demographically complete preindustrial samples (Weiss 1973). Records from the Arkansas and Alabama museum expeditions indicate that a larger number of subadults were in fact discovered, but no skeletal material was recovered from the majority.

A second clear bias in the series is toward selection of cranial rather than postcranial skeletal elements (Table 6.1). Of the 228 individuals, 72 percent are represented almost exclusively by crania and/or mandibles. Many of these also include a few upper cervical vertebrae, apparently fortuitous inclusions in the matrix that covered the skull base. Only 41 individuals, 18 percent of the series, are well represented by both cranial and postcranial remains.

The third bias proceeds from the differential recovery of skeletal material

Skeletal Elements	N	MIDDLE NODENA % represented by elements	N	UPPER NODENA % represented by elements
Cranium	55	79.7	111	69.8
Manible	57	82.6	114	71.1
Teeth	64	92.8	123	77.4
Thoracic Girdle	2	2.9	8	5.0
Cervical Vertebrae	17	24.6	33	20.8
Thoracic and/or Lumbar Vertebrae	5	7.2	8	5.0
Upper Limb	7	10.1	28	17.6
Pelvis	2	2.9	6	3.8
Lower Limb	14	20.3	40	25.0
Ribs	3	4.3	7	4.4
Hands	2	2.9	4	2.5
Feet	5	7.2	8	5.0

Table 6-1. Percentages of Nodena Individuals Represented by Cranial and Postcranial Remains

from the total number of burials discovered. Of the 326 burials recorded by the University of Arkansas Museum at Upper Nodena, skeletal materials were recovered from 26 of the first 50 burials (52 percent), 23 of the second 100 burials (23 percent), 17 of the third 100 (17 percent), and 6 of the fourth 100 (6 percent). The same trend is clear in the Alabama museum series: 40 of the 96 individuals represented (42 percent of the available sample) were taken from the first 200 burials (20 percent of the total excavated). No systematic variations in preservation within the areas excavated were observed within the series that would account for this pattern of differential representation.

The postexcavation history of the Alabama museum collection exacerbated these biases. The majority of skeletal material discovered was offered to Dr. Hampson to augment his own extensive collections, but no extant records indicate how much material he actually incorporated into the exhibits in his private museum (Dan F. Morse, personal communication). Hampson was eventually forced by lack of curation space to rebury the largest part of his collection, retaining only a few crania and pathological long bones that make up the collection at the Hampson Museum.

The magnitude of these identified biases precludes certain types of analyses that would be desirable: the construction of life tables and mortality curves, assessment of subadult health and mortality, and analysis of the impact of morbid conditions upon mortality and quality of life within the population. With respect to the latter category, inferences must be drawn with caution from the present material, given Dr. Hampson's encouragement to other excavators to retain obviously pathological specimens at the expense of less interesting nonpathological material.

Despite these biases, the series does provide a wealth of data on cranial, mandibular, and dental metrics; morphology; and pathology. Whereas only 94 individuals are represented by postcranial remains, 160 present crania for observation and 190 include dentitions in good condition. Cranial and mandibular metric and nonmetric data are available from between 80 and 120 adults, depending upon the features selected. Although analysis of dental morphology was not a part of the present study, the relatively unworn state of the majority of teeth available invites future research into this dimension of Nodena skeletal biology.

Demography

Age estimates for subadults are based upon observations of dental calcification and eruption, degree of epiphyseal closure, and long-bone lengths

(Bass, 1979; Krogman 1973; Ubelaker 1978). Few individuals presented post-cranial features of doumented reliability for assessment of adult age (for example, the public symphysis and sacroiliac articular surface), and estimates for the majority of adults were based, of necessity, upon observations of endocranial suture closure and macroscopic dental wear (Krogman 1973; Ubelaker 1978). Three broad adult-age categories were constructed: "18 to 30 years," "older than 30 years," and "unaged adult." Almost half of the series, 103 cases, was assigned to the first category (Table 6-2). Graphic comparison (Figure 6-1) of the demographic profile of the Nodena series with the profiles of two other prehistoric Native American skeletal series considered to be demographically complete shows clearly the serious underrepre-

		Infants 0-2	Children 3-11	Adolescents 12-17	Younger Adults 18-30	Older Adults +30	Unaged Adults	TOTALS
UPPER NODENA								
Females	N	00.0	00.0	00.0	48.0	27.0	09.0	84.0
	%	00.0	00.0	00.0	57.1	32.1	10.8	100.0
Males	N	00.0	00.0	00.0	28.0	22.0	09.0	59.0
	%	00.0	00.0	00.0	47.5	37.3	154.3	100.0
Indeterminate	N	02.0	03.0	00.0	02.0	05.0	04.0	16.0
	%	12.5	18.8	00.0	12.5	31.2	25.0	100.0
MIDDLE NODENA								
Females	N	00.0	00.0	00.0	14.0	14.0	0.0	28.0
	%	00.0	00.0	00.0	50.0	50.0	0.0	100.0
Males	N	00.0	00.0	00.0	7.0	13.0	1.0	21.0
	%	00.0	00.0	00.0	33.3	61.9	4.8	100.0
Indeterminate	N	00.0	05.0	00.0	4.0	07.0	4.0	20.0
	%	00.0	20.0	00.0	25.0	35.0	20.0	100.0
TOTALS								
Females	N	00.0	00.0	00.0	62.0	41.0	09.0	112.0
	%	00.0	00.0	00.0	55.4	36.6	08.0	100.0
Males	N	00.0	00.0	00.0	35.0	35 .0	10.0	80.0
	%	00.0	00.0	00.0	44.3	43.0	12.7	100.0
Indeterminate	N	00.0	00.0	00.0	06.0	12.0	08.0	36.0
	%	00.0	00.0	00.0	16.7	33.3	22.2	100.0
GRAND TOTALS	N	02.0	08.0	00.0	103.0	88.0	27.0	228.0
	%	00.9	03.5	00.0	45.2	38.6	11.7	100.0

Table 6-2. Demographic Profile of the Nodena Skeletal Series

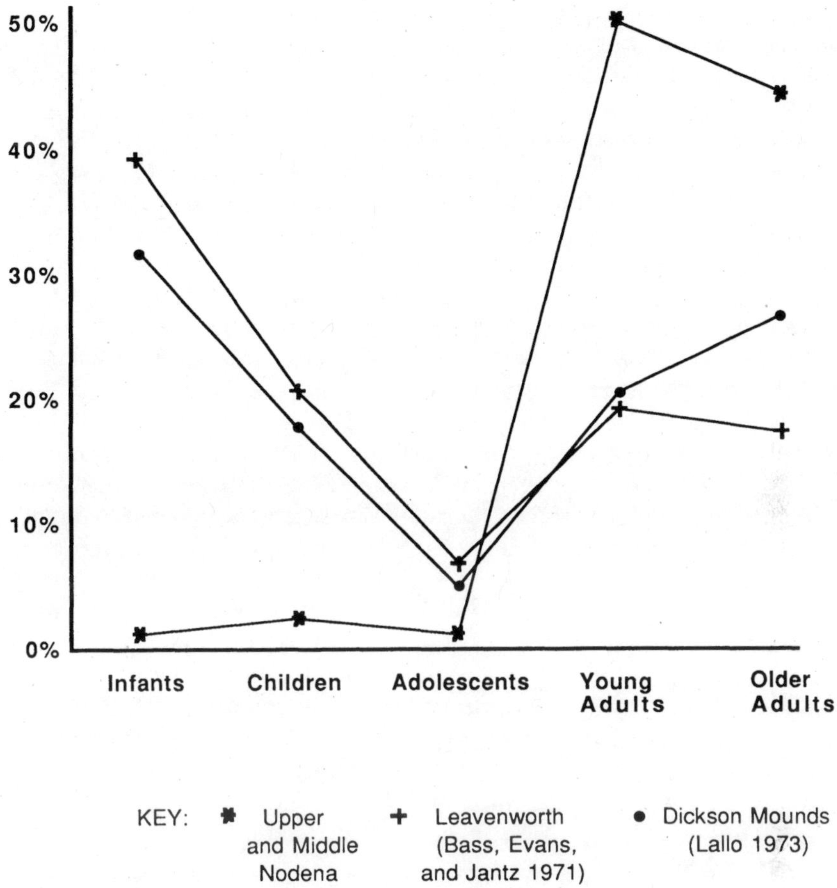

Figure 6-1. Comparison of Demographic Profiles of Three Amerindian Population Samples

sentation of subadults, particularly of infants and young children, and the corresponding overrepresentation of young adults.

Estimation of sex for adults relied upon accepted features of cranial, mandibular, and postcranial metrics and morphology, including pubic shape and sciatic notch width, maximum diameters of the humerus and femur heads and midshaft circumference and anterior-posterior diameter of the femur, robusticity of the mastoid region, brow, orbit and chin, and lateral diameter of the mandible head (Acsadi and Nemeskeri 1970; Bass 1979; Black 1978; Krogman 1973; Phenice 1967; Ubelaker 1978). Each set of observations was seriated by a system designed to quantify osseous expression of secondary sex characteristics within specific human populations (Acsadi and Nemeskeri 1970). For every adult, each feature was assigned a rating on a five-step scale (−2 ultrafeminine, −1 feminine, 0 indeterminate, +1 masculine, +2 ultramasculine). These scores were assigned weights indicative of their relative value for assessing sex; for example, pelvic features were weighted as "2," and the less reliable postcranial metrics were weighted as "1." The summed products of each rating multiplied by the appropriate weight were then divided by the summed weights to produce a total score for each case, for example, −1.5 for an adult unambiguously feminine. Individuals with scores ranging from −.5 to +.5 were classified as "indeterminate."

The initial assessment produced a sex ratio of 70 (males/females × 100) heavily biased in favor of females. Sex ratios ranging from 81 to 108 have been reported by the normal range is considered to lie between 90 and 105 (Teitelbaum 1972). A second assessment was undertaken via discriminant function analysis (using SPSS subprogram DISCRIMINANT), which employed cranial, mandibular, and dental metric data not considered in the initial estimates (Nie et al. 1975). In the first stage of this analysis, functions were derived from a core sample of adults that presented optimal combinations of reliable cranial and postcranial data. (See Power 1989b for a full description of this procedure.) The functions were then applied to all other previously sexed adults (excluding those included in the derivation). The initial sex estimates were confirmed in 90 percent of these cases, supporting the original impression of serious demographic bias with respect to sex. An additional check was provided by the sex estimates made by Dr. Georg Neumann of the 49 adults represented solely by crania in the Hampson Museum collection (Morse 1973b). My estimates matched Neumann's estimates in 94 percent of the cases.

During the Alabama museum excavations at Upper Nodena, Dr. Hampson stated that all but two of the 316 individuals in Mound C were adult males. Morse (1973b:71) suggested that such an unusual burial situation might represent a mass interment of warriors slain in conflict with a neighboring chiefdom. Demographic analysis of the 18 skeletal individuals available in the Alabama museum collection from the Mound C area did not support Hampson's field estimates. Four adult females and two children were identified, along with seven males and five adults of indeterminate sex.

Skeletal Metrics and Morphology

The brachycephalic character of the Nodena population was often exaggerated in females by artificial deformation. The majority of females was classified as hyperbrachycranic, according to conventional standards (Bass 1979:63), as were a few males, although most of the latter were simply brachycranic. One adult male (HM 887) from the Hampson Museum presents a striking contrast to this pattern. Only the cranial vault is present, and the precise location of its discovery at Middle Nodena is unknown. Georg Neumann had identified this individual as white rather than Amerindian when he examined the Hampson Museum crania in 1970 (Morse 1973b:60). Its very long, low, narrow vault contrasts sharply both visually and metrically with the typical broad, rounded, high-vaulted Nodena cranium. This specimen also presents the only unequivocal evidence of nonaccidental trauma in the series: an inscription on the vault, dated 1898, notes that a "black, drill-shaped arrow point" was found embedded just anterior to the right sphenotemporal suture. It is indeed unfortunate that this particular individual is so incompletely preserved, as he may represent an early European intruder into the Nodena homeland.

Although postcranial data were less abundant than cranial and dental data, enough were available to permit characterization of the people of Nodena as a moderately tall and robust population, consistent with early European observations (Bourne 1904:2:140). Cranial and postcranial sexual dimorphism were clearly evident but not exaggerated. Average female stature was estimated at 160 cm and average male stature at 171 cm, based upon maximum femur length applied to standard formulas (Trotter and Gleser 1952.

Skeletal Pathology

Four categories of skeletal pathology were examined in the Nodena series: trauma, degenerative osteoarthritis and vertebral osteophytosis, bony evidence of anemia (porotic hyperostosis and cribra orbitalia), and evidence of infectious disease. The first two reflect physical stresses experienced by the population, mediated by cultural factors such as the division of labor by age and sex. The third monitors a particular nutritional problem, iron-deficiency anemia, which has been widely reported in prehistoric agricultural populations. In the final category, analysis of the anatomical and demographic patterning of lesions of infectious etiology serves two purposes: identification of specific pathological syndromes, and assessment of the population's response to prevalent disease stress.

Few examples of traumatic injury were observed. Two individuals from Upper Nodena displayed bony callus at the sites of old fractures, the first on the distal right tibia of an adult male, and the second marking a greenstick fracture of an adult female's distal left radius and ulna. In the latter case, the fracture may have promoted the osteomyelitis noted in her left femur. Another adult from that portion of the site, represented only by a left femur presenting a normal appearance and a right femur markedly bowed laterally and shortened in overall length, may have experienced a severe leg fracture in childhood that healed with a significant deformity. The only cranial injury was the projectile point wound in the cranium of an adult male of notably non-Indian appearance, as described above.

Of the 43 adults from Upper Nodena with observable long bones, three males displayed osteoarthritic lipping of joint surfaces: the head of the humerus, the olecranon fossa of the ulna, and the patellar facet of the tibia. No similar joint involvement was noted for the 16 adults from Middle Nodena represented postcranially. Osteophyte formation on lumbar vertebrae was recorded for one adult from each locale, an unusually low prevalence undoubtedly influenced by the rare presence of postcervical vertebrae. Three of the 13 adults so represented displayed ovoid or vermiculate grooves known as Schmorl's nodes (Schmorl and Junghanns 1971) in thoracic or lumbar vertebral plates.

One hundred seventy-one crania were examined for cribra orbitalia and porotic hyperostosis, pathological conditions that represent response by bony hemopoietic tissues to anemia of nutritional or other origins (Steinbock 1976). One of the five children displayed the characteristic orbital lesions. Among adults, 14 percent of females and 3 percent of males were affected. This pattern of differing prevalence by sex parallels trends reported in other series (Cybulski 1977; El Najjar et al. 1976; Lallo 1973; Milner 1982; Powell 1983, 1988). The adult prevalence of 9 percent is considerably lower than those reported for southwestern Anasazi (El Najjar et al. 1976) and northwestern coast (Cybulski 1977) samples, yet it is higher than rates noted for earlier Mississippian samples from Dickson Mounds (Lallo 1973), Kane Mounds (Milner 1982), and Lubbub Creek and Moundville (Powell 1983, 1988). The contemporaneous Dallas focus of Toqua in eastern Tennessee displayed a higher prevalence rate: 32 percent in subadults aged 10 years or younger, and 18 percent in older individuals (Parham and Scott 1980), as did several Protohistoric samples from the Alabama River region (Hill 1981b). Anemia apparently became an increasingly common health problem for some populations in very Late Mississippian times, perhaps due to the impact of introduced Old World pathogens upon populations already stressed by heavy intestinal parasite loads and iron-low diets.

The true prevalence of this pathological condition at Nodena cannot be thoroughly assessed in the present collections because of the paucity of subadults and the previously noted possibility that pathological specimens may be overrepresented. The indicative lesions in adult crania should be interpreted as evidence of childhood experience with anemia rather than of recent experience in adulthood (Stuart-Macadam 1985). The presence of several adult cases in the Nodena series, however, does suggest that a higher subadult prevalence probably existed.

Of the 43 Upper Nodena individuals with postcranial remains, 37 (86.0 percent) revealed osseous response to inflammation, probably of infectious origin. Nine of the 16 comparable Middle Nodena individuals (56.3 percent) were similarly affected. In both series the tibia was most frequently involved, followed by the femur and the fibula (Table 6-3), with upper arm bone involvement considerably less common. Similar patterns of skeletal involvement have been reported from other southeastern prehistoric series (Lallo 1973; Larsen 1982; Milner 1981; Powell 1983, 1988). The majority of cases displayed localized patches of mild to moderate periostitis, well-healed at death, most typically affecting the anterior crests and lateral aspects of central and/or distal tibia shafts. Several tibiae exhibit marked shaft expansion of the form known as "sabre tibia."

Four young adults from Upper Nodena and one from Middle Nodena probably died from blood poisoning (septicemia), as suggested by osteomyelitis of the lower limb bones. In one case, exogenous pathogens may have been introduced into the bloodstream by a traumatic injury that fractured the young woman's left arm, only partially healed at her death (as noted above). No associated trauma was evident in the other cases, but in modern clinical samples 60 to 70 percent of cases of suppurative osteomyelitis result from endogenous staphylococcal infection of soft tissues (Robbins and Cotran 1980).

Focal lytic lesions are rare, observed only in crania. Nine adults from Upper Nodena (but none from Middle Nodena) display small circular or oval lesions in the outer table of the frontal or parietal bones, remodeled prior to death in all but one case. The five cases that included postcranial remains all exhibited mild to moderate periostitis of tibia shafts, of the form described above. The femurs in two of these cases display osteitis and/or osteomyelitis active around the time of death in addition to the milder, more extensive periostitis.

Two additional cases from Upper Nodena exhibit more severe focal lytic destruction of cranial bones. One young female was affected by a pathological condition causing progressive erosion of the nasal margins, septum, and palate. The smooth sclerotic borders of the lesions suggest a chronic or intermittently quiescent infection in which periods of destructive osteoclastic activity alternated with periods of osteoblastic remodeling. Invasion of the process into the maxillary sinuses and adjacent inferior orbit walls probably contributed to her early death (Morse 1973b:54). A second young fe-

Element	Stage 1	Stage 2	Stage 3	Percentage Affected Bone X	Percentage of all Bones Affected
Tibia					
Middle Nodena	3 (27.3%)	03 (27.3%)	5 (45.3%)	11/19 (57.9%)	11/23 (47.8%)
Upper Nodena	31 (56.4%)	20 (36.4%)	4 (7.3%)	55/61 (90.2%)	55/94 (58.5%)
Fibula					
Middle Nodena	1 (100%)	0	0	1/6 (16.7%)	1/23 (4.3%)
Upper Nodena	11 (57.9%)	8 (42.1%)	0	19/42 (45.2%)	19/94 (20.2%)
Femur					
Middle Nodena	3 (42.9%)	1 (14.3%)	3 (42.9%)	7/20 (35.0%)	7/23 (30.4%)
Upper Nodena	6 (40.0%)	3 (20.0%)	6 (40.0%)	15/71 (21.1%)	15/94 (16.0%)
Humerus					
Middle Nodena	0	1 (50.0%)	1 (50.0%)	2/9 (22.2%)	2/23 (8.7%)
Upper Nodena	0	2 (100%)	0	2/44 (4.5%)	2/94 (2.1%)
Radius					
Middle Nodena	0	0	1 (100%)	1/7 (14.3%)	1/23 (4.3%)
Upper Nodena	2 (66.7%)	1 (33.3%)	0	3/27 (11.1%)	3/94 (3.2%)
Ulna					
Middle Nodena	0	0	1 (100%)	1/6 (16.7%)	1/23 (4.3%)
Upper Nodena	0	0	0	0	0
Total Affected Elements				Percentages of All Observed Long Bones	
Middle Nodena	7 (30.4%)	5 (21.7%)	11(47.8%)		23 23/67 (34.3%)
Upper Nodena	50 (53.2%)	34 (36.2%)	10 (10.6%)		94 94/275 (34.2%)

Table 6-3. Prevalence of Periostitis in Long Bones

male displays numerous focal lytic posterior cranial vault lesions of the distinctive "stellate" configuration associated with the gummatous skin ulcers of treponemal infection (Figure 6-2) (Hackett 1976; Ortner and Putschar 1981; Steinbock 1976).

Speculation concerning the presence of "syphilis" in prehistoric south-

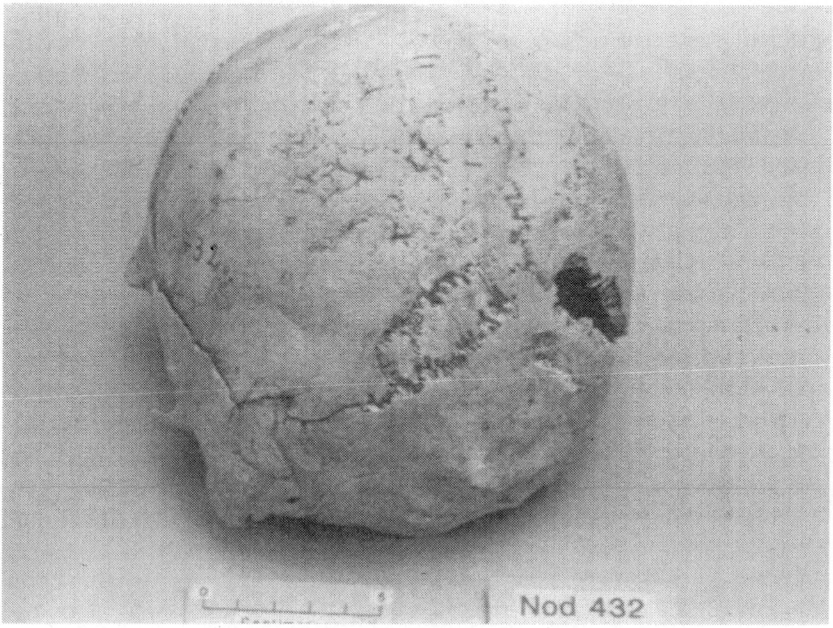

Figure 6-2. Healing Caries Sicca in Adult Female Cranium from Upper Nodena (courtesy of Alabama State Museum of Natural History)

eastern populations began with Joseph Jones's (1876) diagnosis of pathological bones in stone box burials from the Cumberland Plateau in central Tennessee. Similar diagnoses have subsequently been made in skeletal material from Alabama (Haltom and Shands 1938; Powell 1988), Florida (Bullen 1971), Georgia (Powell 1989a), Kentucky (Cassidy 1972), Louisiana (Robbins 1978), and Illinois (Cook 1976). Several Early Historic accounts of typical southeastern aboriginal illnesses include descriptions of syndromes that may represent endemic treponemal infections similar to modern yaws or nonvenereal syphilis (Swanton 1946). Lawson, writing in 1709 of the Indians of North and South Carolina, described "a sort of Rheumatism or Burning of the Limbs, which tortures them grievously, at which time their Legs are so hot, that they employ the young People continually to pour Water down them," and also, "another Distemper, which is, in some respects, like the Pox [venereal syphilis], but is attended with no gonorrhea. This not seldom bereaves them of their Nose" (Lawson 1709:223).

The first condition is reminiscent of the ostealgia (bone pain) of late secondary and tertiary yaws and endemic syphilis, while the second and third features of the aboriginal disease correspond to the absence of urethral discharge and the nasal-palatal destruction characteristic of those modern treponemal syndromes (Hackett 1951). Lawson also noted (1709:18) that the Santee of South Carolina made a clear distinction between a nonvenereal disease that produced skin ulcers and occasional bone destruction that had existed in their realm "for many ages" and the similar but venereally transmitted "Pox" recently introduced by Spanish soldiers and English traders.

The appearance in the Nodena skeletal series of a high prevalence of periostitis affecting primarily the shafts of the lower limb bones and a lower prevalence of small lytic lesions of the cranial vault (occasionally observed together in the same individual) suggests the presence of an endemic treponemal disease. The well-remodeled appearance of the majority of these lesions indicates that in most cases the infection had flourished and subsided some considerable time before death. The absence of any apparent mortality effect in cases not also affected by suppurative osteomyelitis suggests an advanced degree of mutual adaptation between the pathogen and host populations. The presence of two cases that display examples of "classic" treponemal lytic lesions of the nasal-palatal region and the cranial vault strengthens this diagnosis. The lack of subadult material in the Nodena skeletal series precludes the analysis of lesion distribution in that segment of the population, as was possible for Cook (1976) and Powell (1988, 1989a) in their paleoepidemiological investigations of prehistoric endemic treponematosis in Illinois, Alabama, and Georgia. Evidence for dental malformations characteristic of congenitally transmitted treponemal infection in Precolumbian New World populations is still under examination (Della C. Cook, personal communication). However, critical environmental and behavioral constraints upon endemic maintenance of nonvenereal treponematosis (Hackett 1951; Hudson 1958) were in all probability as well satisfied at Nodena as at Moundville in west-central Alabama, the Georgia coast, and in the Lower Illinois Valley.

Dental Pathology

The dental data from Nodena constitute perhaps the single most valuable body of information provided by the series because of the magnitude of the sample and the good quality of preservation. Observations focused upon six features: macroscopic occlusal molar wear, dental caries, antemortem tooth loss, calcified dental plaque, linear enamel hypoplasia, and odontometrics. Their analysis addressed questions of dietary composition, texture, and adequacy, with each issue considered through multiple lines of evidence. Two of the features, molar occlusal wear and odontometrics, also proved valuable in estimations of adult age and sex for demographic analysis, as noted above.

From Middle Nodena, 63 individuals provided 913 teeth for examination. Upper Nodena was represented by 156 dentitions with 1,856 teeth. From this original sample, 13 dentitions with 124 teeth were excluded from the final analysis because of problematical provenience, leaving 143 dentitions with 1,732 teeth.

Dental wear at Nodena was generally light, with only moderate dentine exposure even in older adults. Stone grinders for food preparation are not present in Nodena phase artifact assemblages (Morse and Morse 1983). It seems probable that wooden mortars were used instead for the preparation of plant foods, here as elsewhere in the Mississippian Southeast (Hudson 1976). The slow rate of enamel erosion offered protection against mechanical invasion of the pulp cavity by pathogenic oral bacteria. Yet it also contributed to carious destruction of crown enamel by preserving the complex surface morphology of posterior teeth that trapped food particles, thus promoting bacterial activity.

Table 6-4 presents comparative data on molar occlusal wear at Upper and Middle Nodena and six additional prehistoric southeastern population samples, quantified by the method developed by Scott (1979a). The markedly rapid rate of crown enamel destruction in the preagricultural Indian Knoll sample from Kentucky may be attributed to the use of stone grinders for food preparation (Scott 1979b). Shell Bluff and White Springs in Mississippi (Rose 1981) represent groups undergoing the transition from collected to cultivated subsistence, with a diet intermediate between Indian Knoll and the later populations. The remaining three samples are drawn from late prehistoric agricultural populations at the Campbell site in Missouri (Scott 1979b), Hardin Village in Kentucky (Scott 1979b), and the Alabama Mississippian community at Lubbub (Powell 1983). The Nodena series clearly resembles the later three more closely than the preceeding three samples. This pattern parallels biologically the archaeologically documented similarities in dietary composition and food preparation techniques among the later sites.

Carious lesions are widely prevalent, affecting more than three-quarters of the Nodena dentitions (Table 6-5). The lesions most frequently appear as small cavities located in pits and fissures of the occlusal surfaces of the posterior teeth (Tables 6-6 and 6-7). The Nodena sample displays an average of 3.5 carious lesions and 2.6 carious teeth per dentition, prevalence rates comparable to those reported for other aboriginal agriculturalists in the Southeast and elsewhere (Turner 1979). This pattern is considerably different from that reported for populations less reliant upon cultivated starchy foods, such as the Fourche Maline of Oklahoma (Powell and Rogers 1980). The constant introduction of coarse grit into the Fourche Maline diet from the use of stone grinders produced a rapid rate of crown enamel erosion. This elimination of surface irregularities that trapped food particles reduced the prev-

	M1			M2			M1/M2 Gradient		
<u>Maxilla</u>	N	$\overline{X}$	S.D.	N	$\overline{X}$	S.D.	N	$\overline{X}$	S.D.
Middle Nodena	54	17.9	4.6	42	13.6	3.3	40	4.1	2.6
Upper Nodena	125	17.3	14.3	123	13.5	3.3	111	4.0	2.8
22LO530 [3]	30	26.0	8.3	32	22.0	8.3	--	4.0	-
22IT537 [3]	13	16.6	7.6	17	22.1	6.7	--	4.5	-
Indian Knoll [4]	58	26.7	5.3	58	19.2	5.6	29	7.5	-
Cambell [4]	64	19.3	4.1	64	15.1	2.2	32	4.2	-
Hardin [4]	60	15.4	2.7	60	13.2	2.4	30	2.2	-
Lubbub [5]	25	15.1	2.2	24	13.0	1.5	17	2.3	-
<u>Mandible</u>									
Middle Nodena	57	16.4	3.5	55	13.9	3.0	49	2.7	2.5
Upper Nodena	102	16.5	3.6	101	13.9	3.4	86	2.9	2.0
22LO530	29	25.6	7.2	28	23.1	6.8	--	2.4	-
22IT537	11	28.4	5.9	11	24.0	5.5	--	4.4	-
Indian Knoll	48	27.4	4.3	48	20.1	5.1	24	7.3	-
Campbell	68	17.6	2.8	68	14.6	2.2	34	3.0	-
Hardin	38	17.4	3.8	38	15.1	2.4	19	2.3	-
Lubbub	12	18.0	5.7	18	15.0	4.6	9	3.5	3.7

[1] N teeth [2] N pairs [3] Rose 1981 [4] Scott 1979 [5] Powell 1983

Table 6-4. Molar Occlusal Wear (from Scott 1979a)

alence of pit and fissure carious lesions (Table 6-7) but increased the risk of interproximal lesion formation through promotion of gingival recession (Powell 1985a).

Antemortem tooth loss displayed the same high prevalence and intraoral distribution as carious lesions (Table 6-5). Four out of five teeth lost prematurely were molars. Only 37 percent of observed losses occurred in adults younger than 30 years. The two conditions, dental caries and antemortem tooth loss, should be viewed as two progressive stages of the same infectious process. This is not to claim that all antemortem loss results from carious attack, but the latter condition is responsible for the majority of cases of the former condition in modern clinical populations (Menaker 1980).

SAMPLE	% of Carious Individuals	$\bar{X}$ Lesions per Individual	$\bar{X}$ Lesions per Tooth	% of Carious Teeth	$\bar{X}$ Number Carious Teeth per Individual	$\bar{X}$ Number Teeth Carious or Lost Ante-Mortem
Middle Nodena	76.9% (40/52)	2.73 (142/52)	0.18 (142/783)	15.2% (119/783)	2.29 (119/52)	4.00 (208/52)
Upper Nodena	77.7% (80/103)	3.89 (401/103)	0.26 (401/1548)	19.8% (307/1548)	2.98 (307/103)	4.15 (427/103)
Lubbub[1]	50.0% (15/30)	2.17 (65/30)	0.14 (65/472)	12.5% (59/472)	1.97 (59/30)	3.10 (93/30)
Caddo[2]	90.6% (48/53)	7.60 (403/53)	0.43 (403/928)	25.3% (235/928)	4.43 (238/53)	5.89 (312/53)
Fourche Maline[2]	30.4% (21/69)	1.12 (77/69)	0.09 (77/874)	00.1% (68/874)	0.99 (68/69)	1.68 (116/69)

[1] Powell 1983 [2] Powell and Rogers 1980

Table 6-5. Prevalence of Dental Caries and Antemortem Tooth Loss in Four Prehistoric Amerindian Populations

The presence of moderate to heavy calcified dental plaque on many dental surfaces suggests a diet that would have promoted an alkaline oral environment (Hillson 1979). This feature, considered together with the caries rate and the light occlusal wear, lends support to the idea that dried maize prepared as hominy by soaking in highly alkaline lye water was a dietary staple at Nodena, as in other similar communities (Hudson 1976).

TOOTH TYPE	N TEETH	% OF TEETH IN SAMPLE	% OF TEETH IN COMPLETE ARCADE	% OF ALL CARIES ON THIS TOOTH TYPE	$\bar{X}$ NUMBER OF CARIES ON THIS TOOTH TYPE
INCISORS			25.0%		
Middle Nodena	121	15 5%		0 8% (1/142)	01 (1/121)
Upper Nodena	240	15 5%		0 0% (0/401)	00 (0/240)
Lubbub[1]	93	19 7%		1 5% (1/65)	01 (1/93)
Caddo[2]	196	21 1%		4 2% (17/403)	09 (17/196)
Fourche Maline[2]	156	17 8%		0 0% (0/77)	00 (0/156)
CANINES			12.5%		
Middle Nodena	101	12.9%		1 5% (2/401)	02 (2/101)
Upper Nodena	198	12 8%		1 5% (6/401)	03 (6/198)
Lubbub	81	17 2%		3 1% (2/65)	02 (2/81)
Caddo	126	13.6%		3 0% (12/403)	.10 (12/126)
Fourche Maline	114	13 0%		1 3% (1/77)	01 (1/114)
PREMOLARS			25.0%		
Middle Nodena	221	28 2%		07 0% (10/142)	05 (10/221)
Upper Nodena	449	29 0%		10 7% (43/401)	.10 (43/449)
Lubbub	116	24 6%		13 8% (9/65)	08 (9/116)
Caddo	272	29 3%		16 6% (67/403)	25 (67/272)
Fourche Maline	189	21 6%		10 4% (8/77)	.04 (8/189)
MOLARS			37.5%		
Middle Nodena	340	43 4%		90 7% (129/142)	38 (129/340)
Upper Nodena	661	12 7%		87.5% (351/401)	.53 (351/661)
Lubbub	182	38.6%		81 5% (53/65)	29 (53/182)
Caddo	334	36 0%		76 2% (307/403)	91 (307/334)
Fourche Maline	415	47 5%		88 3% (68/77)	14 (58/415)

[1] Powell 1983 [2] Powell and Rogers 1980

Table 6-6. Caries Prevalence by Tooth Type

LOCATION ON TOOTH	Middle Nodena	Upper Nodena	Lubbub[1]	Caddo[2]	Fourche Maline[2]	22LO530[3]	22IT537[3]
Pit and Fissure	72.5% (103/142)	74.1% (297/401)	44.6% (29/65)	69.0% (278/403)	54.5% (42/77)	45.0% -	35.0% -
Interproximal	21.7% (31/142)	12.7% (51/401)	32.3% (21/65)	15.9% (64/403)	41.6% (32/77)	30.0% -	45.0% -
Cervical	04.2% (6/142)	10.5% (42/401)	16.9% (11/65)	09.7% (39/403)	02.6% (2/77)	15.0% -	05.2% -
Smooth Surface	00.8% (1/142)	01.5% (6/401)	06.2% (4/65)	04.7% (19/403)	01.3% (1/77)	10.0% -	15.0% -
Root	00.8% (1/142)	01.2% (5/401)	00.0% (0/65)	00.7% (3/403)	00.0% (1/77)	00.0% -	00.0% -

[1] Powell 1983 [2] Powell and Rogers 1980 [3] Rose 1981

Table 6-7. Caries Prevalence by Loci

All permanent incisors and canines in the Nodena dentitions were examined macroscopically for the presence of linear enamel hypoplasia. Fewer than 15 percent of the cases displayed no significant enamel disruption. Following the procedure outlined by Goodman and co-workers (1980), the developmental age of the individual at the formation of each lesion was calculated. Only certain instances—those in which synchronous lesions were observed in two or more teeth within a dentition—were scored as evidence of systemic growth disruption. The majority of growth arrest episodes occurred during the third and fourth years of postnatal life (Figure 6-3), an age at which children in traditional agricultural populations frequently experience health problems due to increased nutritional requirements, weanling diarrhea, and increased exposure to intestinal parasites (Scrimshaw et al. 1968). This same chronological peak has been reported for prehistoric agriculturalists in Alabama (Powell 1983, 1988), Illinois (Goodman et al. 1980; Milner 1982) and Arkansas (Rose and Marks 1985).

Summary

The Nodena skeletal series provides a large body of data on the biological dimensions of a large Late Mississippian population located in one of the most beneficent regions of the Mississippian heartland. Although identified biases in sample integrity preclude some types of analysis (e.g., the construction of life tables for investigations of age- and sex-specific mortality), the available age and sex estimates will prove valuable in future mortuary analysis of the Nodena burials. The series also offers opportunities for the

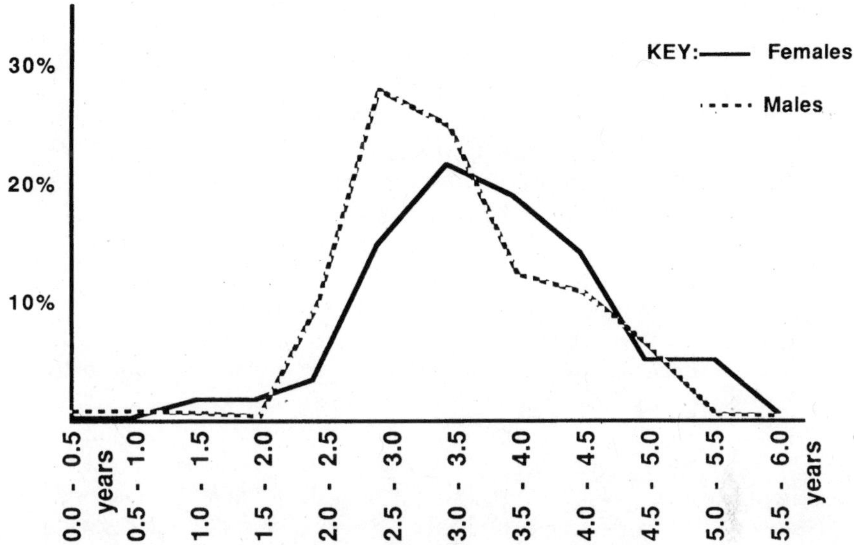

Figure 6-3. Percentage of Enamel Hypoplasia Lesions Formed at Given Postnatal Developmental Ages for Upper and Middle Nodena

investigation of other important features, including metric and morphological characteristics, dental wear and pathology related to diet and food preparation techniques, assessment of nonspecific systemic stress during the developmental years, and the identification of specific pathological conditions.

Cranial and mandibular metric and nonmetric data from between 80 and 120 Nodena adults have been employed for descriptive purposes in this study and elsewhere (Wolf 1976) for the analysis of intra- and extraregional biological distance. Cranial, mandibular, and dental metric data were utilized for the construction of sets of regional-specific discriminant functions designed to assist with sex estimation of incomplete skeletal individuals (Powell 1989b). The only individual in the series who presents a distinctly non-Indian appearance, an adult male represented by an incomplete cranium, also bears the only definite evidence of nonaccidental trauma: a projectile point wound in the right temple.

Quantification of rates of molar occlusal wear and of the prevalence of caries, antemortem tooth loss, and calcified dental plaque have delineated the biological impact upon the dentition of the Nodena diet, high in carbohydrates and soft in texture. Observation of age-associated patterns of linear enamel hypoplasia have revealed peak periods of childhood systemic stress that correspond to those reported for other Mississippian populations. The relatively unworn state of the majority of teeth invites future research into the dimensions of Nodena crown morphology, for the purposes of population characterization and analysis of biological distance.

It is unfortunate indeed that the prevalence rates observed in the series for trauma, degenerative osteoarthritis, vertebral osteophytosis, cribra orbitalia, porotic hyperstosis, and skeletal evidence of infection cannot be considered absolutely indicative of the true prevalence of these pathological conditions in the living Nodena population. However, the case of cranial trauma noted above suggests that at least some of the contacts between the people of Nodena and the non-Indian invaders of their territory were not amicable. As regards infectious disease experience, identification has been made of an endemic treponematosis, probably nonvenereal in nature, whose presence has been documented in other prehistoric populations of the southeastern United States. The modern endemic treponemal syndromes yaws and nonvenereal syphilis typically produce skeletal morbidity in a high proportion of adults but have relatively little impact upon mortality. The age-progressive pattern of skeletal involvement observed in the Nodena series, with no discernible mortality effect save in cases of pyogenic osteomyelitis, suggests that the aboriginal disease followed a similar course.

The people of Nodena and other towns along the Mississippi River were participants in a complex cultural tradition characterized by sophisticated community organization, the creation of well-designed and beautiful artifacts for utilitarian and ceremonial needs, a richly developed system of cosmology dazzlingly depicted in a variety of artistic expressions, a subsistence regimen that incorporated a wide variety of cultivated and collected foods, and a system of intraregional alliances between towns and villages that surprised the de Soto expedition with rapid and efficient mobilization of both personnel and material goods. Our scientific studies of the social and biological dimensions of their lives are far from being completed, and the information presented in this volume should inspire further investigations of their impressive civilization.

Acknowledgments

I would like to thank the Arkansas Archeological Survey for sponsoring my research on the people of Nodena. The directors of the University of Arkansas Museum, the Hampson Museum, and the Alabama State Museum of Natural History generously gave permission for presentation of the data included in this paper. The photograph in Figure 6-2 was taken by Polly Futato, Office of Archaeological Research, The University of Alabama. In my work on this project I have benefited in countless ways from the advice, encouragement, and hospitality of Dan and Phyllis Morse, Martha A. Rolingson, Ann M. Early, and Frank F. Schambach of the Arkansas Archeological Survey, Jerome C. Rose of the Department of Anthropology, University of Arkansas, and Kenneth R. Turner, Laboratory for Human Osteology, The University of Alabama. I have incorporated many of their suggestions into the final report; my errors, however, are all my own.

7

The Parkin Site and The Parkin Phase

Phyllis A. Morse

The Parkin site is located on the east bank of the St. Francis River in Cross County, Arkansas (Morse 1981:1), immediately below the confluence of the Tyronza River. The site measures 17 acres (6.9 ha) and is bordered by a wide ditch on the northern, eastern, and southern edges. A flat-topped temple mound 6.5 m high is on the central western edge, overlooking the St. Francis River. The site was built up with occupational debris and rises 3 m above the surrounding floodplain. An extension of the temple mound 1.5 m high is present on its southern end. Six smaller mounds, ranging from 0.5 to 1 m in height, were mapped in 1940 (Phillips et al. 1951:331). They were located near the temple mound. A plaza ceremonial area is located in the center of the village.

The village plan of the Parkin site is not unique. In fact, Phillips, Ford, and Griffin (1951:329) named a "St. Francis–type" village site classification after their initial survey and mapping of the Lower Mississippi Valley in 1940. St. Francis–type sites were planned villages, laid out as rectangles. They had an unusually deep concentration of refuse. Ditches surrounding these sites emphasized their height, and the concentration of refuse within the ditched area indicated the former presence of stockades. Houses at St. Francis–type sites were grouped around the site edges, leaving a low plaza in the center. Twenty-one sites were placed in this classification by the Lower Mississippi Survey team. Fourteen of these had dominant temple mounds on the western edge.

Parkin phase sites have very similar percentages of certain pottery types (Phillips 1970:931). These percentages differ enough from those of other nearby site clusters to be defined as "valid culture-historical units" (Phillips 1970:930). According to Phillips, a very large percentage of Parkin phase pottery is Mississippi Plain, *var. Neeley's Ferry*. The next most frequently found types are Barton Incised, *var. Togo* (formerly called Parkin Punctated), *Barton*, and *Kent*, and Old Town Red. Small percentages of Fortune Noded, Vernon Paul Applique, Avenue Polychrome, and Nodena Red and White are present. Bell Plain and Carson Red on Buff are very seldom found at Parkin phase sites (Phillips 1970:932). A later analysis of 15,926 sherds from surface collections at the Parkin site show this same pattern of pottery type percentages, as Phillips had previously defined (Morse 1981:Figure 7).

Complete vessels from the Parkin site are most frequently found as burial furniture. These vessels are often utilitarian cooking pots rather than the more elaborate ceremonial ware found in Kent, Nodena, or Walls phase burials. In fact, C. B. Moore used such adjectives as "lop-sided," "insufficiently fired," "trivial decoration," "inferior surface," "rude and scanty" (decoration), and "great monotony" (1910:260) to describe the pottery he found along the St. Francis River. However, Moore depicted well-made kneeling human figurines, cat-serpent effigies, fish-shaped bowls, red and white striped water bottles, and human-head bottles from Parkin phase sites. Jars are the most frequently found pottery shape, with food bowls and bottles less common. Five head pots are known to be from Parkin itself, as well as one dog effigy "teapot" (Morse 1981:27; Dr. James Cherry, personal communication).

Pottery disks, made from sherds, are a common Parkin phase artifact. Leaf-shaped Nodena points are the representative projectile point style. Other stone tools include Mill Creek chert stone hoes, small end scrapers, and chert hammerstones, bifaces, and debitage. Bone and antler awls, punches, needles, fishhooks, and scrapers are present at Parkin. Shell beads made from Gulf Coast conch shell include the Protohistoric square shell button variety (Putnam 1882:85).

Sites attributed to the Parkin phase have been excavated at least since 1879, when Edward Curtis collected archaeological specimens along the St. Francis River for the Peabody Museum of Harvard University (Putnam 1881:18). Curtis sent back over 900 complete pottery vessels from the Rose Mound, Fortune, Halcomb, and Stanley Mound [Parkin] sites. Curtis was very enthusiastic about the potential of this area, claiming that 10,000 (complete) pieces of pottery could be obtained in six months with five good hands (Morse 1978:9).

The next recorded individuals who made extensive collections from Parkin phase sites were "Captain" C. W. Riggs (Brose 1980; Riggs 1893) and Wilfred Hall (Griffin 1981). Riggs took his entire family with him to dig along the St. Francis. He brought back barrels of pottery to Cincinnati, including a human-head effigy from near the mouth of the Tyronza (Riggs 1893:10–11) that was probably from the Parkin site. Riggs kept a diary and wrote articles for newspapers. He even took photographs in the field for these articles. He published a price list and sold arrow points, pots, and "vases" to collectors and museums. The Riggs collection was exhibited at the World's Columbian Exposition in Chicago in 1893 and portions of it were sold to the Chicago Field Museum and Buffalo Museum of Science.

Captain Hall supplied various patrons with archaeological specimens. His most notable collection was acquired by the Putnam Museum in Davenport, Iowa (Morse 1986b). This collection was not kept separate by site and is therefore not as useful as it would otherwise have been to scholars.

The Smithsonian Institution, Bureau of Ethnology, Division of Mound Exploration was the next significant archaeological survey to visit the Parkin phase area (Thomas 1894). The first accurate mapping of sites was accomplished, and all specimens were cataloged and sent to Washington. The Parkin phase Miller Mounds and Taylors Shanty sites were tested by Edward Palmer for the Smithsonian during this project.

Clarence B. Moore took his private steamboat the *Gopher of Philadelphia* up the St. Francis in 1910 (Moore 1910). He brought his own crew with him and dug such Parkin phase sites as Rose Mound, Parkin, Neeley's Ferry, Vernon Paul (Jones and Borum) Twist, and Fortune. Moore believed that the Parkin site was already dug out and did not spend much time there. His major excavation was at the Rose Mound, where he spent 12 days and found 587 vessels in 207 burials (Moore 1910:278).

The University of Arkansas Museum, directed by Samuel Dellinger, sent excavators to northeastern Arkansas in the 1930s (Hoffman 1981). Large collections from the Neeley's Ferry and Hazel sites were sent back to Fayetteville. Pottery from these sites has been used as data for various theses (Green 1974; Zinke 1975).

Philip Phillips of Harvard University, James B. Griffin of the University of Michigan, and James Ford of Louisiana State University began their monumental survey of the Lower Mississippi Alluvial Valley in 1940 (Phillips et al. 1951). This survey is the beginning of modern archaeological research in the area. No longer was the emphasis on acquiring complete pottery specimens for museum exhibits. Instead, the research team gathered surface collections from 385 sites, mapped all of these sites, analyzed the ceramics, and conducted small test excavations. Their major research problem was the origin and development of Mississippian societies. The Rose Mound site was tested during this project (Phillips et al. 1951:284–90).

The Arkansas Archeological Society, an amateur organization, dug at the Parkin site under professional supervision in 1965 (Davis 1966). They found evidence of house construction, a possible palisaded wall, four burials, a plaza area relatively clean of artifacts, and a complex series of lenses in the apron area of the large mound. A University of Arkansas field school was held at Parkin the next summer (Klinger 1977). A large charred posthole was excavated in the top of the mound, as well as a series of test pits in the mound apron, an oval structure, and 10 burials.

University of Arkansas field schools were held at the Hazel site in 1964 and in 1969 (Davis 1973). Highway salvage excavations were conducted there by the university and later by the Arkansas Archeological Survey (Morse and Smith 1973). Morse excavated 76 features, including 62 burials, in an 11-day period at Hazel. Most features were worked out quickly in between the movements of the road-building machinery. Evidence of at least six rows of houses was recorded. Two large pits were also excavated.

The Parkin Archeological Project of 1978–79 was funded by a special grant from the Arkansas legislature as part of a study to determine whether the Parkin site should become a state park (Morse 1981). Research began with a literature and archival search to bring together all excavation notes, records, and reports that dealt with both Parkin and related sites. A contour map of the site was drawn by a professional engineering firm (Figure 7-1). Sites that were thought to be related to Parkin were visited, and surface collections were made. The sizes of these sites were measured, using soil probes to determine whether midden was intact rather than scattered from land leveling. The edges of the midden areas were flagged and a rolatape measuring wheel was then used to determine acreage.

Sites determined to be in the Parkin phase because of their diagnostic artifacts were then compared by drawing 1-km catchment circles around each site and using modern soil maps to determine how much land suited for hoe agriculture was present within this circle. An intensive surface survey was done in a 1-km circle around the Parkin site. One test excavation was carried out at a Parkin phase site during the project time period.

The Parkin project attempted to place the Parkin site within its temporal and topographical boundaries. One aspect of this research that has continued is the probability that the first European historical contact in the Central Mississippi Valley visited and described the Parkin site in 1541, when the de Soto expedition came through this region (Hudson 1985:3–12; Morse and Morse 1983:305–15).

The intensive surface survey conducted in a 1-km area around the Parkin site produced 129 separate site locations (Morse 1981:33–43). Many of these were isolated finds such as single projectile points. A deliberate attempt was made to discover small Parkin phase farmstead sites, which were thought to be present surrounding the large Parkin village. Only evidence of Middle Mississippian period farmsteads was discovered, as well as Late Archaic, Woodland, and undiagnostic lithic and fire-cracked rock loci. Not one

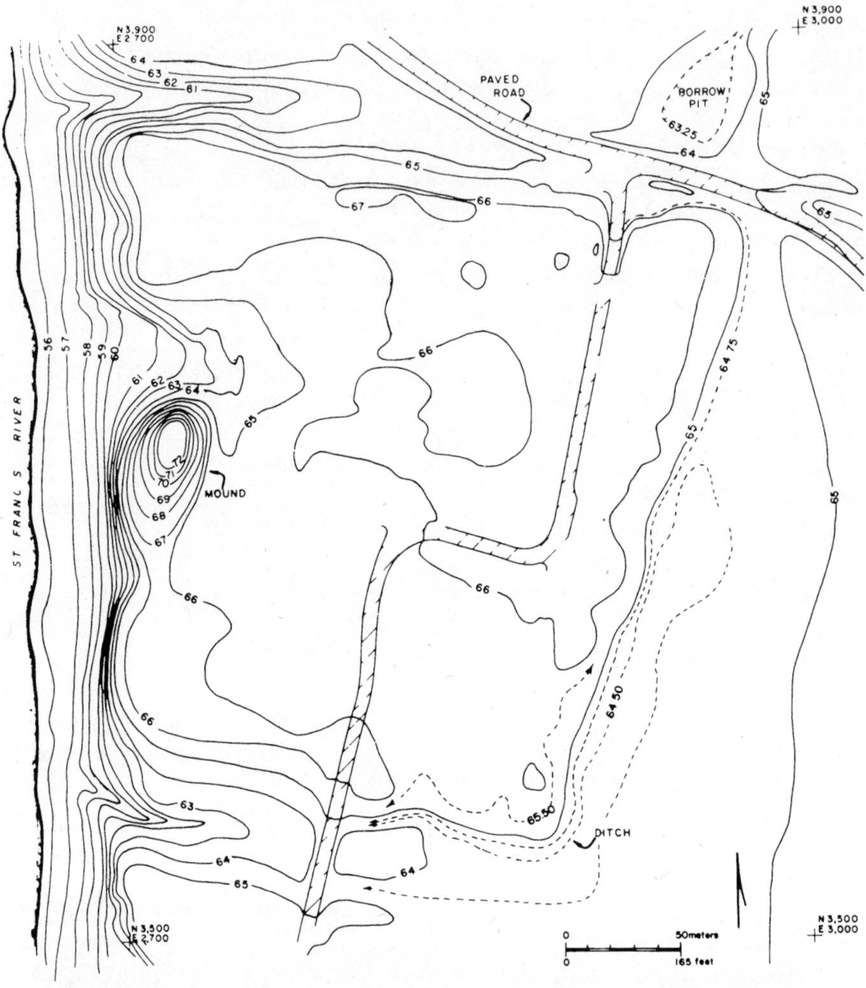

Figure 7-1. Contour Map of the Parkin Site

diagnostic sherd of Parkin Punctated or Barton Incised pottery was found on the farmsteads located. One isolated find of an end scraper across the St. Francis River on the western bank was the sole Parkin phase artifact found outside the village area.

This patterning of Protohistoric period artifacts that occurred mainly within the large Parkin village site was unusual and unexpected. When intensive areal surface surveys are conducted, usually small farmsteads are found that are associated with earlier larger Mississippian ceremonial centers (Smith 1978a). The Parkin project areal surface survey was expanded in a conscious effort to locate any Parkin phase farmsteads, and none were found. It then became necessary to explain their absence.

The Parkin phase "St. Francis–type" sites had obvious ditches around them when observed by Phillips, Ford, and Griffin. The ditch around Parkin itself was still in an excellent state of preservation in 1978, although farming had obscured many of those at other sites. It was concluded that the inhabitants of Parkin phase sites had to reside within fortified walls at night, venturing out during the day to tend to their nearby agricultural fields. People may have been concentrated or nucleated for protection. This concentration of population would increase in some areas and leave others uninhabited.

An examination of the other 20 sites classified as belonging to the Parkin phase gives us further information on the settlement system (Morse 1981:45–59). The Parkin site is by far the largest, measuring 6.9 ha (17.1 acres). Four other sites measure between 2.7 and 3.7 ha in area. These are Neeley's Ferry, Rose Mound, Glover, and Barton Ranch. Three of these sites are almost the same distance from Parkin, about 8 km (Figure 7-2). Barton Ranch is almost 20 km away along the Tyronza River.

Ten sites are included in the size category of 1.5 to 2.4 ha fortified village. Six other sites measured 0.3 to 0.8 ha in size. In general, Parkin phase sites are ca. 4 km apart and are spread along the St. Francis and Tyronza rivers. This pattern is similar to the linear stream pattern recorded for Formative villages in Mesoamerica (Flannery 1976:180).

The Parkin site is located in the most strategic location, below the junction of the two rivers. It could control the major means of transportation, travel by canoe. Major trails would also tend to follow the rivers. The paramount chief would probably live in this largest village. Other large villages probably were governed by lesser chiefs.

A study of the soil types available for agriculture in a 1-km circle around each Parkin phase site was done to see if any significant conclusions about conscious choice of site location could be reached. Parkin phase people were agricultural, growing corn, beans, and squash. Besides the ethnobotanical specimens, large Mill Creek hoes and hoe chips are commonly found on these sites. It was assumed that the Parkin phase villages would logically be located near expanses of land most conducive to hoe agriculture. One-km circles around each site were compared to each other, to measure their potential productivity (Morse 1981:82). A measurement of how many metric tons of corn could be grown per year was derived after measuring and comparing the soil types around each village. A positive correla-

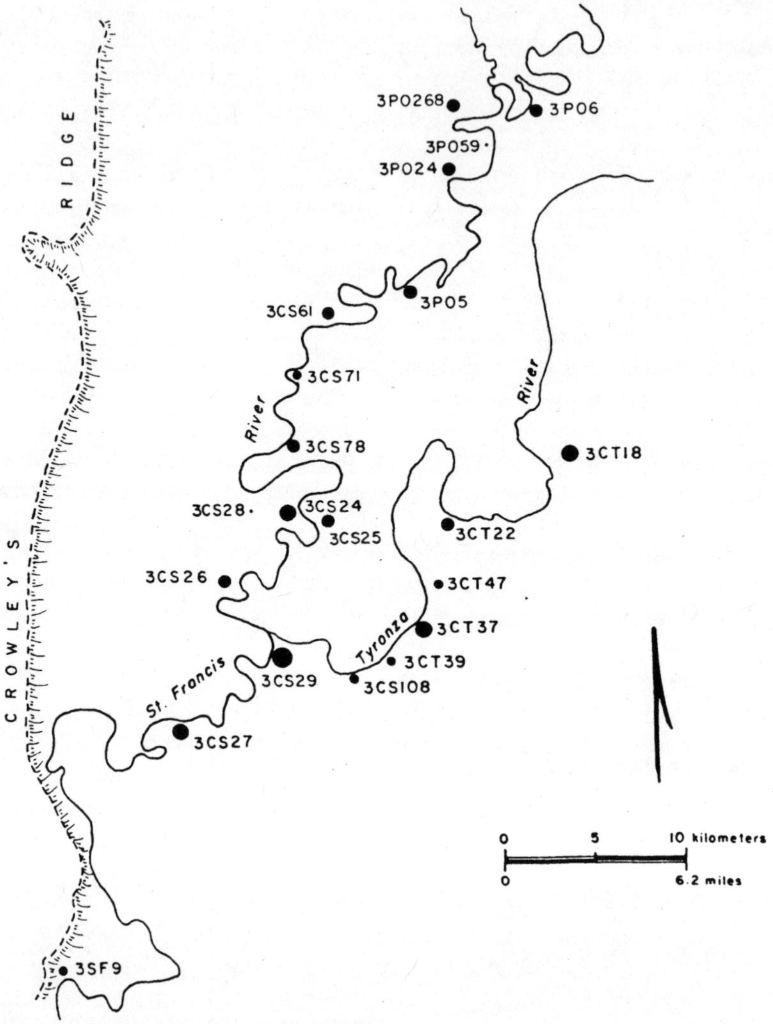

Figure 7-2. Distribution of Parkin Phase Sites

tion occurred between site size and the total arable hectares of land around each Parkin phase village, with the larger sites being near the most productive areas (Morse 1981:87).

The major exception was the Parkin site, which could not be self-sufficient in farming using this comparison. It is by far the largest village but is placed in a relatively agriculturally unfavorable area. The strategic location of the Parkin site must have been more important than mere agricultural productivity. In many chiefdom societies, tribute is regularly paid from smaller to larger villages. Ceremonies and feasts were conducted on a regular basis, using this tribute. The paramount chief would collect more of this tribute than the others. The residents of the Parkin site must have collected and stored goods for chiefly consumption, rather than raising all their food themselves.

Two artifacts of European origin that date to the sixteenth century were found at the Parkin site. A small brass bell was excavated that was associated with a child burial (Davis 1966:11). Three Barton Incised, *var. Kent*, and one Vernon Paul Applique, *var. Vernon Paul*, pottery vessels were also found associated with this burial. The type of bell has been named the Clarksdale bell (Brain 1975b:132). Clarksdale bells were made of sheet brass. The halves were joined by an obvious seam, which was folded and soldered. A wide brass loop was attached to the top. Later bells have a wire loop attachment. The Parkin bell has a shell bead as a replacement for the clapper. Both little bells and glass beads were part of the core gift kit carried by the early Spanish explorers in the Southeast (Brain 1975b:130).

A seven-layered glass chevron bead was the other sixteenth-century European object found at Parkin (Klinger 1977:Figure 19). The bead has a green center and layers of white, green, white, red, white, and blue that were cut into facets at the ends to show all seven colors. The central green color, the number of layers of glass, and the faceting are all typical of beads made before 1580 (Smith 1977:15). The Parkin bead was found during a University of Arkansas field school excavation on the surface of the Parkin site.

The only carbon-14 dates available for the Parkin phase are derived from charcoal collected in 1933 by the University of Arkansas Museum at the Hazel site (Zinke 1975:37, 56). Four dates from the upper levels of Trench 5 are: A.D. 1350 ± 90; A.D. 1450 ± 80; A.D. 1580 ± 70; and A.D. 1580 ± 180. Three dates from cluster burials are A.D. 1260 ± 70; A.D. 1260 ± 70; and A.D. 1480 ± 70. Using the date corrections recommended by Damon, Ferguson, Long, and Wallace (1974), the Trench 5 dates become A.D. 1330 ± 102; A.D. 1421 ± 94; A.D. 1540 ± 86; and A.D. 1540 ± 191 (Morse 1980:Table 7-3). The Damon et al. correction for cluster burial dates changes them to A.D. 1249 ± 83; A.D. 1450 ± 86 (Morse 1980:Table 7-3).

The Hazel site contains both Middle and Late Mississippian components (Morse and Smith 1973). Two of the cluster burial dates may actually be based on Middle Mississippian charcoal. However, it is obvious that the Parkin phase component at the Hazel site generally dates up to Late Period Mississippian, A.D. 1350–1650, as well as to the de Soto time line, 1539–43.

The Parkin Phase as the Province of Casqui

The accounts left by members of the de Soto expedition of 1539–43 are the only known descriptions of the powerful chiefdoms once present in the Central Mississippi Valley. Scholars have been trying to trace the exact route of the expedition for centuries, using information from these accounts and correlating it with known topography and archaeological sites. The most recent hypotheses are the "official" Swanton route (1939), the Phillips route (Phillips et al. 1951), the Brain, Toth, and Rodriguez-Buckingham route (1974), and the most recent proposals of Morse and Morse (1983) and Hudson (1985).

Barring any new discoveries in the Archives of the Indies, the four historical accounts available are those of Biedma, Ranjel, and Fidalgo of Elvas (in Bourne 1904), all of whom were actually with de Soto during the trip, and Garcilaso de la Vega, who recorded the reminiscences of several survivors (Garcilaso 1951). These accounts vary greatly in length and in detail. They also vary in style, from the crisp and businesslike Biedma to the colorful and adventurous Garcilaso. The number of pages concerning the province of Casqui and its political relationships with other provinces is meager, but reveals a great deal when combined with present anthropological knowledge of chiefdoms and how they operate.

The Swanton route was reconstructed for the 400th anniversary of the de Soto expedition and was the official congressional commission synthesis. Swanton covered the entire five years of the route and used local experts where they were available. Unfortunately, the archaeological knowledge available at the time was not precise and many sites chosen by Swanton as locations were not of the correct time period. Swanton chose Sunflower Landing, Mississippi, as the crossing point where de Soto entered Arkansas, down by Helena. He located Casqui west of Crowley's Ridge, and determined that the White River was the River of the Casqui.

Philip Phillips revised the Mississippi and Arkansas portions of the de Soto route, proposing a crossing point near Commerce Landing, Mississippi, which was considerably upstream (Phillips et al. 1951:347–91). Phillips used the revised dating and expanded archaeological knowledge of the Lower Mississippi Valley Survey when proposing his alternative routes. Phillips located the province and site of Casqui at or near Parkin.

Brain, Toth, and Rodriguez-Buckingham (1974), in a new interpretation of the de Soto route, chose the area of Friars Point, Mississippi, as the critical crossing area. This was ca. 25 miles (40 km) north of Sunflower Landing. They then placed Casqui near the Starkley site on the Lower St. Francis River and identified the Belle Meade site in the very southern end of Crittenden County as Pacaha.

Brain, Toth, and Rodriguez-Buckingham applied an "ethnohistoric archaeology" approach to the route of de Soto, using evidence from many disciplines including history, prehistory, and ethnology. They evaluated the historic records in light of new ethnographic and archaeological data, emphasizing the increase in archaeological knowledge since the Swanton and Phillips, Ford, and Griffin interpretations of the route.

Their new archaeological information consisted mainly of the refined classification of pottery that Phillips published in 1970, with types and varieties that were also regional culture-historical time markers. Those refined varieties of pottery that defined the de Soto expedition time period of the mid-sixteenth century consist of 20 diagnostic types, listed in their Table 6-1 and pictures in their Figure 6-2. Sites that were occupied in 1541 by definition contained several of these pottery types. Sites without these pottery types were excluded from consideration as possibly being in existence during the de Soto entrada (Brain et al. 1974:235).

While trying to reconstruct the route of de Soto in western Mississippi and northeastern Arkansas, Brian et al. made the following statements, which dismiss the possibility of either the Parkin or Nodena phases being seen by de Soto:

> The Commerce Landing hypothesis receives a severe setback when the route finally winds its way to Casqui. As discussed as a possible Quiguate candidate for the Memphis hypothesis, the Parkin/Rose Mound/Neeley's Ferry group *simply does not seem to have had a substantial occupation as late as 1541. Diagnostic ceramic markers are rare and the sites probably represent a slightly earlier settlement of the St. Francis Basin.* . . . These sites *lack the appropriate ceramic markers* and were probably settled at the same time as the main occupation at the Parkin group . . . about a century before de Soto. (Brain et al. 1974:247; emphasis added)

Thus, Brain et al. first list the 20 diagnostic pottery types that were indicative of the de Soto time line and then simply make the claim that these types and varieties were not present at Parkin and Nodena sites.

However, an examination of Phillips's 1970 Lower Yazoo Report in which these pottery types and varieties are defined—the very data they used to dismiss phases as too early—totally disproves the above statements. Phillips (1970) refined his types and varieties based on the earlier Phillips, Ford, and Griffin collections. Phillips pictured a typical vessel for each of these varieties, and these illustrations were reproduced in Brain et al. as their Figure 6-2. A search was made through the literature and through copies of Phillips's photograph files to ascertain exactly where the pottery used to illustrate these types, the archetypes, had been found. Of the 17 vessels that were pictured as the archetype for specific varieties, four are from the Parkin phase sites of Rose Mound, Vernon Paul, and Cummings (Table 7-1). Five others are from the Nodena phase sites of Upper Nodena, Wapanocca, Pecan Point, and Bradley. The very vessel used to illustrate the cover of their report, a kneeling human male effigy, is from the Parkin phase site of Vernon Paul, formerly known as the Jones and Borum Place (Moore 1910:Plate xx). The majority of the names of the diagnostic 1541 time line pottery varieties were even derived from Nodena, Parkin, and Walls phase sites.

It is obvious from this evidence alone that both the Nodena and Parkin phases were in existence in 1541, using Brain, Toth, and Rodriguez-Buckingham's own criteria. All their other arguments, supporting a de Soto route beginning with a crossing at Friars Point, are based on the nonexistence of the Parkin and Nodena phases in 1541. Once this mistaken interpretation has been corrected, it is time to reconsider the role of the Parkin site and phase in retracing the route of de Soto.

Morse (1981:61–72) concluded that the Parkin site was the capital of Casqui after studying the sizes and locations of other Parkins phase sites, the artifacts and their current dating, and the relationship of the Parkin phase sites to those known for the Kent, Nodena, and Walls phases. Further research while compiling a general summary of the whole prehistoric sequence in the Central Mississippi Valley (Morse and Morse 1983) has only reinforced this conclusion. In particular, our current knowledge of the patterning of Protohistoric sites dating to the de Soto time line as only existing in certain restricted areas of prime agricultural soils correlates with the historical accounts very neatly.

Charles Hudson has been retracing the entire de Soto route recently in cooperation with archaeologists throughout the Southeast (Hudson 1985). He briefly summarized his ideas on the Arkansas portion, choosing the area near Lake Cormorant, Mississippi, as the crossing point. This choice of where de Soto crossed the Mississippi River is obviously crucial to any later discussion of the rest of the route. The area where the Spaniards stopped and built their barges was near three villages of the Quizquiz, 1 league apart from each other. The Quizquiz were tributary to the "lord of Pacaha" (Biedma 1922:25). It took almost a month to build the four barges, which had to carry 60 men and five or six horses at a time. Every day during this construction the Spaniards were harassed by up to 250 canoes of warriors from the province of Pacaha (Ranjel 1904:138).

The Mississippi River was crossed June 18, 1541, early in the morning. No raiders molested the Spaniards. The barges were then brought 1.5 leagues upstream to a village of the Aquixo, where the metal was retrieved. Several towns of the Aquixo were seen, but the inhabitants had fled. A few captives offered to lead de Soto to the land of the Casqui

BRAIN	TYPE	PHILLIPS 1970	SOURCE	TYPE SITE	PHASE
a	Fortune*	p. 83	Brown 1926 Figure 269	Walls	Walls
b	Harris*	p. 151	Phillips Negative 3484	Upper Nodena	Nodena
c	Parkin*	p. 151	Phillips et al. 1951 Figure 94a	Rose	Parkin
d	Ranch*		Phillips et al. 1951 Figure 99g	Beck	Walls
e	Vernon Paul*	p. 167	Griffin 1952b Figure 124h	Vernon Paul	Parkin
f	Kent	p. 46	Phillips Negative 3357	Wapanocca	Nodena
g	Rhodes*	p. 157	Moore 1911 Figure 38	Rhodes	Walls
h	Owens	p. 149	Phillips Negative 3022	Beck	Walls
i	Nodena*	p. 141	Moore 1911 Plate xxxv	Pecan Point	Nodena
j	Avenue	p. 41	Phillips Negative 2040	Pecan Point	Nodena
k	Walls*	p. 169	Moore 1911 Figure 50	Bradley	Nodena
l	Carson*	p. 62	Moore 1910 Plate xiv	Cummings	Parkin
m	Horn Lake*	p. 157	Phillips et al. 1951 Figure 98b	Beck	Walls
n	Bell*	p. 58	Moore 1911 Figure 30	Rhodes	Walls
o	Blanchard	p. 105	?	?	?
p	Chickasawba*	p. 135	Brown 1926 Figure 300	Walls	Walls
q	Beaverdam	p. 146	Moore 1910 Plate xx	Vernon Paul	Parkin

*Name based on Nodena, Parkin, or Walls phase sites.

Table 7-1. Diagnostic Types Used by Brain et al. 1974

Instead of choosing to go directly to Pacaha, whose warriors had been harassing them, the Spaniards may have deliberately chosen to go first and visit the Casqui, the enemies of Pacaha. To reach Casqui, they passed through "the worst tract for swamps and water that they had found in all Florida" (Ranjel 1904:138).

The province of Casqui was reached when the Spaniards came to the "summit of some high hills, where they beheld a town of 400 houses, situated beside a river larger than the Guadalquivir at Cordova" (Garcilaso 1951:430). They were cordially received and fed by the Casqui. The route to the main village of the Casqui was described as 7 leagues upriver. The land was described as high and more level. "The greater part of the way lying through fields thickly set with great towns, two or three of them to be seen from one" (Fidalgo of Elvas 1904:118).

The main town of the Casqui contained the dwelling of their chief. This house was on one side of the town on a high mound. Ten to 12 other large structures for his wives and servants were nearby. The Spanish camped in a nearby orchard and received presents of food and clothing.

It is very unusual for all four historical accounts to describe the same incident, so the following event has unusual significance. There had been a drought, so the Casqui asked de Soto to help make it rain. The Spaniards built a great cross and placed it on top of a mound overlooking the River of the Casqui. Fifteen to twenty thousand individuals witnessed the placing of the cross from the opposite side of the river. The Spanish priests conducted a suitable ceremony. It fortunately did rain the next day. A cane fence was then placed around the cross by the Indians.

The Casqui joined with de Soto to invade the province of Pacaha. In fact, the enmity between the groups was of long duration. "For many centuries Casquin . . . had waged war on the lords of Pacaha . . . since these lords were more powerful in both vassals and lands and were still pushing Casquin into a corner, almost to the point of surrender" (Garcilaso 1951:435). Casqui amassed 5,000 warriors and 3,000 bearers to invade Pacaha.

They followed along the River of the Casqui and spent the night in a Casqui village. They then had to cross another swamp, where a bridge was built for the Spaniards. The Pacaha did not expect the invasion and fled to an island retreat. Then the Casqui sacked the town, destroying a temple that held the bones of the ancestors of the Pacaha chief. Heads of Casqui were displayed on points of lances at the temple doors. Many Casqui captives were released.

The Casqui displayed cowardice while confronting the Pacaha at their island fortification. The Pacaha chief reportedly said: "Remember well what you are doing and what you already have done, for when these strangers have departed, then we shall see what manner of men you are in battle" (Garcilaso 1951:442). De Soto then made peace with the chief of the Pacaha and stayed with them for almost a month. Casqui returned and made peace with both. A banquet was held, and there was a disagreement between the two chiefs as to who should sit at the right hand of de Soto. Pacaha claimed precedence: "You know I am a greater lord than you, and of more honorable parents and grandparents" (Ranjel 1904:144). Casqui agreed that Pacaha's forebears were greater than his but that his greater age should give him precedence. De Soto chose Pacaha to sit on his right. He also accepted the gift of a daughter of Casqui and of a wife and a sister of Pacaha. De Soto sent several expeditions out to explore the country. Finding no gold, he then returned briefly to Casqui and went further downstream, to the province of Quiguate.

In our most recent reconstructions of the de Soto route, the area of Walls, Mississippi (north of Commerce Landing) (Charles Hudson and David H. Dye, personal communication), is proposed as the most logical staging area where the barges were built. Three Protohistoric sites, Irby, Lake Cormorant, and Norfolk, date to the proper time period in this location. These sites are geographically located according to the de Soto accounts. Across from Walls, Mississippi, are such sites as Belle Meade and Beck, which were located on a large oxbow lake near the Mississippi. These could be where the de Soto barges were pulled and taken apart, in the province of Aquixo. The archaeological phase is known as "Walls West" or Belle Meade.

The "worst swamp" is placed in the backwater area between the meander belts and the levees of the Mississippi and St. Francis rivers. The high hills that greeted the Spaniards emerging out of the swamp could be Crowley's Ridge, an erosional remnant that bisects northeastern Arkansas from the Missouri border to Helena, Arkansas. The Parkin phase Big Eddy site is located right on the edge of Crowley's Ridge and beside the St. Francis River (Morse 1981:Figure 24). Big Eddy dates to this Protohistoric period, with two diagnostic head pots known to be from this site. The River of the Casqui at this point is described to be similar to the Guadalquivir River at Cordova. Jeffrey Mitchem of the University of Florida measured the length of a Roman bridge across the Guadalquivir River at Cordova as ca. 750 feet (227 m). Right at the eddy of the St. Francis, this comparison is not too far off. The River of the Casqui was also described as a tributary of the great River of Pacaha, the Mississippi, and the St. Francis River is such a tributary.

The Spaniards followed the River of the Casqui up to their major village. Sites such as the Rose Mound could be representative of the large towns seen along the way. In fact, the Parkin phase settlement system, with larger villages spaced ca. 4 km apart and smaller ones in between, is described by the de Soto accounts. The site of Parkin certainly fits neatly with the description of the capital, with a high mound on one edge of the village, beside the river. The house of the Casqui chief could have been placed on the summit of this mound and the residences of his wives and retainers on the southern mound apron or extension. This mound could also have been the site of the raising of the great cross, the first Christian religious ceremony held in Arkansas. The Parkin mound is placed right on the edge of the St. Francis River, where the thousands of onlookers could see the ceremony while standing on the opposite bank.

The joined armies of de Soto and Casqui again followed the River of the Casqui towards Pacaha. The Tyronza River, which joins the St. Francis half a mile above the Parkin site, could be considered as an extension of the River of the Casqui and leads back east toward the Mississippi River, the territory of Pacaha. Several villages of the Parkin phase are located along this river, and a site such as Barton Ranch could have been the frontier village occupied by the armies for one night. Barton Ranch is located on the edge of another swampy region, which could serve as the geographical boundary between two warring groups.

The enmity between Casqui and Pacaha had existed "for many centuries." Pacaha had already taken over the provinces of Aquixo and Quizquiz, located further south along the Great River. The Casqui were even afraid to leave their own borders.

The Parkin phase settlement system, composed of fortified large and small villages and a lack of farmsteads, is probably a reaction to this constant warfare. It is more efficient to place small family farmsteads in the fields to better work and protect the crops, as was the case during Middle Mississippian times (Price 1978; Smith 1978b). However, this dispersed farmstead pattern would have to be abandoned when warfare and raiding were a frequent occurrence. Having all inhabitants within fortified village walls at night creates the densely occupied villages typical of the "St. Francis–type" sites noted by Phillips, Ford, and Griffin (1951).

Pacaha has been interpreted as the Nodena phase, while the Quizquiz and Aquixo are identified as the eastern and western Walls phases (Morse and Morse 1983:313). It is interesting to note that the artifacts of the Nodena and Walls phases are similar, especially when compared to those of the Parkin phase. In fact, they were once lumped together as "Walls–Pecan Point" (Griffin 1952a). The province of Casqui, the Parkin phase, would have not had much contact with these other groups and would have grown more isolated in material culture. This seems to have been the case, particularly when one compares the burial furniture of the Parkin phase with that of the Nodena and Walls phases.

The historical accounts of the interactions between the chiefs of Casqui and Pacaha with the Pacaha claim of "greater forebears" fit neatly with what is known about the politics of chiefdoms. The ranking of lineages is normal in chiefdoms, with the top lineage supplying both political and religious leadership (Sahlins 1968). All other lineages are subordinate to the ranking one and are also ranked as higher or lower to each other. The Casqui chief was already admittedly in a lineage subordinate to that of the Pacaha chief. These societies were probably both matrilineal.

Pacaha had already incorporated several other chiefdoms under one chief and was still expanding in power. The future of Casqui, once the Spaniards left, must not have been too promising.

What happened once de Soto left the area can only be learned through archaeological excavation. The political disruptions must have been immediate. The consumption of any food surplus by the de Soto entourage also would have serious consequences. But one cause of social change, the introduction of European diseases, probably had severe and long-lasting effects on all the societies contacted by the Spaniards. The spread of measles, smallpox, and influenza in "virgin-soil" populations can be rapid and devastating (Milner 1980). When the populations are packed into fortified towns, the chances of acquiring these diseases are even higher. Over 90 percent of the Mandan tribe died of smallpox in a six-month period after the disease was introduced in 1837 (Trimble 1986). Similar disease patterns could have occurred in many areas of the Southeast (Dobyns 1983).

A study of 14 skeletons from the Parkin site showed that two had lesions suggesting tuberculosis (Murray 1989). The presence of tuberculosis among the American Indian is thought to be due to European contact (Morse 1967). The proper excavation and study of skeletal series from several Protohistoric sites may produce more evidence of such diseases.

The next historical account of European contact in the Central Mississippi Valley was not until 130 years later, when Joliet and Marquette came down the Mississippi by canoe (Marquette 1900). They described no villages in the area once occupied by the Nodena and Walls phases. In fact, they saw no occupation below the Ohio River until they reached the Quapaw near the mouth of the Arkansas. The flourishing societies with towns of 400 houses, thousands of warriors, and great canoes full of archers were gone. The spread of European disease is the most logical explanation for this disappearance.

The site of Parkin is currently being purchased by the Archeological Conservancy and is slated to become an Arkansas state park. Once the site and its surrounding acreage are acquired, a development similar to that at Toltec Mounds State Park (Rolingson 1982) will probably take place. A museum with an archaeologist on the staff will plan exhibits and conduct research at Parkin and surrounding sites. More will be learned about the Protohistoric period before all the evidence is lost to intensive farming and treasure seeking.

8

The Walls Phase and Its Neighbors

Gerald P. Smith

Within the Mississippi Valley at least four major cultural and geographic subdivisions of the Mississippian tradition can be distinguished. The major point of distinction to date is ceramic complexes that have required different ceramic typologies in each region for nearly all types except the plainware. Contrasts in a wide range of other aspects of culture are also coming into view as work continues but are not yet as obvious or well defined as the ceramic discontinuities. These will be discussed in more detail in the respective sections of this paper; here a summary of the major ceramic discontinuities must suffice to set the stage.

In the South, from the mouth of the Mississippi northward to the vicinity of the mouth of the Arkansas River, is the territory encompassed by the Coles Creek–Plaquemine–Natchezan cultures and here termed the Southern Delta Tradition. It can be characterized ceramically by such major types as the former Addis and Coles Creek Plain types accompanied by Coles Creek, Hardy, and Greenhouse Incised; French Fork Incised; Chevalier Stamped; Fatherland Incised; Bayou Goula Incised; Manchac Incised; Dupree Incised; Plaquemine Brushed; and the Anna–L'Eau Noir—Evangeline Engraved/Incised group. From about the mouth of the Arkansas northward to the vicinity of the Tennessee-Kentucky border is the area included by what is here referred to as the Northern Delta Tradition. Vessel forms include jars, bowls, bottles, and a variety of effigies (Figures 8-1, 8-2a, b). The major types here include Neeley's Ferry and Bell Plain, Parkin Punctated (Fig. 8-2c), Barton Incised (Figure 8-2d), Ranch Incised (Figure 8-3b), Rhodes Incised (Figure 8-3c), Kent Incised (Figure 8-3d, e), Fortune Noded, Nodena Red and White, Avenue Polychrome, Old Town Red, Walls Engraved (Figure 8-3a), and Hull Engraved. In the earlier complexes Coles Creek types are conspicuous by

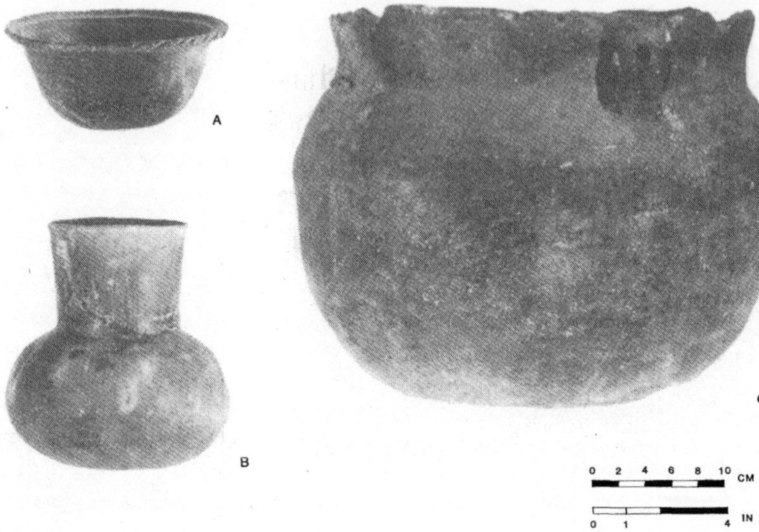

Figure 8-1. Basic Vessel Forms, Northern Delta Regional Tradition

their absence; Larto Red, Evansville Punctate, and "Mazique Incised" are the main decorated types involved. Phillips (1970:38–40) has revised Mazique into Alligator Incised, which works well on the Sunflower but seems to exclude a variant to the north that is essentially Barton Incised on Baytown paste and appears to be later than Alligator but does not fit Manchac particularly well either.

From the Tennessee-Kentucky border northward perhaps as far as Ste. Genevieve, Missouri, up the Ohio into southwestern Indiana and central Kentucky, and up the Tennessee and Cumberland Rivers into central and eastern Tennessee, northern Alabama, and northern Georgia is an oversized unit here referred to as the Cairo Lowland Tradition. The major types here are Neeley's Ferry Plain, Matthews/Moundville/Carthage Incised, Beckwith Incised, O'Byam Incised, Wickliffe Incised, Kimmswick Fabric Impressed, O'Byam Engraved, Old Town Red, and the Angel-Sikeston-Nashville Negative Painted group. North of the Cairo Lowland region is the area centered on Cahokia, in which Powell Plain, St. Clair Plain, Ramey Incised, Tibbets Bean Pot, Monks Mound Red, Cahokia Red, and Cahokia Cord Marked are the major ceramic elements.

The Walls phase is the last prehistoric phase of the Mississippian culture in the immediate vicinity of Memphis on the east side of the Mississippi

River. It is one of numerous such units definable along the Lower Mississippi River and its major tributaries during the Mississippian cultural period. Most of these units are part of local sequences of phases with relatively stable territories through the entire period. Such territories will be referred to here as "districts," following the usage of Lehmer and Caldwell (1966:512).

David Clarke (1968:358–388) cites analysis of ethnographic data, which suggest that in many cases configuration of traits and trait frequencies closely correspond to tribal territories. Clarke worked with the extensive ethnographic data collected during the 1938–43 Culture Element Distribution studies in California and adjacent areas conducted by Alfred L. Kroeber and his associates. Driver (1961:213) notes that "material objects show the same reliability as the entire body of data," thus lending strong support to the potential validity of archaeological inference of cultural units from material remains alone. Kroeber (1953:425–29), in an appendix to Driver's (1939:297–433) report on northwest California notes not only contrast among tribal units but also sharply contrasting groupings within the study area. Internal intercommunity correlations within tribes normally exceed .90 but drop as low as .67 between tribes in the same cluster and below .43 between

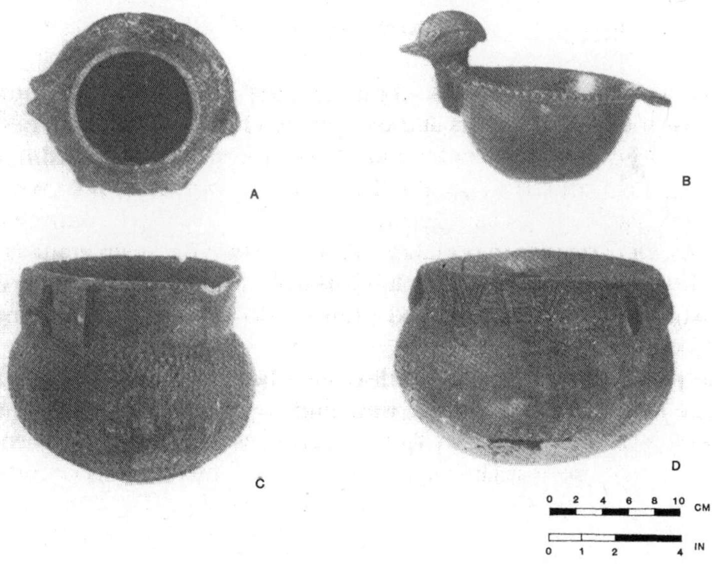

Figure 8-2. Basic Vessel Forms, Northern Delta Regional Tradition: a. fish effigy bowl; b. wood duck effigy bowl; c. Parkin Punctated jar; d. Barton Incised jar

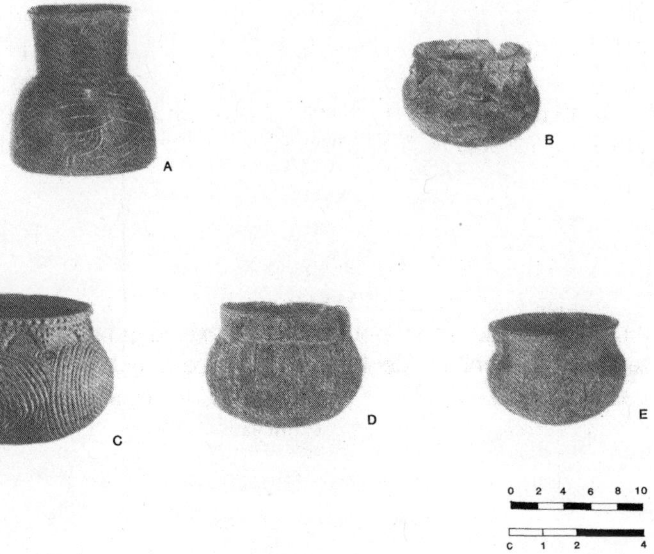

Figure 8-3. Distinguishing Walls Phase Ceramic Types: a. Walls Engraved; b. Ranch Incised; c. Rhodes Incised; d,e. Kent Incised

members of different clusters. Combining the results of these studies implies that sociocultural units at the tribal (or chiefdom) level can be distinguished by contrasting configurations of material culture within larger groupings, not a particularly novel idea. Further implied, however, is that frequency patterns of shared elements may be crucial to making such distinctions. Of particular importance in this regard is the steep gradient of frequencies at cultural boundaries. It is this latter implication that is here taken as a working hypothesis, applied primarily to ceramics, in defining local phases of Mississippian culture.

The particular districts to be discussed here include Walls and its immediate neighbors to the south, west, and north, including the Commerce, Horseshoe Lake, Nodena, and Tipton districts as well as others nearby (Figure 8-4). I will briefly relate the Nodena phase to the rest without further detail inasmuch as that is covered by Dan Morse in chap. 5. As defined for present purposes, the Walls district includes sites in Shelby County, Tennessee, and DeSoto County, Mississippi; the Commerce district includes sites in Tunica and Coahoma counties, Mississippi; the Horseshoe Lake district includes sites in Crittenden County, Arkansas; the Nodena district includes sites in Crittenden and Mississippi counties, Arkansas; and the Tipton dis-

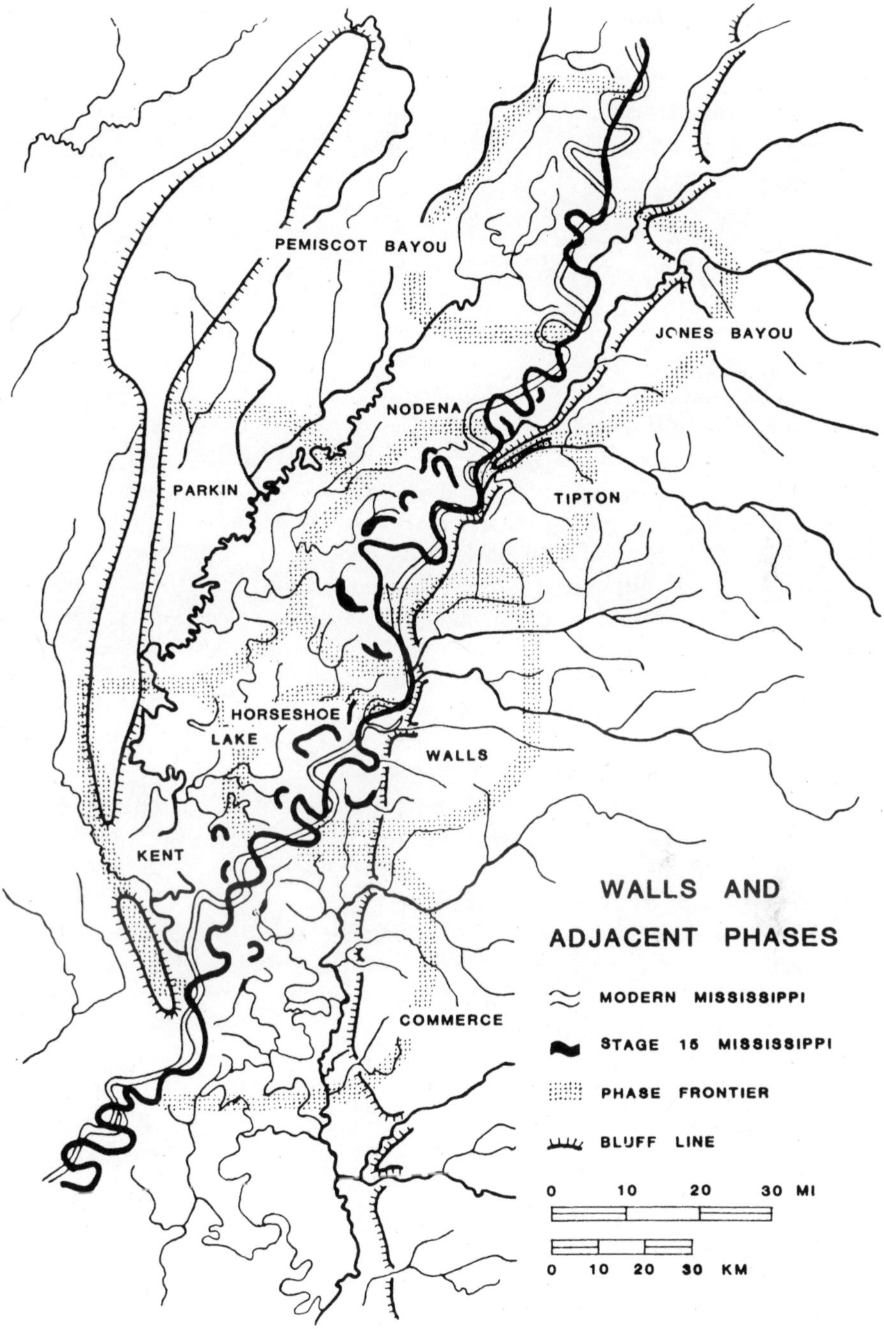

Figure 8-4. Walls and Adjacent Phases

trict includes sites in Lauderdale, Tipton, and Shelby counties, Tennessee. The districts and phases and are most readily distinguished by their contrasting frequency patterns of ceramic types (Figure 8-5), but other contrasts are also important.

Walls District

Walls Phase

The Walls phase is one of the most widely known of the phases cited in this study and also includes some of the most extensive excavation data from the area. Sites with single-platform mounds include Chucalissa, site 40SY28, Cheatham, Irby, Lake Cormorant, Norfolk, Walls, and Woodlyn (Figure 8–6). Numerous smaller village sites are scattered along natural levees in the Mississippi River floodplain and along the top of the bluffs in Shelby County. Hunting camps occur up the various creeks and rivers as much as 20 miles (32 km) into the loess hills. Mid-nineteenth-century maps of the area near the old bridges at Memphis indicate a major site with numerous mounds. The site was incorporated into Fort Pickering during this period and most traces of it were obliterated. Two of the mounds on this site still survive in Desoto Park, and test excavations conducted by Weaver (1982) in 1981 indicate the presence of a Walls phase component. This site may well have been the primary center for the Walls phase, but more research will be required on the site remnant in order to make a better determination of this possibility. Of the mounds on the other sites, only those at Chucalissa, site 40SY28, Cheatham, and Woodlyn still survive.

The Walls ceramic complex includes large globular jars, short-necked water bottles with globular to ovate bodies, and a variety of bowl forms (Figures 8-1, 8-2 a,b). Vessel decoration is primarily in the form of incised, engraved, or punctated motifs and the addition of applique features or appendages to the basic vessel forms. Jars are usually made of Neeley's Ferry paste and water bottles of Bell paste. Bowls occur in both, but Bell is predominant. The basic ceramic type frequency configuration for the phase consists of 55 to 75 percent Bell Plain, 15 to 35 percent Neeley's Ferry Plain, 5 to 10 percent Parkin Punctated, 1 to 3 percent Barton Incised and Old Town Red Filmed, and a consistent presence of Kent Incised, Rhodes Incised, Ranch Incised, Walls Engraved, and Hull Engraved in the 0.2 to 2 percent frequency range for each.

The 1847 map of Fort Pickering suggests that this site had its main mound facing eastward across a plaza defined by at least three other mounds

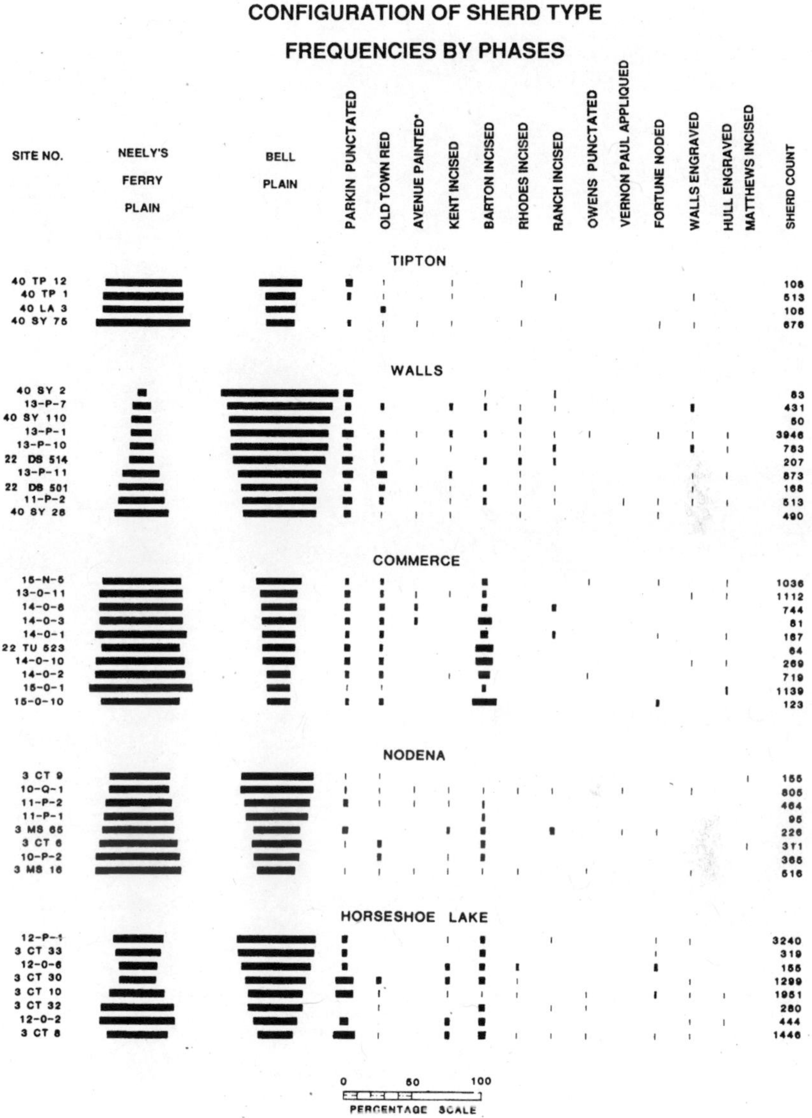

*"Avenue Painted" follows the Phillips, Ford, and Griffin 1951 convention of combining Avenue Polychrome, Nodena Red and White, and Carson Red on Buff in order to maintain comparability with the earlier data indicated by Harvard site designations.

Figure 8-5. Configuration of Sherd Type Frequencies by Phases

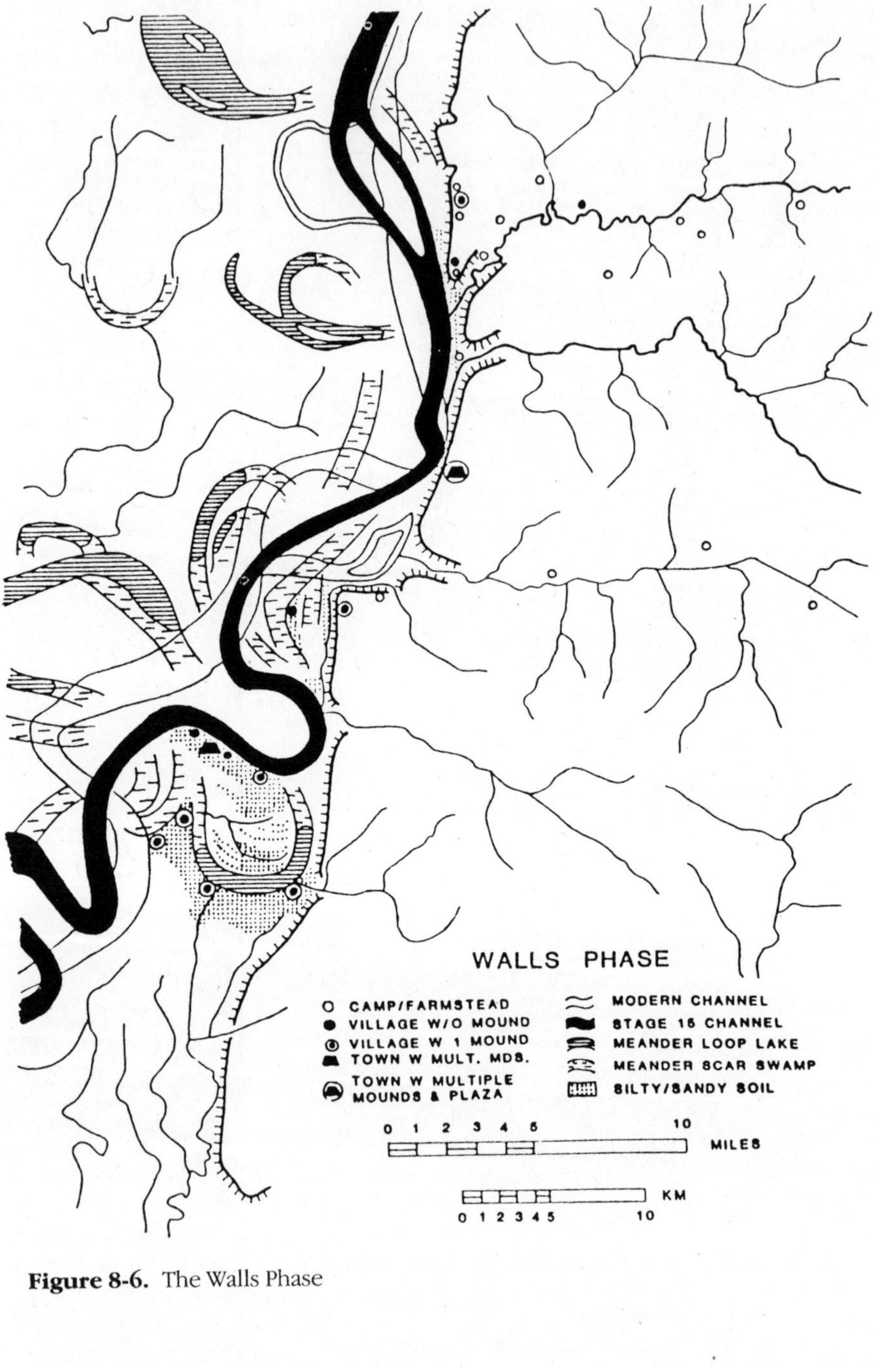

Figure 8-6. The Walls Phase

(Weaver 1982). An additional three mounds along the bluff may have been part of the same complex. This site was at least partly under urban development by 1850, incorporated into the massive Fort Pickering of 1862 vintage, then reurbanized shortly afterward. Today only the main mound, severely damaged by a Civil War artillery emplacement, and the second largest mound survive in recognizable form. Chucalissa, Cheatham, and Woodlyn, the only secondary centers surviving in adequate condition, all have mounds facing south across their town plaza. Brown (1926:122) notes that the mound at Cheatham had several small mounds around it. The main mound at that site is so close to the old river channel on the north side of the site that mounds to the north seem unlikely. At Chucalissa there is an earlier mound at the southwestern corner of the plaza and a series of Walls phase house mounds around the plaza. These sites all have additional occupation midden beyond the plaza and cover areas on the order of 6 to 10 acres (2.5 to 4 ha).

Only six smaller occupation sites without mounds have been recorded. These are on the order of 1 to 3 acres (0.4 to 1.2 ha). Most are in the Mississippi River floodplain or on the bluff overlooking it, but one, site 40SY33 (Smith and Willis 1951), was 4 miles (6.7 km) up the Loosahatchie Valley. Numerous Late Mississippian hunting camps, identifiable only by the presence of Nodena and small triangular arrow points, are scattered up the tributary stream drainages as far as 20 miles (33 km) from the bluffs.

The settlement pattern indicated for the sites at and south of Chucalissa includes the placement of "permanent" sites on or adjacent to very fine sandy loam soils along meander loop lakes or meander scar swamps. The DeSoto Park site, the postulated paramount site, overlooked a large floodplain area that has since been reworked and largely reoccupied by the Mississippi. A similar situation also applies at site 40SY28, but the floodplain area appears to have been quite small throughout most of the recent geological history of that vicinity. The time of the channel change in the Walls-Cheatham sector from active channel to meander loop lake is unknown relative to the Walls occupations there.

Overall the verifiable settlement pattern suggests use of natural levee soils in the Mississippi River floodplain for agriculture and the uplands for hunting. Secondary centers in the floodplain are placed at 2- to 3-mile (3.2- to 4.8-km) intervals with the exception of Norfolk and Woodlyn. These two sites are virtually within view of each other, less than 1 mile (1.6 km) apart. Such close proximity raises the possibility that both sites were left by the same community, living first at one location and then moving to the other. The apparent wider separation of the bluff-top sites may result either from irregularities in the environment or site obliteration by the city of Memphis prior to any survey work. Hamlets in the southern portion of the territory appear as outliers to secondary centers where fine sandy loam soils cover extensive or discontinuous areas. The northern hamlet at site 40SY110 may well be an extension of this pattern, but site 40SY33 seems most likely to have served as a base for exploitation of upland resources in the Loosahatchie Valley.

Data from the secondary center at Chucalissa provide the most extensive excavated information from the phase (Beaudoin 1953; Blake n.d., 1965; Dye 1976; Lahren and Berryman 1984; Nash 1972; Robinson 1977; Smith 1975; Smith 1969). Further site plan detail from this site includes the use of paired structures on the main platform mound and the existence of a belt of overlapping single-stage house mounds around the plaza. Burial areas appear to have been scattered throughout the site in association with house sequences. Coextensive variations in several aspects of the data suggest the existence of at least three basic status categories in the community.

The largest structure excavated was on the western half of the main mound and was ca. 50 feet (15 m) square. Its roof was largely carried on interior cypress posts 1.0 to 1.5 feet (30 to 45 cm) in diameter set in a rectangular pattern with paired posts at the midpoint of each long side. The floor was lowered into the mound summit ca. 0.8 foot (20 cm), with the wall posts set inside the sunken area at its very edge. A shallow "wall stub" trench provided seating for the base of the wattle and daub to seal out rainfall trying to run under the wall. Extensive wear and repair of the floor area around the hearth was noted. The hearth itself was roughly centered in the front half of the interior post pattern. Several large storage pits were located between the interior support posts and the outer wall. These pits were 8 to 12 feet (2.5 to 3.5 ml) long, ca. 3 feet (1 m) wide, and up to 6 feet (2 m) deep at the deepest end. The deep ends were 2.5 to 3 feet (.75 to 1 m) in diameter and lined with strips of bark, probably cypress. Household debris (including sherds; charred nut hulls, corn, and beans; bones; and lithic debris) suggests a residential use for the building. The extensive insect damage present in the wood charcoal and the lack of artifacts apparently burned in a fire suggest that the structure was stripped of usable items and deliberately demolished by fire after it had deteriorated beyond repair. Preliminary tests indicate the existence of a somewhat smaller structure on the eastern side of the mound, set much deeper (ca. 3 feet or 1 m) into its summit. These structures have the highest incidence of decorated and suspected trade ceramics of any area on the site. The storage pit walls reveal multiple mound construction stages averaging ca. 1.5 feet (0.5 m) in thickness with two to three structures in sequence on each surface before addition of the next stage. A single burial was found on the mound, intrusive from the surface and possibly interred after abandonment of the site.

House mounds around the plaza are on the order of 30 to 40 feet (9 to 12

m) square at the base and 1.5 to 2.5 feet (0.5 to 0.75 m) in height. The houses on them are on the order of 18 to 22 feet (5.5 to 6.7 m) square. Two to four structures appear to have been built in sequence on each mound before its use was discontinued. Each mound was apparently then replaced by a new, often partly overlapping mound of the same approximate size rather than receiving a new construction stage. Small trash dumps sometimes occur on the faces of these mounds.

Burials occur in clusters both in the circumplaza zone and in the area beyond, but they cannot yet be associated with specific house sequences. Subfloor burials do not occur. Grave goods tend to be more plentiful in the circumplaza zone, with notable variation among some burial clusters. For example, one burial cluster in this zone had mostly high-quality, possibly even specialist-made, pottery while the adjacent cluster had mostly ordinary household ware but also included all the conch-shell jewelry from the site.

Beyond the circumplaza zone houses approximate 15 to 20 feet (4.5 to 6 m) square. This area has the lowest incidence of decorated and trade ceramics as well as the lowest incidence of grave goods on the site. The outer margin of this sector of the site is poorly defined, mainly due to severe sheet and gully erosion around and into the area.

Floral and faunal studies of Chucalissa materials have been carried out by Blake (n.d., 1965), Parmalee (1960), and Smith (1975). Corn was the overwhelming majority plant food represented, but beans, sunflower, persimmon, hickory, acorn, and butternut were also represented (Blake 1965). Hickory and persimmon were the main wild species present. Plant food samples are too small for meaningful intrasite comparisons. The most useful faunal analysis are those carried out by Bruce Smith on water-screened samples from the village area behind the mound and from the large structure on top of the mound. The sample from the mound is much smaller than the village area sample and thus must be viewed with caution regarding the lowest frequency species. Deer is by far the overwhelming majority of both samples by any measure. When viewed in terms of pounds of meat per species, as calculated from the indicated minimum number of individuals, the importance of most secondary species is far greater in the mound structure sample (See Table 8-1). This contrast is most apparent in the case of turtles, birds, and squirrels. Notably missing from the mound structure sample are opossum and bear. Both species yield very fatty meat and in historic times were often rendered down for oil or lard rather than eaten. Any status differentiation present here appears to be in the greater frequency of secondary species rather than in absolute presence or absence of particular species. David Stevenson (personal communication) has recently noted an extremely high incidence of apparently diet-related cranial osteoporosis in burials from beyond the circumplaza zone.

CLASS/SPECIES	VILLAGE UNIT 6 LBS. MEAT	%	STRUCTURE 1 (MAIN MOUND) LBS. MEAT	%
FISH	?	?	20.5	5.6
AMPHIBIANS AND REPTILES				
Snapping Turtle	22.5	0.2	7.5	2.1
Terrapin	32.7	0.3	2.0	0.5
Other	27.1	0.2	3.0	0.8
Subtotal	82.3	0.7	12.5	3.4
BIRDS				
Canada Goose	22.0	0.2	11.0	3.0
Blue/Snow Goose	8.0	0.1	-	-
Mallard/Black Duck	8.0	0.1	2.0	0.5
Wood Duck	4.0	< 0.1	1.0	0.3
Merganser	5.0	< 0.1	-	-
Turkey	178.5	1.4	17.0	4.7
Other	21.3	0.2	2.0	0.5
Subtotal	246.8	2.0	33.0	9.0
MAMMALS				
White-tailed Deer	10,192.0	83.4	245.1	67.0
Elk	350.0	2.9	-	-
Black Bear	420.0	3.4	-	-
Mountain Lion	60.0	0.5	-	-
Bobcat	30.0	0.2	-	-
Raccoon	296.0	2.4	8.5	2.3
Opossum	85.0	0.7	-	-
Rabbit	96.5	0.8	5.0	1.4
Fox Squirrel	72.0	0.6	10.5	2.9
Beaver	80.0	0.7	20.0	5.5
Other	214.5	1.8	10.6	2.9
Subtotal	11,896.0	97.4	299.7	82.0
TOTAL	12,225.1	100.1	365.7	100.0

Table 8-1. Species Composition of Potential Meat Supply at Chucalissa (Based on data from Smith 1975:157-164)

Boxtown Phase

The Boxtown phase is defined primarily on the basis of its component in Stratum III at Chucalissa, but it is also represented at Walls and other sites in the district (Figure 8-7). Its ceramic complex consists primarily of Neeley's Ferry Plain, *var. Neeley's Ferry,* and Boxtown (55–75 percent), with Bell Plain as the secondary plainware (20–35 percent) and Parkin Punctated (1–5 percent) Owens Punctated, Barton Incised (2–8 percent), and Old Town Red as the primary decorated types. Low-rimmed jars with slightly everted rims and poorly defined necks are the primary vessel form (Figure 8-8a). Loop, wide loop, and strap handles are present on these jars, but none of the arcaded or decorative series.

Houses are constructed with their walls oriented approximately to the cardinal directions. Construction is apparently of the open-corner wall-trench type, but many have wall trenches only slightly wider than the post diameters and have postmolds penetrating well below the base of the apparent wall trench. The houses are roughly square and range from 10 to 16 feet (3 to 4.8 m) on a side. A single, much larger structure was partly exposed by Charles Nash's excavations in Unit 6 at Chucalissa; its full size and use remain unknown.

Review of the carbon dating for the component by Charles McNutt (personal communication) with respect to recent reanalysis of the associated ceramics indicates a mean data of ca. A.D. 1340 within a likely A.D. 1250–1400 span.

Boxtown settlement patterns generally parallel those of the Walls phase, but no definitive mound association determinations are yet possible. Apparent village components are present at Walls, site 22DS506, Chucalissa, and sites 40SY5, 40SY11, 40SY37, 40SY71, and 40SY215. Scattered camps or homesteads are found along the Loosahatchie River and Nonconnah Creek. The lower reaches of Wolf River and the area between Nonconnah Creek and Loosahatchie River were not effectively surveyed before urbanization and thus the lack of recorded sites is not reflective of past settlement. Data from Phillips, Ford, and Griffin's Cut B at Walls (1951:245–47) may be regarded as confirming the Chucalissa stratigraphic placement of the phase as precedent to the Walls phase.

Commerce District

The Commerce phase includes a series of sites in Tunica and northernmost Coahoma counties in Mississippi (Figure 8-9). Most of the sites are ad-

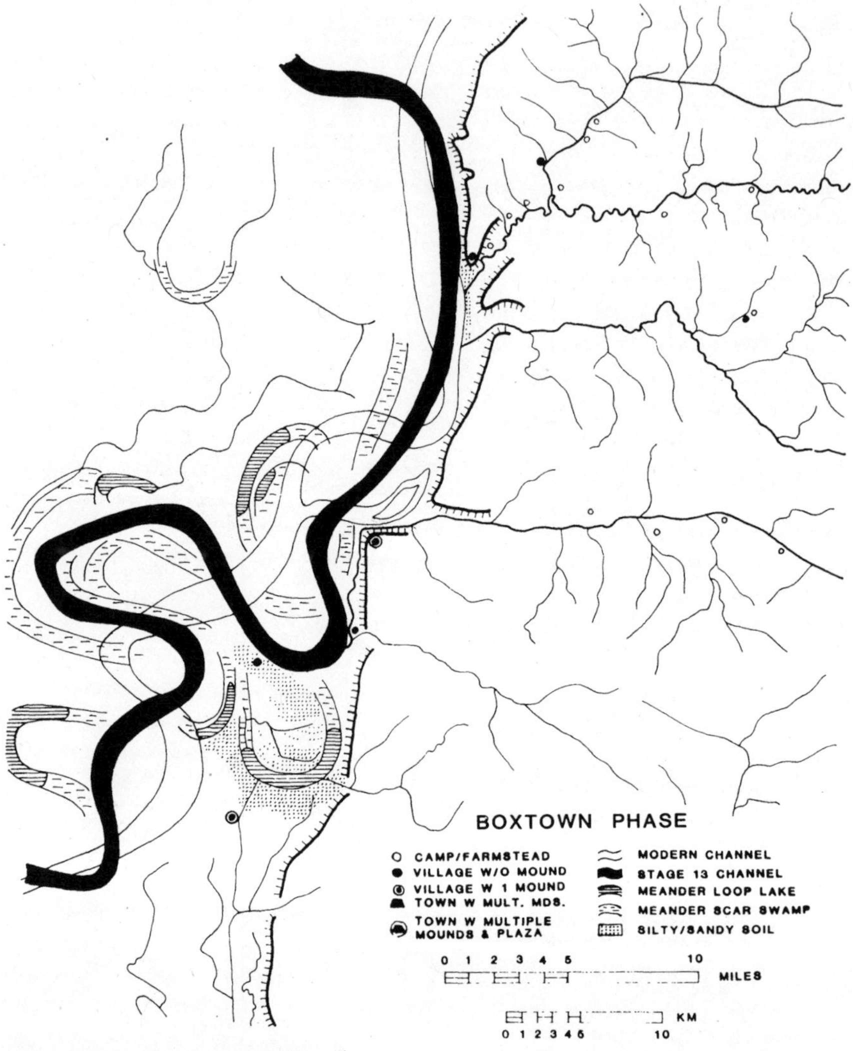

Figure 8-7. The Boxtown Phase

jacent to old Mississippi River meanders that should have been either meander loop lakes or swamps at the time of site occupation. The phase is characterized by a ceramic complex that includes 40 to 60 percent Neeley's Ferry Plain, 20 to 40 percent Bell Plain, 2 to 5 percent Parkin Punctated, 5 to 11 percent Barton Incised, 1 to 3 percent Old Town Red, and up to 1 percent combined multicolored painted types. Other types are very rare to absent, thus forming a sharp contrast with the Walls phase complex immediately to the north as well as the Horseshoe Lake and other phases across the Mississippi River. Vessel forms consist of the same basic jar, bowl, and bottle forms as in the Walls and other adjacent phases.

Dundee appears to have been the primary center for the phase, with secondary centers at Evansville, Beaverdam, Owens, West, and Parchman. Wilford, Salomon, Canon, and Commerce appear to represent local centers on the frontiers of the territory. Test excavations conducted at Wilford by the Mississippi Archaeological Survey (Connaway and McGahey 1970:11–12) have provided the only excavation data from sites for the phase. Some salvage work has been possible on other Mississippian sites in the area, such as Flowers No. 3 (Connaway 1981:50–55), whose affiliation with this phase or an earlier or later one remains unclear. Several houses were excavated at each site. Those at Wilford are virtually unique in the archaeological record and appear to be the remains of large structures with floors raised well above the ground. The houses at Flowers No. 3 were the open-cornered wall-trench structures commonly found on Mississippian sites throughout the extent of the culture.

Figure 8-8. Vessel Forms Characteristic of Specific Phases: a. Boxtown phase; b. Walls phase; c, d. Horseshoe Lake phase

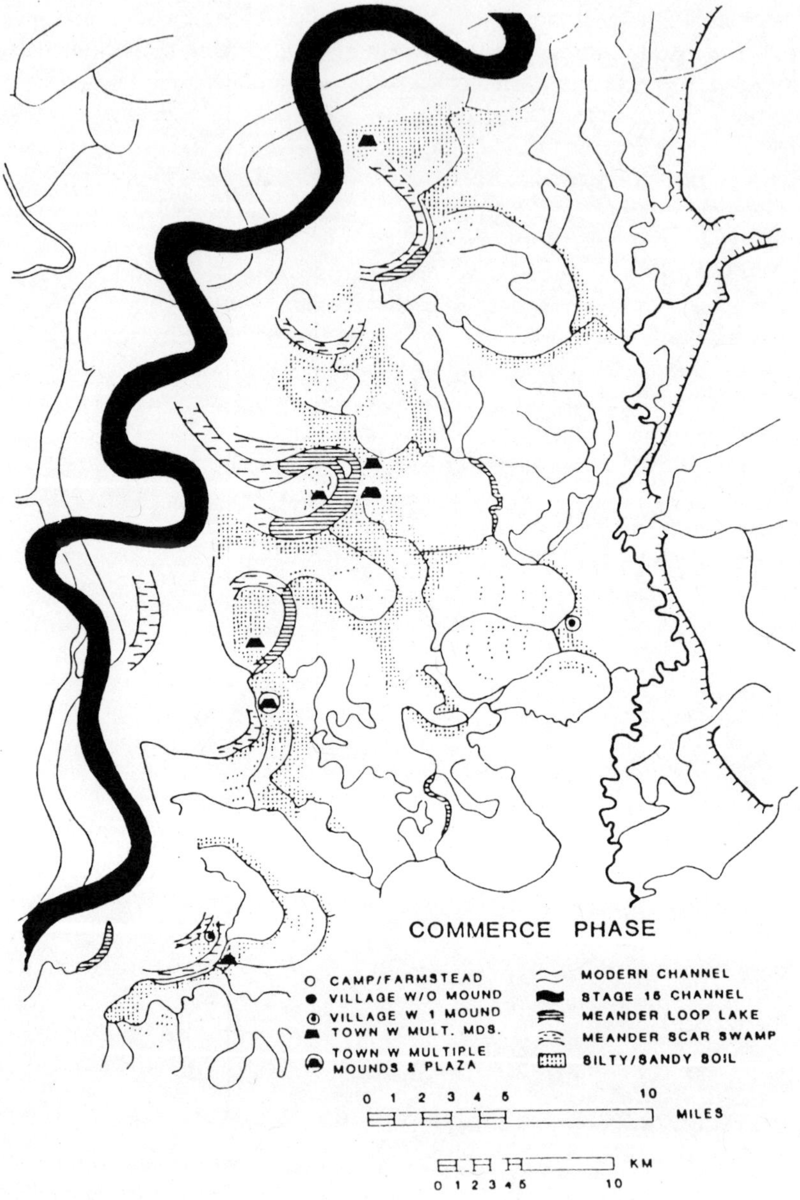

Figure 8-9. The Commerce Phase

Patterns of subsistence and sociopolitical organization remain unknown beyond the most general level of inferences possible from the partial settlement pattern on record. Placement of the major sites and population centers on the active Mississippi River meander belt implies more use of riverine and swamp resources than found in the Walls phase with its relatively easy access to the loess uplands. Canon may represent a base for the exploitation of upland resources, but the uplands in the area are too poorly known for useful discussion beyond noting that occasional small Mississippian components are known to occur there.

All the sites noted as centers with at least one mound are interconnected by bayous, thus providing easy travel throughout the territory without recourse to the Mississippi River. This connection would have been an important factor inasmuch as the population and political centers were otherwise separated by large areas of backswamp. While the territory covered is quite extensive, approximating 50 miles (80 km) across, most of the area was backswamp prior to modern drainage canals and of little or no agricultural or residential use. Dundee appears to have been the paramount site by virtue of its remnant large mound and plaza complex (not cut by a railroad) and large size, but the area around Beaverdam Lake may well have had a larger population if all the apparent secondary sites there were occupied simultaneously.

Tipton District

Tipton Phase

The Tipton phase has a ceramic complex with very little decorated ware included, itself a contrast with Walls phase. Parkin Punctated, Old Town Red, and Kent Incised are the only types consistently present, the only sporadic occurrence of Barton Incised providing the best ceramic contrast with the Nodena phase across the Mississippi River. Lack of Kent Incised in the phase to the north, along with far more Parkin Punctated and Old Town Red in that phase, provides a contrast in that direction. Test pits have been excavated in two sites of the phase, sites 40SY75 and 40TP1, with inconclusive results other than confirming the presence of midden deposits at both sites. The phase area extends at least 6 miles (10 km) up the Hatchie Valley (Figure 8-10), with the Fort Pillow ridge as the apparent boundary between it and the Jones Bayou phase to the north. Tipton phase sites that now or formerly had mounds include sites 40LA3, 40TP1, and 40SY75, each with a single platform mound. No potential major center has yet been located, but the removal of

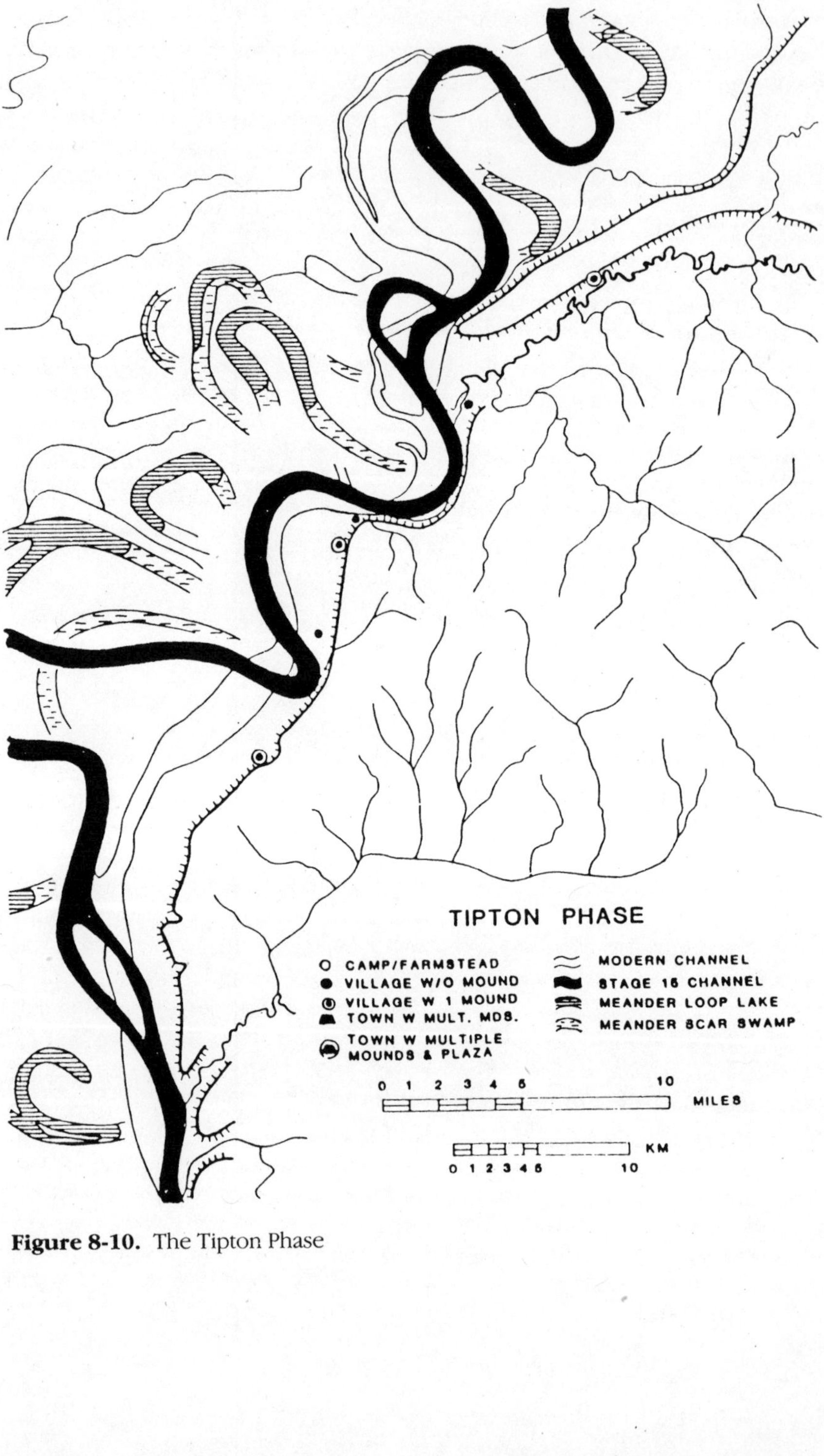

Figure 8-10. The Tipton Phase

most of the potential huge floodplain zone west of site 40SY75 and other areas by more recent Mississippi River channels may well have removed any such site.

Jones Bayou District

Jones Bayou Phase

The Jones Bayou phase is the main Late Mississippian phase of the district. Most of the sites are on natural levees of Channels 14 and 15, with others on the bluff at points where creeks enter the delta and along earlier channels probably occupied by bayous or cutoff lakes at the time of site occupation (Figure 8-11). Ceramics from these sites include 30 to 40 percent Bell Plain, 50 to 60 percent Neeley's Ferry Plain, and the balance composed of various decorated types, with Ranch Incised, Vernon Paul Applique, Nodena Red and White, Avenue Polychrome, Hollywood White, and Jones Bayou Brushed present at the 0.3- to 0.7-percent frequency level.

Neeley's Ferry Plain jars are the dominant vessel form, with flared-rim jars composing two-thirds of the category and cylindrical-necked jars the remaining third. Bell Plain bowls are of secondary importance. Three-fourths of the bowls are flat-bottomed, flared-wall vessels, and one-fourth are round-bottomed cylindrical vessels; water bottles with narrow lip flanges are known but were apparently rare.

Small platform mounds with the corners oriented to the cardinal directions are present at sites 40LA2, 40LA4, 40LA7, and 40LA17. The mound at site 40LA19 is somewhat anomalous for this phase in having the sides of the mounds oriented to the cardinal directions and a possible double ramp on the south side.

The lithic complex includes isosceles triangular arrow points, Nodena points, and chipped and polished celts. Site 40LA11 appears to have been used primarily for heat treatment of the tan chert gravels locally available from the bluff just below the loess. The site is on the crest of the bluff and is heavily littered with fired clay, fire-shattered gravel, discarded heat-treated pebble cores, and unfinished projectile points. No midden was apparent and potsherds were present in only small quantity by comparison with heat-treated stone, suggesting that the site served as a seasonal camp rather than a permanent village.

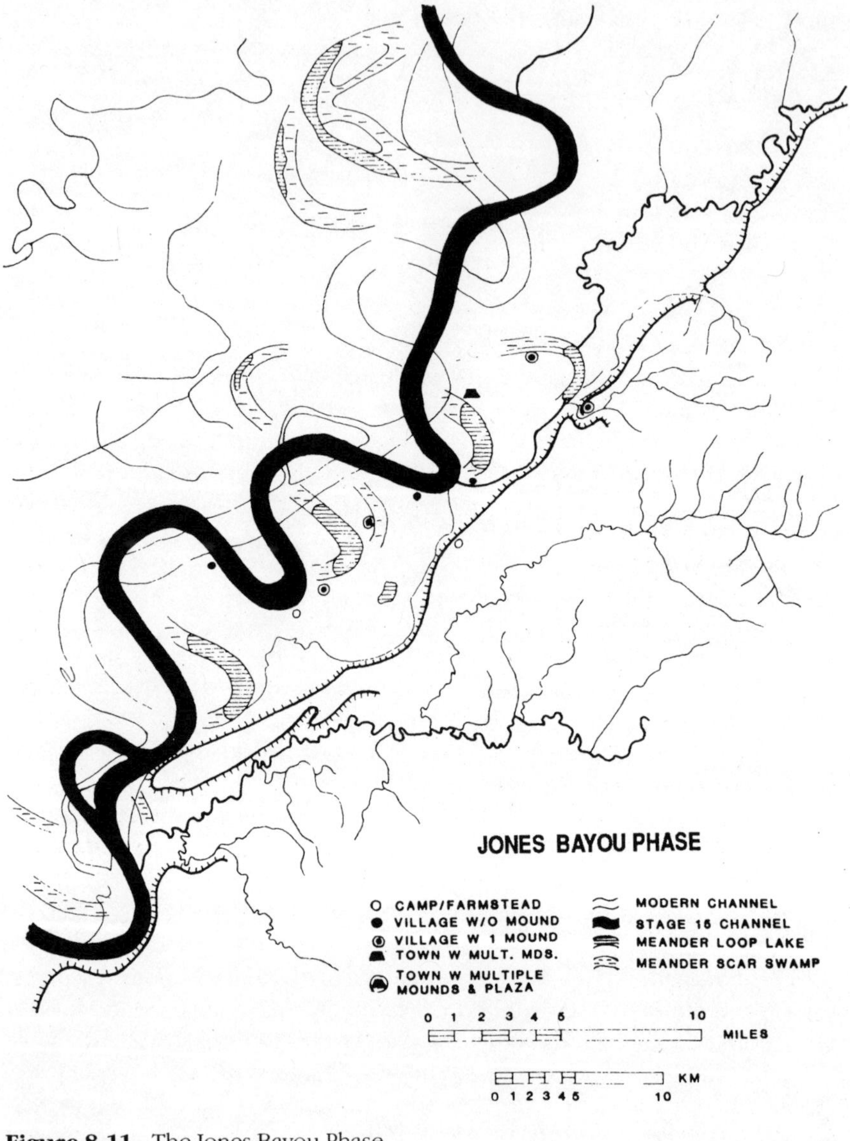

Figure 8-11. The Jones Bayou Phase

Lower St. Francis District

Early Kent Phase

The phase here termed the Early Kent phase includes the Lower St. Francis River drainage of Lee and southern St. Francis counties, roughly near and south of Forrest City, Arkansas (Figure 8-12). Part of this territory is included by Phillips (1970:Figure 447) in his Kent Phase territory and part of it is in his Parkin phase territory. The actual area as defined here corresponds closely to that shown by Morse and Morse (1983:Figure 12.1), although the Morses do not discuss their version of the phase in detail.

The ceramic complex of the phase consists primarily of 70–85 percent Neeley's Ferry Plain and 5–12 percent Bell Plain accompanied by 2–16 percent Parkin Punctated, 2–15 percent Old Town Red, 2–10 percent Barton Incised, up to 1 percent Kent Incised, up to 1 percent multicolored painted wares, and occasional other late decorated types such as Rhodes Incised, Ranch Incised, Fortune Noded, and Walls Engraved. Parkin Punctated frequencies in the 5–10 percent range and Old Town Red frequencies in the 5–10 percent range are usual. Vessel forms include jars, bowls, and water bottles, with a wide range of effigy forms included.

Sites of particular apparent importance include Big Eddy, a large St. Francis–type site with a mound and large plaza; Clay Hill, with two mounds reported; Mound Cemetery, with two mounds; and Castile Landing, formerly a large St. Francis–type site with a stepped rectangular mound, and possibly others. Sudan is a site with multiple mounds reported and is a reputed source of many ceramic vessels from the time period, but there are little firm data about the site. Davis, Starkley, Moore, and Riverside appear to be secondary centers, with a single mound at each. Castile Landing may well have been the primary center, but it has been largely destroyed by an early St. Francis River flood-control project; likewise, Big Eddy has been essentially strip-mined for its pots and Clay Hill nearly destroyed by overlapping potholes dug in recent years. Clay Hill appears to have been one of the latest sites in the phase by virtue of its relatively high Bell Plain frequency; it has also produced early European trade items.

Late Kent Phase

Four closely grouped sites in the southern part of the territory have ceramic complexes that are within the Early Kent phase range, so far as their decorated ware frequency configuration is concerned, but have sharply

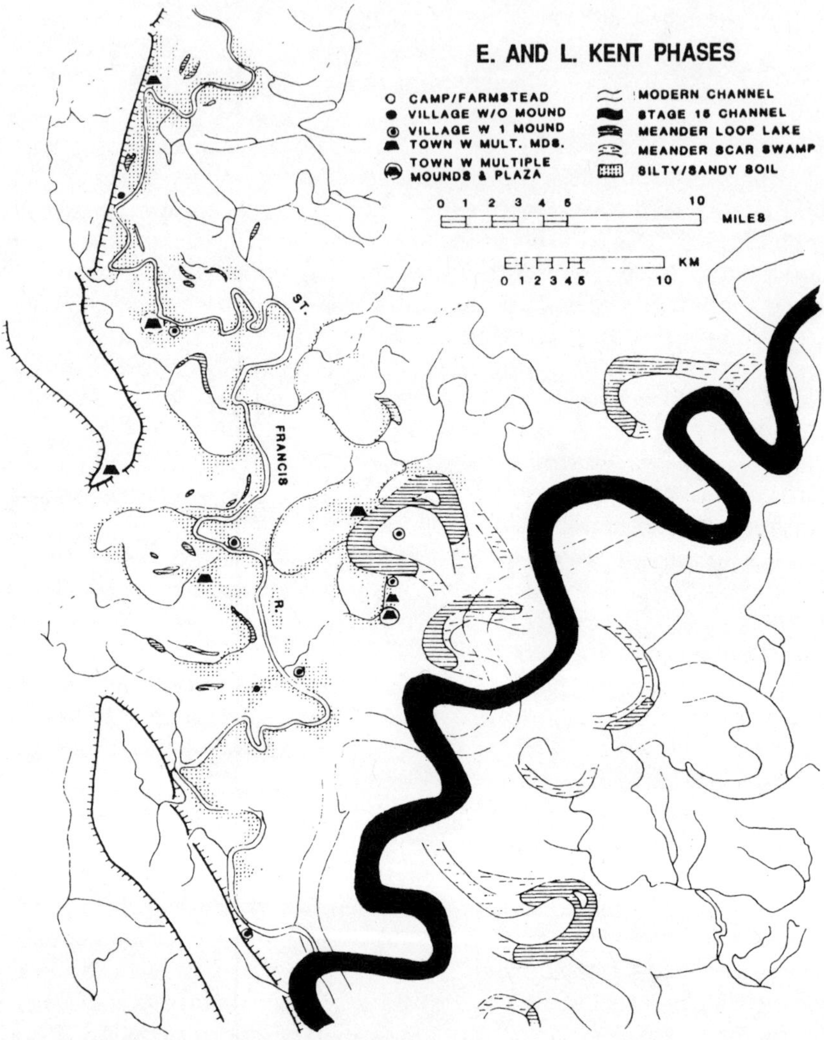

Figure 8-12. The Early and Late Kent Phases

higher Bell Plain frequencies. Bell Plain in these collections, from Applewhite, Kent, Jerusalem Church, and Murdock, ranges from 25 percent at Applewhite to 55 percent at Murdock. Three of the collections are rather small for reliability (less than 100 sherds), but the pattern is shared in the huge collection from Kent (30 percent Bell Plain) and is therefore of potential significance. Kent is a large site with two mounds and a plaza, potentially the primary center, while Applewhite and Murdock have one each, and one is reported from Jerusalem Church. Kent has produced a few glass beads from a burial in the top of the main mound. Unfortunately here, as at Clay Hill, the field information is too poor to permit adequate assessment of the relationship of the materials involved to the main site occupation. An apparent disruption of the established pattern at Clay Hill, in which its relatively high Bell Plain frequency for its phase is accompanied by increased Neeley's Ferry and decreased decorated type frequencies, may also be related to cultural disruptions accompanying European contact.

Horseshoe Lake District

Horseshoe Lake Phase

The Horseshoe Lake phase as defined here consists primarily of the trans-Mississippi portion of the old Walls–Pecan Point phase and perhaps the northeastern frontier of Phillips's Kent phase (Figure 8-13). Its ceramic complex contrasts with Walls in having notably more Parkin Punctated, Kent Incised, Barton Incised, and Fortune Noded but much less Old Town Red. The Kent phase complex to its south has far more Old Town Red and approximately the same amounts of Parkin Punctated, Barton Incised, and Kent Incised but virtually none of the other incised types normally present in Walls and Horseshoe Lake contexts. Vessel forms include the usual local range and an additional form unique to the phase; a deep, unrestricted jar that is known so far only in Neeley's Ferry Plain ware (Figure 8-8c) and a high-shouldered, flat-bottomed jar with a sharply defined neck (Figure 8-8d).

The settlement pattern follows the association of silty/sandy natural levee soils with recently abandoned meander loops. This pattern is emphasized in the shift of settlement area from that of the previous Cramor phase locations (Figure 8-14) into meander loop lake locations left behind when the Stage 12 and 13 channels in the area were abandoned by the Mississippi River. Pouncy, Beck, and Belle Meade are all large sites that had multiple major mounds around a plaza and are possible candidates for the paramount site of the phase. Belle Meade is within sight of Beck and has a midden whose basal materials correspond with the top of the Beck deposits, making it a possible relocation of the same community. Secondary sites include Mound Place, Young, and Rhodes (Brackenseik). Edmondson may also have been a secondary site but has been too badly damaged by the modern town on top of it to permit this determination now. Smaller villages and hamlets should exist but are not yet recorded.

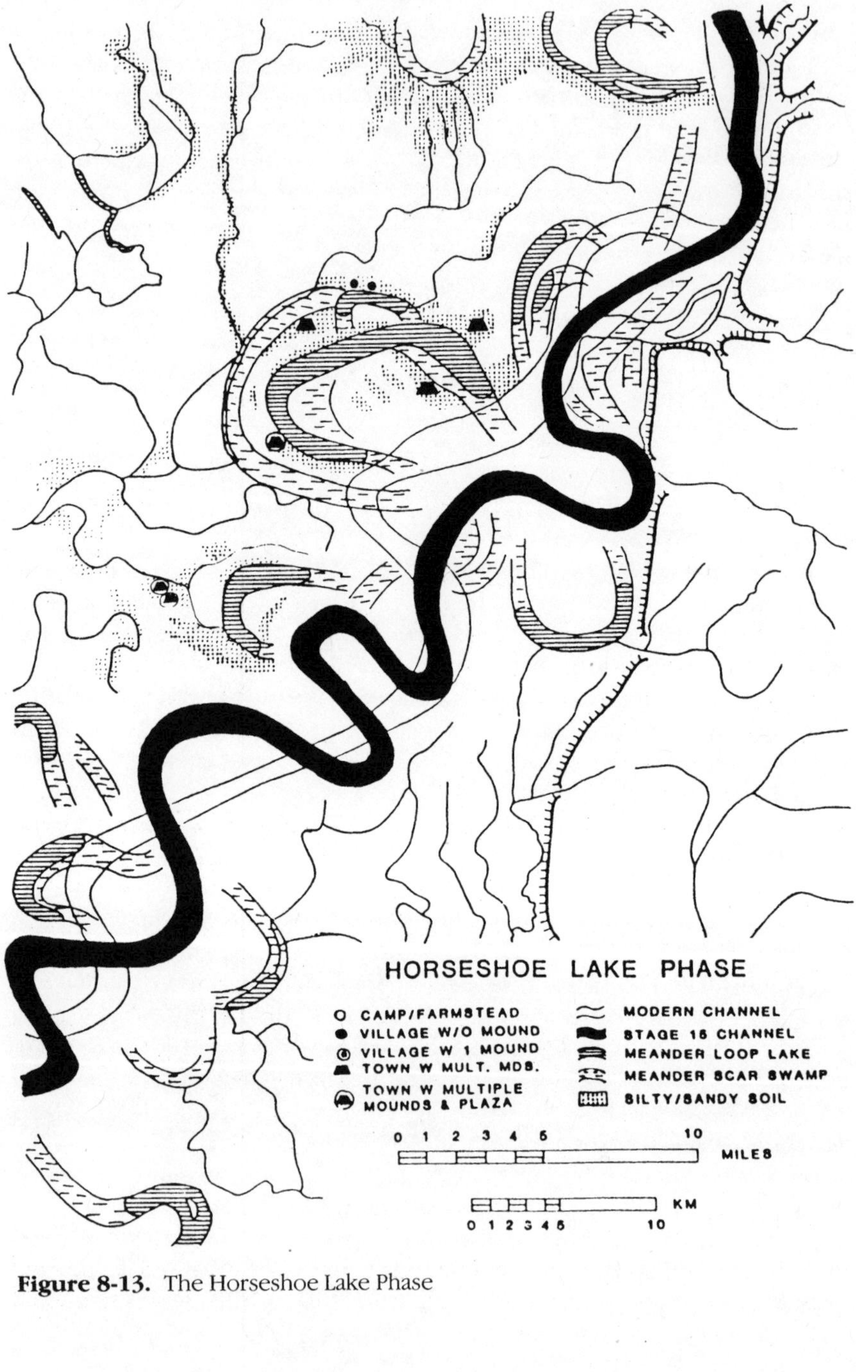

Figure 8-13. The Horseshoe Lake Phase

Test pits excavated at Beck and Belle Meade by Terry Childs, a Memphis amateur to whom I served as a part-time adviser, constitute the only such archaeological work done on these sites. The excavations at Belle Meade were essentially strata cuts in deep midden, primarily revealing structure floors and occasional potted-out burials. Additional work there consisted of making a site map and salvaging a few burials washing out of the bare ground around the barns. The work at Beck began with a zoned surface collection, to be followed by strata cuts. The main work there included a strata cut enlarged to trace several burned individual post houses, which turned out to have been in partly overlapping shallow pits. These structures were on the order of 16 to 20 feet (4.8 to 6 m) on a side with rounded corners. Data analysis from this work is still in progress. Additional work on this phase was begun by David H. Dye in 1987 at the Belle Meade site.

Cramor Phase

Most of the sites in the Cramor phase are in the western portion of the district and include one site (Cramor) with large mounds and a plaza, a second site (Nickel) with multiple mounds, four sites with single mounds (Turner, Nickey, New Bethel, and Barrett), and at least four smaller non-mound sites. Site locations are along bayous on natural levees, usually along old Mississippi River meander scars (Figure 8-14). In many cases the meander scars may well have still had lakes or permanent swamps in them.

The ceramic complex consists of 60 to 80 percent Neeley's Ferry Plain, 5 to 20 percent Bell Plain, 5 to 7 percent Barton Incised, 1 to 12 percent Parkin Punctated, and a consistent 1 percent Old Town Red. Kent Incised, Fortune Noded, and multicolored painted types are normally present in collections over 500 sherds. Notably absent are such late types as Rhodes Incised, Ranch Incised, and Walls Engraved.

Lack of excavation data prevents effective discussion of architecture, subsistence, or lithic and bone tool complexes for the phase. Salvage work by the Arkansas Archeological Survey at Barrett revealed that its platform mound was constructed in multiple stages, but the mound was destroyed before further data could be collected.

Nodena Phase

Because the Nodena phase has been covered extensively in Dan Morse's recent publications (Morse 1973b; Morse and Morse 1983) and is discussed

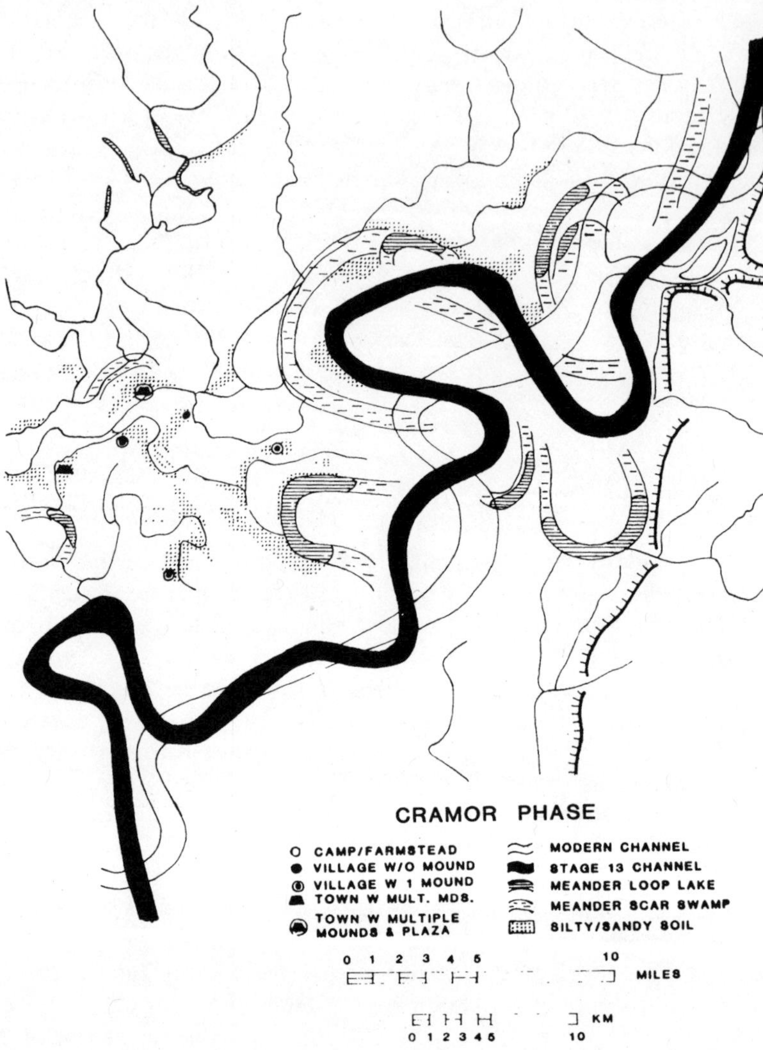

Figure 8-14. The Cramor Phase

in chap. 5, I will only point out some of its salient contrasts with adjacent phases in this particular model. Despite its fame for painted mortuary pottery, the Nodena ceramic complex as a whole is quite low in all its decorated ware frequencies. These sites are characterized by the consistent low-frequency presence of Parkin Punctated, Old Town Red, Kent Incised, and Barton Incised. Most other types in the area are represented but only on a sporadic basis from site to site.

The settlement pattern follows the association of silty to sandy natural levee soils (Morse 1973b:78) on old natural levees or point bars with old meander scar lakes or swamps. Potential primary centers are at Shawnee Village and Upper Nodena (Morse 1973b:74), with secondary centers at Sherman, Turnage, Pecan Point, Nettle Ridge, and site 3MS75 within the Nodena phase territory as defined here (Figure 8-15). The northern area of sites around Blytheville is here considered a separate phase on the basis of its consistently very weak showing or absence of Barton and Kent Incised, and a consistent presence there of Ranch and Matthews Incised along with somewhat higher frequencies of Parkin Punctated and Old Town Red. Nodena phase also includes several small villages and numerous hamlets without mounds, presumably outliers of their respective local centers. Subsistence data from Upper Nodena (Blake and Cutler 1979) note corn and beans as cultigens whose remains have survived, along with hickory, persimmon, walnut, pecan, wild cherry, hazel, and pawpaw as wild plant remains.

Extensive excavations at Upper Nodena indicate that it was a planned town with rectangular houses laid out in rows within the fortified zone. Test work at a Nodena phase hamlet near Bradley (Perino 1966) also revealed rectangular houses, these with individual-post construction. Perino's (1966) excavations at Banks Village have provided a report with extensive data on lithics, burials, ceramics, houses, and subsistence, but the whole makes a rather enigmatic combination of early and late characteristics. Perhaps the best interpretation now would be that the long-necked water bottles, short-necked jars, and perhaps the wall-trench houses belong to a pre-Nodena occupation and the short-necked water bottles, higher-necked jars, Manly Punctated, and Walls Engraved vessels, Nodena projectile points, and perhaps the individual-post houses belong to a Nodena phase occupation. Such data as square profiles and materials from the various midden levels would be crucial to resolving this problem and thus clarifying the other data for better use.

Pemiscot Bayou District

The Pemiscot Bayou phase is the initial Late Mississippian phase of the district. Components are known from site 23PM13, site 23PM43, and Kinfolk Ridge (Figure 8-16). The ceramic complex is dominated by Neeley's Ferry Plain, but Bell Plain increases rapidly from about 5 percent at Kinfolk Ridge to nearly 30 percent at site 23PM13. Parkin Punctated and Old Town Red are

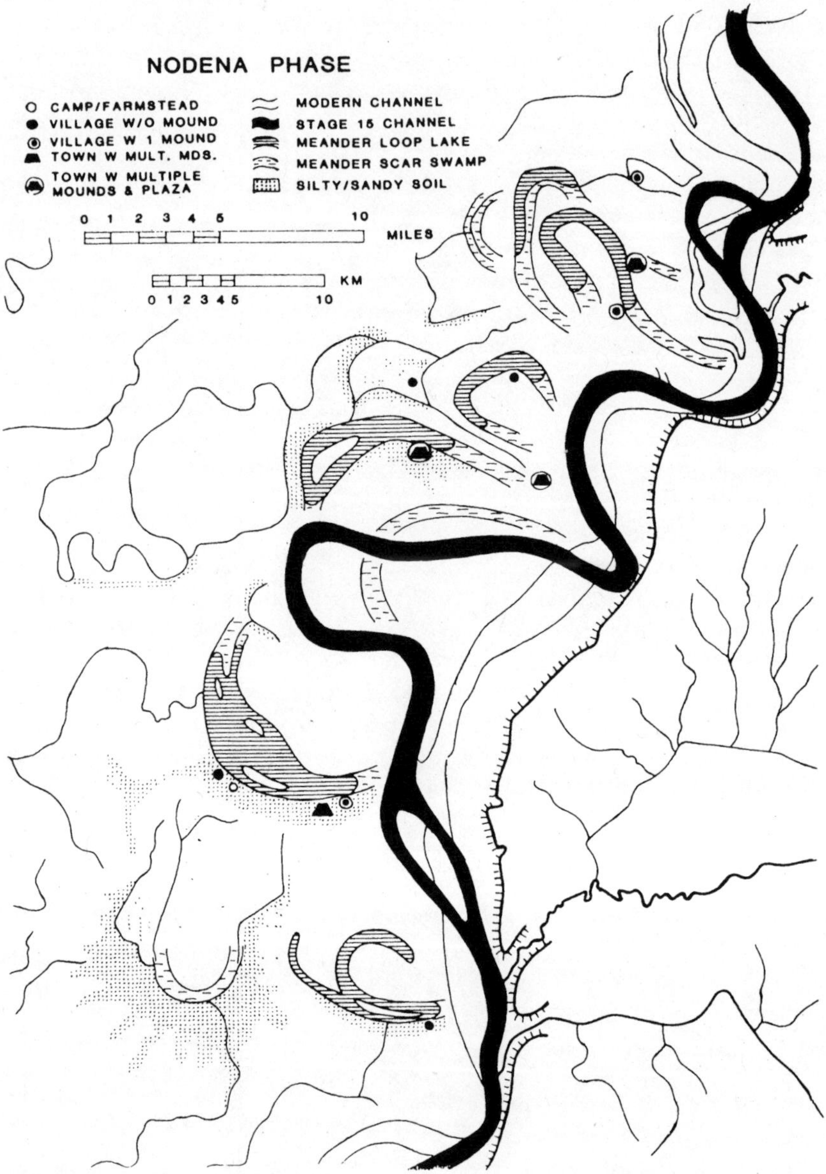

Figure 8-15. The Nodena Phase

apparent local decorated types, while Matthews Incised on Neeley's Ferry paste is present as a probable trade ware at site 23PM43 and Kinfolk Ridge. The basal midden at site 23PM13 included a Pemiscot Bayou component, providing stratigraphic evidence that this phase is antecedent to the Campbell phase, as indicated by seriation. Williams (1954:189) notes that Kinfolk Ridge appears to be a single component site with two mounds, one of which might be a burial mound.

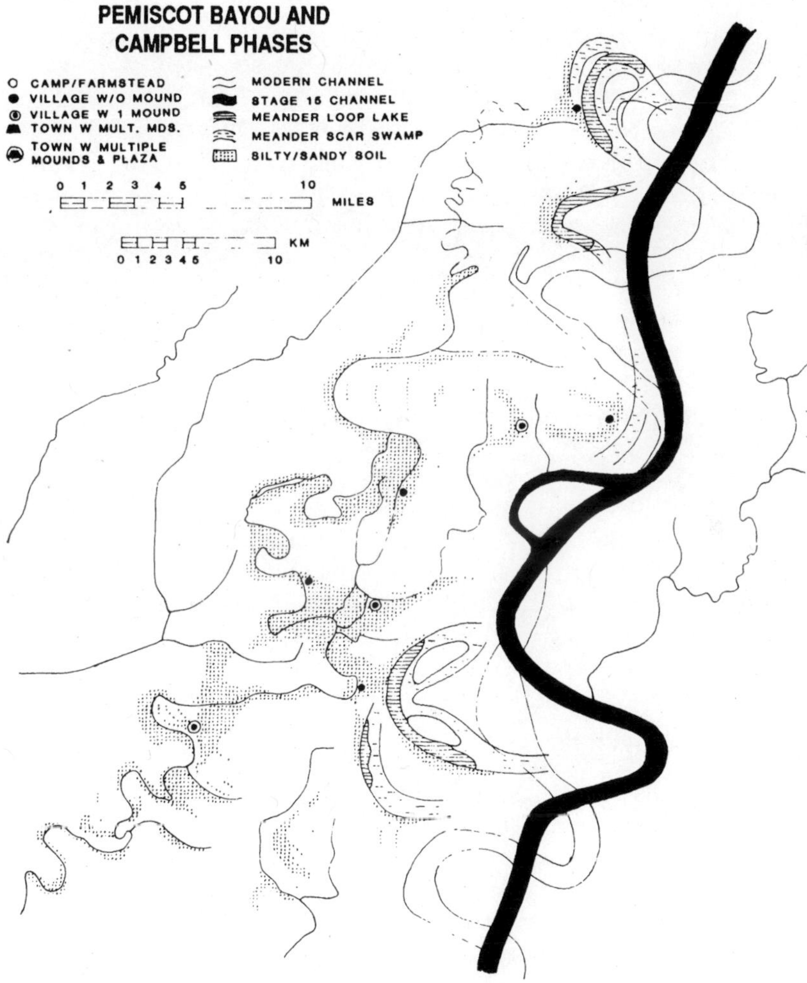

Figure 8-16. The Pemiscot Bayou and Campbell Phases

Campbell Phase

The Campbell phase is one of the better known phases in the region. Major sites include Campbell, Cagle Lake, McCoy, and site 23PM2. Portions of two cemeteries have been excavated at Campbell (Chapman and Anderson 1955), a house excavated at Cagle Lake (Williams 1968:106–16), and a detailed map showing house and cemetery locations made at McCoy (Williams 1968:100–105).

The ceramic complex shows continuity of the trends developing in Pemiscot Bayou, with the addition of a wide variety of decorated types, most of them apparently introduced from the south. The decorated types include Nodena Red and White, Kent Incised, Barton Incised, Rhodes Incised, Ranch Incised, Fortune Noded, Walls Engraved, Campbell Punctate, Matthews Incised on Neeley's Ferry paste, Hollywood White Filmed, and Carson Red on Buff. Vessel forms include low- to flared-rimmed jars; water bottles with large globular bodies and relatively small cylindrical necks; water bottles with globular bodies and flared necks, often with flared annular bases; flared-wall and subhemispherical bowls; rim effigy bowls including tail-riding effigies; effigy water bottles; and head effigy vessels. While quite similar to the Nodena ceramic complex, that of the Campbell phase is distinguishable by its higher frequency of Parkin Punctated, Old Town Red, and Campbell Punctate, much lower frequency of Kent and Barton Incised, and distinctive bottle forms.

The lithic assemblage includes Nodena points, snub-nosed end scrapers, bipointed perforators, ground-stone celts, chipped and polished adzes and chisels, and ground-stone discoidals. Bone artifacts include bone pins, split bird-bone awls, antler tine flakers, and possible arrow points. Shell items include disk beads, conch columella beads, and mushroom-shaped earplugs.

The Campbell site has a medium-sized temple mound facing a rectangular plaza directly east of it; Cagle Lake, a small temple mound; and McCoy, a small temple mound facing an ovate plaza to the east. The McCoy site map (Williams 1968:Figure 24) indicates rectangular houses arranged in a horseshoe pattern with the open end to the south-southwest and a cemetery across the open end of the horseshoe. Scattered burials are also present among the houses. The houses have their corners oriented to the cardinal directions and their axes running north-northeast/south-southwest. House dimensions average 12 by 18 feet (3.6 by 5.4 m) (Williams 1968:102).

At Campbell one cemetery was north of the plaza, apparently surrounded by occupation area, and the other was southeast of the plaza. A single burial, possibly a shaman (Chapman and Anderson 1955:77), was near the northeast corner of the plaza. In the north cemetery males tended to be oriented to the east and females to the north. In the southeast cemetery orientations were generally to the north or east, but the sexual distinction was not present. Chapman and Anderson (1955:114) note that, of the effigy bowls, only bats were found in the southwest cemetery and fish and birds in the north cemetery. They suggest that this effigy distribution might be indicative of clans making up moiety divisions. The presence of European trade goods in the form of blue glass beads and reportedly other items at Campbell suggests that this phase continued into the seventeenth century

The de Soto Problem

Examination of the areas of the districts and phases discussed in this and other papers, the apparent district sequences, Fiske's (1944) and Saucier's (1974) river channel studies, and the de Soto expedition accounts suggest possibilities for the de Soto expedition route that seem worthy of further investigation. Phillips, Ford, and Griffin (1951:348–91) eventually came to favor a Mississippi River crossing point at or near Commerce Landing. In their hypothesis the sites in the vicinity of Horseshoe Lake are identified as Aquixo, the St. Francis Parkin area sites as Casqui, and the Nodena sites as Pacaha. Their route reconstruction was finally halted by the postulated Fiske Stage 15 shift of the St. Francis River directly from a course entering the Mississippi near present-day Horseshoe Lake to its present location. The effect of this geomorphological reconstruction was thought to place the province of Quiguate either in the area previously assigned as Aquixo or in an area singularly lacking in sites of the proper age. Although the problem of the shifting location of the mouth of the St. Francis River, perhaps a relatively frequent event, still requires resolution, Saucier's more recent work greatly alleviates the need for a firm determination of this factor before archaeological interpretation can proceed.

Application of the new district definitions proposed here and further consideration of geographic possibilities provide support for a slightly modified version of the route proposed by Phillips, Ford, and Griffin (1951). Extensive post-1951 surveys have essentially eliminated the entire Western Lowland from consideration as part of these provinces, while the Parkin phase territory is not matched by any upstream phase of proper age or even remotely proper scale for prehistoric Casqui. The identification of Parkin phase as the remains of Casqui, and Nodena phase as Pacaha, for a host of reasons enumerated by Phillips, Ford, and Griffin (1951:384–89) and others, seems secure. The newly proposed Horseshoe Lake phase encompasses an area postulated as Aquixo by Phillips, Ford, and Griffin (1951) and is distinct from the Kent phase on the Lower St. Francis River, in an area they suggest for Quiguate.

The crucial matter of the northeastern Arkansas route thus seems to be at least under sufficient control to render further consideration of the route east of the Mississippi River a more promising venture. Clearly the most eligible potential crossing point lies somewhere between the mouth of the St. Francis River and the southern edge of the Nodena phase territory. Total failure of the de Soto accounts to mention the 100-foot (30-m) bluffs along the east side of the valley suggests that they somehow missed the experience of scrambling down their nearly vertical face. This consideration implies rather severe constraints on possible routes to the river, such as along the bottom of a stream valley from which the bluffs would not have been apparent. A second key constraint is the implication that the sequence of Quizquiz villages visited lay along a line roughly perpendicular to the course of the Mississippi River, inasmuch as they did not see it or even seem aware of it until they reached the third village. Presuming for the moment that Quiguate controlled the territory near the present mouth of the St. Francis River, Commerce Landing seems a bit too far south and the Walls phase territory becomes of more serious interest.

Quick review of the Commerce phase sites reveals that their alignment is markedly parallel to the Mississippi, thus further diminishing their potential. The known Walls phase site distribution in Tennessee is also linear, mainly along the crest of the bluff without significant occupations apparent along the Loosahatchie River, Wolf River, or Nonconnah Creek, and thus Walls phase appears unlikely as the area visited by the de Soto expedition. Even further north, the Tipton phase sites are also strung out along the crest and base of the bluffs all the way to the Hatchie River, 30 miles (48 km) north of Memphis. Only the poorly known Hatchie River Valley offers even a chance of site alignment that would shift the crossing northward and remain within the various constraints imposed by the related data for both sides of the river.

The Mississippi portion of the territory, however, has a site and geographic configuration of great potential. Not just one but two sets of sites are aligned along the southeast-to-northwest route roughly perpendicular to the general course of the Mississippi, all in the floodplain. One set consists of Dogwood Ridge, Cheatham, and Walls, all at ca. 3-mile (4.8-km) intervals. The approach to this series is down the face of the bluff into and across a swampy meander scar, surely a memorable experience, let alone the view of the countryside obtainable from the crest of the bluff. This series also encounters the river or a recently abandoned meander loop at Cheatham, the second site. This series thus appears rather unpromising.

A second series of sites consists of Irby, Lake Cormorant, and Norfolk/Woodlyn. Irby, the easternmost of the series, is on the natural levee of an early meander loop at the point where Johnson Creek enters the loop from the uplands. The creek valley forms a natural route from the loess hills through the bluff line into the floodplain. A route along the natural levee to Lake Cormorant and Norfolk, sites ca. 3 miles (4.8 km) apart, would bring a traveler to Fiske Stage 15 channel near Norfolk. This channel was at or just beyond the western bank of the present course of the Mississippi in the Norfolk vicinity. A bankside camp along this sector of the Stage 15 channel would thus have been washed away by subsequent river movements.

The Fidalgo of Elvas account, which contains the most detailed description of the crossing, notes that "in passing over they were carried down, so as to land opposite the camp; but, before arriving there, at twice the distance of a stone's cast, the horsemen rode out from the piraguas to an open area of hard and even ground, where they all reached without incident" (1904:115). The landing point thus seems to have been a stable sandbar, whose exposure suggests a moderate water level in the river at the time. Fidalgo of Elvas continues (1904:116): "The Rio Grande being crossed, the Governor marched a league and a half, to a large town of Aquixo. . . . As the town toward which the governor marched was near the river, he sent a captain, with a force he thought sufficient, to take the piraguas up the stream." Should Norfolk be postulated as the crossing point, Pouncy, Brackenseik, and Mound Place, each ca. 5 miles (8 km) from a point opposite Norfolk, appear to be prime candidates for the town mentioned by Fidalgo of Elvas.

Postulation of Stage 15 as the active channel at the time of the de Soto expedition is based partly on Fiske's (1944) estimate of 100 years as the duration of the later stages and partly on apparent aboriginal demographic factors. While apparent site associations with Stage 15 and earlier channels are quite frequent, associations with Stage 16 and later channels are extremely rare. Dobyns (1966:414) has proposed an overall depopulation ratio of 20:1 for the Americas, with nadir population reached at ca. 130 years after contact. He cites the introduction of European diseases into nonresistant populations as the major factor. In the same article he cites examples of disease-induced extinction of some modern South American forest groups within half a century, with most of the population loss occurring in the first decade of contact.

Fidalgo of Elvas (1904:67) notes apparent severe population loss in the vicinity of the earlier Ayllon colony, and Hariot (1955:378–79) notes severe population loss from disease within a few days of English visits to Carolina Algonquian towns in 1585. It thus seems that a working hypothesis that collapse of Mississippian culture in the northern Mississippi River alluvial as a result of rapid, massive population loss from European diseases may be considered plausible pending acquisition of better data. Sufficient forest regrowth for the Marquette and Joliet expedition of 1673 and that of La Salle in 1682 to consider the area a virgin forest with only a few inhabitants provides a temporal reference for this depopulation. A roughly contemporaneous abandonment of most of the sites in the area occupied at the time of the de Soto expedition due to death of the inhabitants appears quite possible. The almost total depopulation of the area of the 1541–42 winter encampment by the winter of 1542–43, as noted by the expedition accounts themselves, provides an example of how swiftly such events could have proceeded.

Summary and Discussion

Review of data from the Walls phase, particularly Chucalissa, and adjacent areas indicates phases definable mainly by internally consistent configurations of ceramic types that appear to exist in the area. Phases antecedent to the adjacent Walls and Horseshoe Lake phases have been defined that are also distinct and foreshadow the major features of the later phases in those respective districts. The spatial subdivisions proposed here vary significantly in some aspects, especially in the proposed Horseshoe Lake district, from those previously proposed by Phillips (1970). While Phillips at many points notes his distrust of frequencies of types and varieties from surface collections, he clearly uses them in addition to the emphasis placed on marker types and modes. One may not like to rely so heavily on surface collections but must in the end face the fact that there is still very little else to work with in the Mississippi Alluvial Valley. I have been able to use some more recent survey and excavation data not available to Phillips and thus to flesh out the ceramic framework with information on other aspects of culture here and there

The use of ceramic frequency configurations is intended to permit use of smaller collections than those necessary to produce enough marker-type or mode specimens for phase assignment. The collections used must be unselected and preferably from a site that has not been severely eroded and/or deeply plowed. It is hoped that a more accurate picture of the sites assignable to phase units results from using the entire body of the ceramic material rather than a relatively small fraction. While both schemes require extensive testing by excavation work, the particular units proposed here at least seem to correspond better with the sociopolitical units briefly described in the de Soto expedition accounts. Intensive modal analysis of ceramics from the various districts, as well as extensive excavation work, would be a major step in more refined definitions of the various phases.

With regard to the de Soto expedition, the best that can be offered here is to pinpoint particular sites for further investigation, none of which has had a great deal of attention except from land leveling and railroad construction crews. The identity of the Parkin and Nodena phases and the St. Francis River with the de Soto account counterparts appears too secure at this point for serious challenge with current data; there simply appear to be no viable alternatives. In this regard I believe the Clarksdale bells at Clarksdale, Mississippi, often cited as support for a more southerly crossing, may be better explained as obtained by visitors to the expedition winter camps somewhere in the vicinity of the mouth of the Arkansas River. This statement does no violence to the crucial geographic and archaeological realities of the Arkansas side of the Mississippi River and retains consistency with both the nature of the bells themselves and the expedition accounts by numerous visitors from each of the Mississippi as well as by local residents, especially during the first winter.

The configuration of ceramic-type frequencies from Mississippian sites in the general vicinity of Memphis has revealed geographic clusters of sites that appear to correlate with historically reported sociopolitical units. There are many problems with the archaeology and the ethnography, both of which require long-term attention. This paper may best be regarded as a more detailed framework for much of that work. Should detailed reconstruction of the de Soto expedition route through the area become possible, then the accounts of the only eyewitnesses to these societies who left written records may become valuable for interpretation of the archaeological data. Certainly such matters as local sequences and chronologies, phase complexes, and eventual reconstructions will require major research efforts.

9

The Vacant Quarter and Other Late Events in the Lower Valley

Stephen Williams

In returning now to Memphis, the City on the Bluffs, I realize that my more than 30-year perspective on its growth and change should help me in my attempt to understand events in the prehistoric past from this same vantage point. Front Street, where in the early 1950s cotton still often lay like early snow in the gutters; Crump Boulevard, which was *the* expressway then; and a very decaying Beale Street was certainly no tourist spot—those were some of my landmarks. Fuller Park, now known as Chucalissa, was just becoming archaeological; and Mud Island was just that. There have been changes in people, places, and even attitudes.

Memphis does provide a very special platform from which to review events in Lower Mississippi Valley archaeology, especially those of the last four or five hundred years, even if I must side with my colleague Jeffrey Brain (1985a, 1985b) and indicate that I believe it unlikely that Hernando de Soto ever shared this specific vantage point. However, de Soto's arrival in this general area remains a very important event, as I will indicate shortly.

The Memphis archaeological locality, as exemplified by its two remaining large mound sites, Chucalissa and the nearly destroyed DeSoto Park Mounds, did participate in the late climax cultures of the valley that we term quite appropriately "Mississippian." This great and well-known cultural tradition, which has its greatest expression on the Mississippi River from near St. Louis south to Vicksburg, has been archaeologically recognized for over one hundred years. James B. Griffin (1985a) has recently chronicled the conceptual development of this tradition. Our interests here will be focused on the last part of this story of cultural birth, maturity, and old age: the "Late Mississippian" as we call it.

However, to understand current views on this topic, we must do a little intellectual history ourselves. Mississippian culture is defined by its large platform mounds (often called "temple mounds," although they sometimes served secular purposes as well, or so we think), its well-known and elaborate ceramic complex (examples come from thousands of burials excavated from nearby sites in Arkansas and Mississippi), its established agricultural base, and its large villages with thatch-roofed houses (Williams 1963). It was quite easily recognized in the late 1930s as the latest culture in the prehistoric sequence as then defined. There were thought to be some possible continuities into the Protohistoric period and to some of the ethnographic tribes of the Southeast (Williams 1981).

A provisional chronology was established in the late thirties, especially following WPA work in the Tennessee Valley and elsewhere in the Southeast. It set the "de Soto time line" (A.D. 1539–43) as a very significant time on which to focus attention (Ford and Willey 1941; Griffin 1946); it was a known anchor point. Attempts to use it directly with the then-known archaeology were made in 1939 by John R. Swanton, of the Smithsonian Institution, who led the research team and published a long but rather unsuccessful tome on tracking de Soto. The historical research was excellent, and the volume has just been reprinted with a new introduction (Brain 1985b), which provides a much-needed update on the archaeological connections and does so in a very useful way.

One of the main events in post–World War II developments in eastern archaeology was the establishment of this historic archaeology as a legitimate field of research. Although some decent work preceded this watershed, as described by Harrington in 1952, growth was very rapid in the next decades. Just focusing on the Lower Valley, one can see that the prewar work at sites like Fatherland in the Natchez area, not to forget Jennings (1941) and the Chickasaw, was followed up very successfully by a number of projects in the 1960s. At about that time I reviewed the status of historical archaeology both generally (Williams 1966) and more particularly in the Lower Valley (Williams 1962). While Neitzel took up the task at Fatherland (Neitzel 1965), following the vicissitudes of the Natchez to their disastrous end in 1730–31, I was chasing the Taensa, across the river in Louisiana (S. Williams 1967).

Even more important, at least from my perspective, the Lower Mississippi Survey (LMS) of the Peabody Museum began in 1971, and over a decade of work took place along the east bank of the Mississippi, from the Yazoo Bluffs on the north to Tunica Bayou on the south, focusing on the late cultures found there. This research was the work mainly of Jeffrey Brain and Ian Brown, my LMS colleagues. Brown (1982, 1985a) ultimately focused on the Natchez, while it became the "Trail of the Tunica" for Brain (1977, 1979).

This Tunica saga, beginning in 1970, was a climactic breakthrough—not the "treasure" itself, but what it showed us very clearly about resident Indian cultures in the Lower Valley at ca. A.D. 1750. Despite the important work at Fatherland, where strangely so few historic materials were recovered from the postwar excavations, we saw for the first time with the Tunica, and Trudeau site material in particular, that aboriginal ceramics were still being made in some quantities and in varieties, in our nomenclature, of types that had existed in much earlier prehistoric times (Brain 1988). Strong continuities did exist—that had been clearly posited by Ford nearly 40 years earlier for the Natchez—but here we found in burial contexts Tunica wares associated with Natchez vessels and even Caddoan pots from the Red River. I had attempted earlier to deal with these Caddoan ceramics (Williams 1964), but the Tunica association spelled out the message loud and clear: here were really Late Mississippian ceramics in the Lower Mississippi Valley with unequivocal historic dates (Brown 1983). That we had not had before.

Of course, this LMS work was not where all the action was. To the north, Jim Price and his program (Price 1978) were getting things in order in southeastern Missouri—nothing very, very late however, or so he felt, except in his southern border area. Dan Morse and his northeastern Arkansas program (Morse and Morse 1983) were where it was at, first working at Upper Nodena (Morse 1973b) and nearby sites (Morse et al. 1974) over on the river; then back to the St. Francis for Parkin (Morse 1981), Hazel, and others that provided lots of materials that certainly came up to the de Soto time line and later. Work by others in Arkansas including Hoffman (1977) with the Quapaw and Jeter (1986) with his Tillar complex added fuel to the fire.

As I have noted elsewhere (Williams 1980), as early as 1953 I had run into some perplexing complexes around Nodena and farther to the north in Missouri with the Campbell site (Chapman and Anderson 1955) and a few of its close relatives. Out of these new data outlined in the preceding paragraphs came what I called the Armorel phase and the Markala horizon (Williams 1980). Obviously these new concepts went hand in hand with the work of Brain and others (Brain et al. 1974) on finally tying down some real data on de Soto and the materials that would help us identify his actual movements. This work was the basic archaeological development of marker types of artifacts such as the Clarksdale bells (Brain 1975b) and the refinement of the glass bead types begun by John Goggin in Florida many years before.

With some now well-dated benchmarks, it was soon realized by some that what had happened in past reconstructions (Griffin 1946, 1952b, 1967), my own included (Williams 1954, 1956, 1963; Williams and Stoltman 1965), was that very Late Mississippian complexes had just been extended (stretched, if you will) to fill the gap from A.D. 1350/1400 to 1600/1650. The facts were that no real evidence existed for such a monolithic view of the whole Southeast.

In late times in parts of the Mississippi Valley, broadly considered, for example at Cahokia, Kincaid, Angel, and even Moundville, really late occupations were scanty to absent at these major mound centers. Here the then-recent work at Cahokia (Fowler and Hall 1972), Mueller's (1978) excavations in the Black Bottom (Kincaid locality), and the work of Peebles (1979, 1981) and later Steponaitis (1983) at Moundville were essential. The very comprehensive Powers phase project in southeastern Missouri (Griffin 1985b:15; Price and Griffin 1979) provided very detailed knowledge of Late Mississippian happenings up to ca. A.D. 1400, but work there and to the south on Fourche Creek by Price (1976) indicated some strong evidence for depopulation.

All these data had a strong impact on me, and I first tested the waters with a paper (Williams 1983) at the Southeastern Archaeological Conference meetings in Lafayette, Louisiana, in 1977 that briefly spelled out the hypothesis that there was a "Vacant Quarter" around the mouth of the Ohio but stretching north on the Mississippi to include Cahokia and to the south, stopping on the Missouri side just north of the Arkansas border. I had important input from some colleagues, especially Jim Price; unfortunately, rather inconvenient delays postponed publication of that paper until 1983. Meanwhile, I followed up with the paper on the Armorel phase in 1978—which contained another reference to the Vacant Quarter Hypothesis. It was published (Williams 1980) before my original paper saw daylight.

Perhaps it would be best to restate that hypothesis here (Figure 9-1). The Vacant Quarter Hypothesis raises the question of the general abandonment of the major Mississippian ceremonial centers in an area centering on the mouth of the Ohio River, on the Mississippi from Cahokia, Illinois, on the north to New Madrid, Missouri, on the south, up the Ohio River to Evansville, Indiana, and up the Cumberland River to the Nashville Basin. The desertion of these large mound sites and their supporting villages apparently took place in the period A.D. 1450–1550. Population relocation rather than wholesale decimation is posited as part of a likely explanation.. The term "vacant" is used; however, it should not be understood to suggest that the area was completely devoid of use by native American peoples who hunted there and made use of other resources as well. The year-round settled villages are no longer there. Cordell (1985:37) has made the same point for the abandoned areas in the southwestern United States. This vacant situation here is also mainly a phenomenon of the alluvial valley areas. The uplands, surrounding the Vacant Quarter, still probably supported a reasonable population of long-term settled groups, but this notion needs more testing.

I might briefly note that this "emptiness" is not exactly a new idea. I suggest that you look at Samuel F. Haven's 1856 volume, *The Archaeology of the*

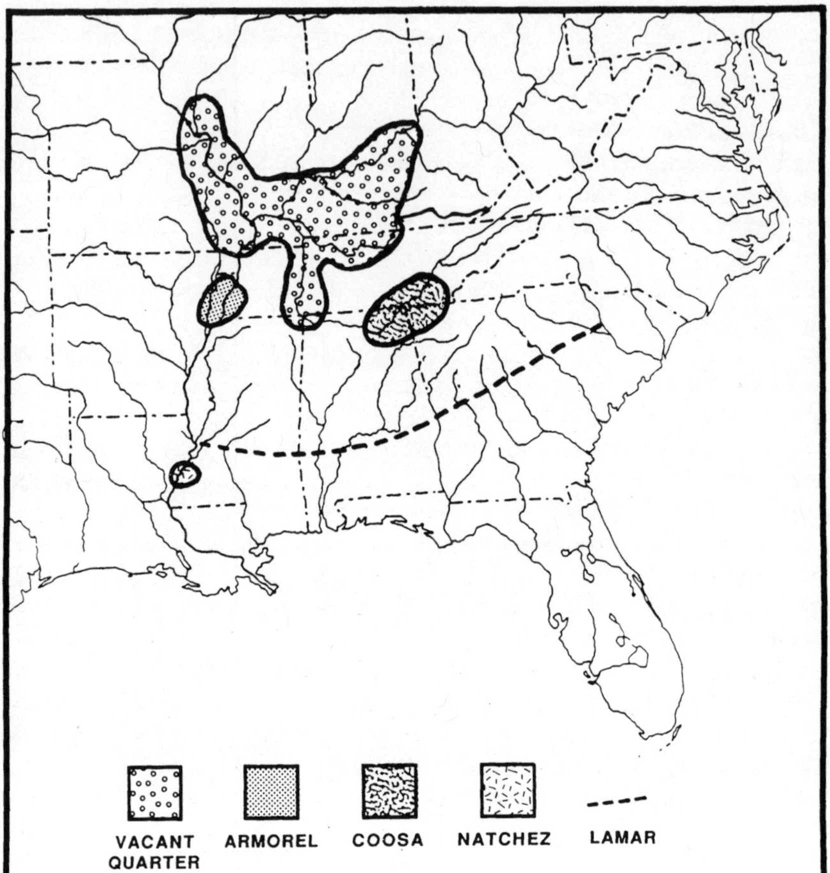

Figure 9-1. The Vacant Quarter: The Southeastern United States ca. A.D. 1500

United States, where in his closing paragraphs (Haven 1856:158) he discussed the lack of permanent Indian occupation along the Ohio, especially the "Dark and Bloody Ground" of Kentucky, and speculated on the possible causes of the population destruction.

However, it does seem to some today, myself included, that it is an idea that needs and deserves thorough testing. I chaired a symposium on the subject at the thirty-ninth annual meeting of the Southeastern Archaeological Conference in 1982, and I found quite general support for the concept. Some recent southeastern volumes, including those by Morse and Morse (1983) on Arkansas, Walthall (1980) on Alabama, the edited volume on Alabama and the borderlands (Badger and Clayton 1985), and even a monograph on the Little Bear Creek Reservoir in northwestern Alabama by Charles McNutt and Guy Weaver (1985) all generally support the notion, some more strongly than others. Bruce Smith's (1986) comprehensive summary article on the Southeast backs the concept to the extent of publishing another map of the vacant area. However, some vocal critics such as R. Barry Lewis (1982, 1986, 1987) have contrary opinions that surely must be heard.

Lewis believes that archaeological phases, especially in southeastern Missouri and adjoining Kentucky, fill the temporal gap until after the time of de Soto (Lewis 1986:Figure 2). The contrary view held by myself and others (Smith 1986:86–59) is that the late (A.D. 1450–1550) ceramic varieties, so well known in the Armorel phase to the south, are clearly absent in that region and are sometimes even found in association with European trade goods. This late phase is marked by a series of new ceramic traits such as applique designs and a totally different lithic complex utilizing new forms and materials. Interestingly, this late phase is clearly seen as post–Late Mississippian in a very small locality in southern Indiana at the mouth of the Wabash River (Green and Munson 1978), and it is not just a slight frequency change in a few specific modes.

What the Vacant Quarter Hypothesis does not mean is swift population decline for all of the Southeast during this time. There is also no reason to suggest a single-cause catastrophe. As Linda Cordell has recently, and with great effect, discussed for the American Southwest (Cordell 1984, 1985), startlingly similar changes occurred in that area at rather the same time period. She discounts any single-cause solution, be it weather (drought), population decline, or warfare. What is clear is that some major changes in the sociopolitical arrangements were taking place. Thus some areas were left almost devoid of occupants while others saw populations packing together—in the northern Rio Grande, for example—at a level never before seen.

The climatic aspect of the southwestern situation has been tempered from a direct causal link with the A.D. 1276–99 drought period, noticed long ago in the tree rings, to a hypothesis of changing precipitation patterns over a wide area; this idea makes sense in the relatively fragile balance of maize agriculture/precipitation amounts that affects so much of the Southwest. It does not seem to serve as a good model for the Southeast, on the other hand (Smith 1986). Griffin (1961) some time ago tried out some correlations between climate and culture change in the East, but not with much success. In terms of other possible causal factors, a study of possible warfare effects in the Southeast was made by Lewis Larson (1972); I am not too sanguine on that possibility having any important impact.

However, in searching for more viable factors, social/political change seems to be a worthwhile area of research in both the Southeast and the Southwest. Cordell (1985:38) has said with regard to the explanations in the northern part of the Southwest, "These observations indicate that the eventual abandonment of villages in the Mesa Verde region may reflect the inability of the social system to sustain dense populations." The question of social changes in the Southeast has been the subject of a recent paper (1986) by Christopher Peebles. In this interesting work, Peebles refers to "Social Devolution" in the area at the time in question here, as seen from the perspective of Moundville. By A.D. 1400 the great mound center had attained its climax, but "within one hundred years of reaching its zenith, Moundville was abandoned" (Peebles 1986:30). Without choosing among four possible causes, including climate, he strongly suggests social and economic (exchange) forces as paramount. The Moundville decline does seem to join the Vacant Quarter syndrome a bit on the late side, but it does join.

Some interesting publications on this time period in the Southwest have appeared recently (Riley 1982; Wilcox 1984; Wilcox and Masse 1981). Their Protohistoric period (A.D. 1450–1700) fits well with the time segment being discussed here. All these works have made quite detailed use of Spanish documents in a broad geographic framework so that large-scale interactions, both intra-and interareal, are discussed, such as Southwest/North Mexico connections and panareal trade networks (Riley 1982:Figure 1). Some suggestive comparisons could be made between the Southwest and the Southeast, including Riley's concept "The Greater Southwest—A Diverse Unity," but here is not the place for that. However, David Wilcox's (1984) detailed discussion of the "Dog Nomads" of the Plains and their trade with the Pueblos, and the reverberation of this trade network over a large area, certainly suggests possible analogies to what was going on in the Southeast at about the same time: the Spiro "Gateway" hypothesis, for example. Would that our southeastern data on trade, especially as seen from Spanish sources, had been as carefully analyzed.

In much of the so-called Central Basin of the Mississippi, one certainly finds that the major ceremonial centers no longer have large populations. In the Vacant Quarter, construction of mounds, added mantles in most cases, is virtually ended. "The center is burned out." Not so for the southern part of the Lower Valley, as previous work by the LMS at the Haynes Bluff site and in the Natchez area has clearly indicated and as the 1985 summer's work in the Boeuf Basin region in northeastern Louisiana has also documented. Indeed, that western edge of the valley, the Ouachita and Bayou Bartholomew, seems to be another late "hot spot" much like the St. Francis to the north, but that story is just now being unraveled. Perhaps one can even speak of a "Post-Classic" afterglow here, as is done in Yucatań.

As I have said, we seem to be looking at population rearrangement rather than real population decline. This statement is not an easy one to make. Good population estimates are difficult to obtain archaeologically, as Griffin (1985a:42) has wryly indicated: the 1818 estimate for Cahokia made by Brackenridge has about the same quality of verification as do interestingly similar estimates for the same site made in the 1970s.

However, as with the Southwest (Cordell 1985:37), we do not see unusual increases in mortuary remains in these late times. Instead the current southwestern view is that of slow depopulation in certain areas such as Mesa Verde and a regrouping in a few rather crowded areas. In the Southeast we have the Vacant Quarter that quite possibly was slowly emptied, possibly with Cahokia leading the way. Then areas like eastern Tennessee, the Lower Valley from northeastern Arkansas to below Natchez, the Oneota homelands to the north and west of the Cahokia/American Bottom center, and the Upper Ohio Fort Ancient region are some of the obvious places where a noticeable population increase seems to have occurred. I do not wish to suggest, however, that there are any signs of large-scale migrations.

Thus there are these strong differences: in some places, such as the Vacant Quarter, things had gone downhill, or were even deathly quiet, when de Soto appeared, while others were surging to new levels. In a broad band across the Southeast, from South Carolina to Biloxi, at and below the Fall Line (the Lamar cultures, Fort Walton phase in Florida, Bottle Creek phase in Alabama), all participated strongly in what I have termed (Williams 1980) the Markala horizon—a widespread series of very specific artifacts. There was an ongoing trade network (despite what Peebles [1986] has said to the contrary) as seen, for example, by the distribution of the definitive pierced celt. Aboriginal materials do move widely (Williams 1980:Figure 1), and soon European trade goods were caught up in the same still-operational network.

The de Soto time period and the geographical area of the entrada have recently been subjected to much needed archaeological scrutiny by my colleague, Jeffrey Brain (1985a, 1985b). Others, especially in the Tennessee/Georgia/Alabama junction area, such as Hudson, Smith, Hally, Polhemus, and DePratter (1985b), are locating historic Coosa and other such mysteries. Meanwhile, Phil Phillips and Jeffrey Brain are studying published sources and museum holdings from the same area, much as they did with the Spiro shells, to get grave-lot associations of shell gorgets and other related artifacts. Using these data, site interconnections are being elaborated by distinguishing three trade horizons: Prehistoric, Protohistoric, and Historic. These studies further strengthen the known extent of the Markala horizon.

The Armorel phase towns (Williams 1980), easily witnessed from these very bluffs, were vibrant cultural centers at one end of the Markala trade routes. They were very much going concerns when de Soto and his Spanish vanguard arrived: up and down the St. Francis River were dozens of thriving towns and villages, grouped into separate demi-Chiefdoms. They met de Soto, and he was theirs, forever.

But turning to less well-known happenings, what of the death and destruction hypothesis to clear the Vacant Quarter? The sweep of disease so well documented for the eighteenth century, and logically hypothesized for the sixteenth and seventeenth centuries as the Europeans brought new viruses as well as new ideas and artifacts into North America, seems like a good explanation. Could a disaster such as a catastrophic epidemic or a series of local disasters account for the vacant ceremonial centers of prehistoric times discussed above? George Milner (1980) discussed the disease hypothesis for historic times with some care. For another view, see Ann Ramenofsky's (1982) work.

Certainly such a pandemic could decimate village-living farmers and their central towns. However, the archaeological evidence for such a disease-caused disaster would be most difficult to recognize inasmuch as all the swift killers like smallpox, plague, and even measles leave no signs on the bones as clues to the cause of death. Also, at times of such widespread devastation there often are not enough well people even to bury the dead, leaving the corpses for natural scavengers, and thus no archaeological traces remain. Disease is a facile answer to population decline, but it is virtually unprovable via archaeology; even data from mass graves may be equivocal (Milner 1980). The European plague of the Middle Ages is best known from documentary evidence, not from funeral monuments.

Other forms of catastrophic impact on these cultures can be invoked—earthquakes such as the New Madrid quake of 1813 could have occurred earlier in the valley and probably did, but they would not seem to be of broad enough scope. Plant diseases attacking the basic maize crops might have created the "sharp edges" of the affected area as we presently map it, but again they are virtually impossible to "see" archaeologically. From this podium one should not forget the late Charles Nash's lighthearted suggestion that perhaps the devastation wrought by a population explosion of passenger pigeons, such as we know to have occurred in the late eighteenth and early nineteenth centuries, might have caused the Late Mississippian decline as a result of massive crop destruction. I've always been fond of that idea.

Our available windows on the past often do not provide us with the clearest vision. What would a plague-ridden village site look like? How different would it appear from one slowly but surely evacuated over a period of some years, as Cordell believes the southwestern pueblos were abandoned? In some few cases, thanks to careful excavation and good fortune (in about equal amounts) at several Powers phase sites (Price and Griffin 1979) in southeastern Missouri, evidence seems to show that the local inhabitants packed up and left, burning their villages behind them at a time just prior to A.D. 1400. Perhaps this action was the beginning of the movements that were to empty that part of the Vacant Quarter in the next century.

So as to the causal aspect, we have ended with a problem without a clear solution. That's not very unusual in archaeology. The evidence seems quite positive in one direction. Change took place. After A.D. 1450–1500, well before de Soto arrived, no more were the great ceremonial centers in the Vacant Quarter the nexus of these chiefdoms or possible pseudostates. These once-grand monumental assemblages were empty: the well-kept earthen pyramids, formerly crowned with sacred temples, grew up in trees and grass, the buildings atop them were either burned or had collapsed, the tall palisaded walls with numerous bastions now lay scattered and rotting on the ground, the birds and animals took back the broad plazas for their own, and the fertile fields were cane covered and vine entangled again. An echoing silence now occurred where throngs of ballplayers had once noisily been challenged in near–life-and-death events.

Elsewhere in the Southeast things were very different; "after-glow" peaks of activity and population would be reached at the very time of the Spaniard's arrival. Perhaps the ceremonial and religious life was not quite so rich; I see this as the time of decadent Southern Cult paraphernalia, for example, but it was not a time of real decline, anymore than it was on the Rio Grande in Pueblo IV times, in what was formerly called "Regressive Pueblo" times. At this time, village populations in the southern part of the Lower Valley, the area that I know best, grew along the secondary rivers, the St. Francis and the Ouachita, and even back from the main Mississippi channel in the Natchez Bluffs area as well.

These crowded villages would have been prime locales for disease spread, but that seems not to have been the prehistoric situation. After de Soto, the story was different. Waves of epidemics would sweep the country, leaving only tattered remnants. Whole tribes were decimated, and by the 1800s the few survivors were destitute refugees in a land they had once owned. In the Lower Valley some Quapaw remained on the Arkansas, a few Tunica and other remnants were at Marksville, some Kadohadacho survivors huddled near Shreveport (Williams 1964), and Chitimacha, Choctaw, and Houma, in small clusters, were far to the south in southern Louisiana. Other Chickasaw and Choctaw in Mississippi were waiting to move west into the now-empty valley

It was change, indeed, as seen from a Memphis perspective. After the Chickasaw and Choctaw moved in, next came the flatboats, then steamboats and the planters. Then in the 1860s came the gunboats. After that the city felt its own taste of disaster, yellow fever. But a new and brighter era was to come at the end of the century. Then Dollar Cotton and the Speed Highway—could Elvis be far behind? So our narrative closes; history catches up with us. Yet there are still more potsherds to collect, more sites to locate, and more hypotheses to test. A Cajun friend, upon seeing archaeologists arrive in his village, used to say, "I see the bone jockeys are back in town." Yes, we are, and we will come back to Memphis again, and soon. There's plenty left to do.

Acknowledgments.

I wish to acknowledge the aid of my LMS colleagues, Jeffrey Brain and Ian Brown, for their corrective help on this version.

10

The Hernando de Soto Expedition: From Mabila to the Mississippi River

Charles Hudson
Marvin T. Smith
Chester B. DePratter

Hernando de Soto's expedition into the southeastern United States was dealt a serious blow in the battle with Tascaluza and his allies at Mabila. De Soto lost men and horses, as well as clothing, food, and equipment. About 150 of his soldiers were wounded, many with multiple wounds (Fidalgo of Elvas 1866:89–90; Ranjel 1922:127). He also lost the pearls that he had taken from the temple at Cofitachequi and that he intended to use as evidence of the wealth that could be obtained in the Southeast. His soldiers were dispirited. They had not yet discovered great wealth or a state-level society, and the Indians at Mabila had been so bellicose it was clear that it would be very costly to subdue the people of the Southeast by force. De Soto was aware that if he followed his plan of going south to rendezvous with the ships that were to meet him at Pensacola Bay, his army could disintegrate, and he would be ruined.

At present, all that can be said about the location of Mabila is that it was in the vicinity of the Lower Cahaba River and perhaps at the site of Old Cahawba (lDS32), which lay on the western side of the mouth of the Cahaba River. Two indirect kinds of evidence argue for our placing Mabila in the vicinity of the Lower Cahaba River (Figure 10-1). One is that this location is consistent with the travel time of de Soto's line of march from Athahachi, whose location near the junction of the Coosa and Tallapoosa rivers is confirmed by evidence from the Tristán de Luna expedition (De Pratter et al. 1985:121; Hudson et al. 1985a). This location for Mabila is also consistent with placing Piachi—the town de Soto visited just before arriving at Mabila—in the vicinity of Durant Bend, where the topography accords well with Ranjel's description. A second indirect kind of evidence is that Mabila was a heavily fortified frontier town, and while ceramics do not always co-

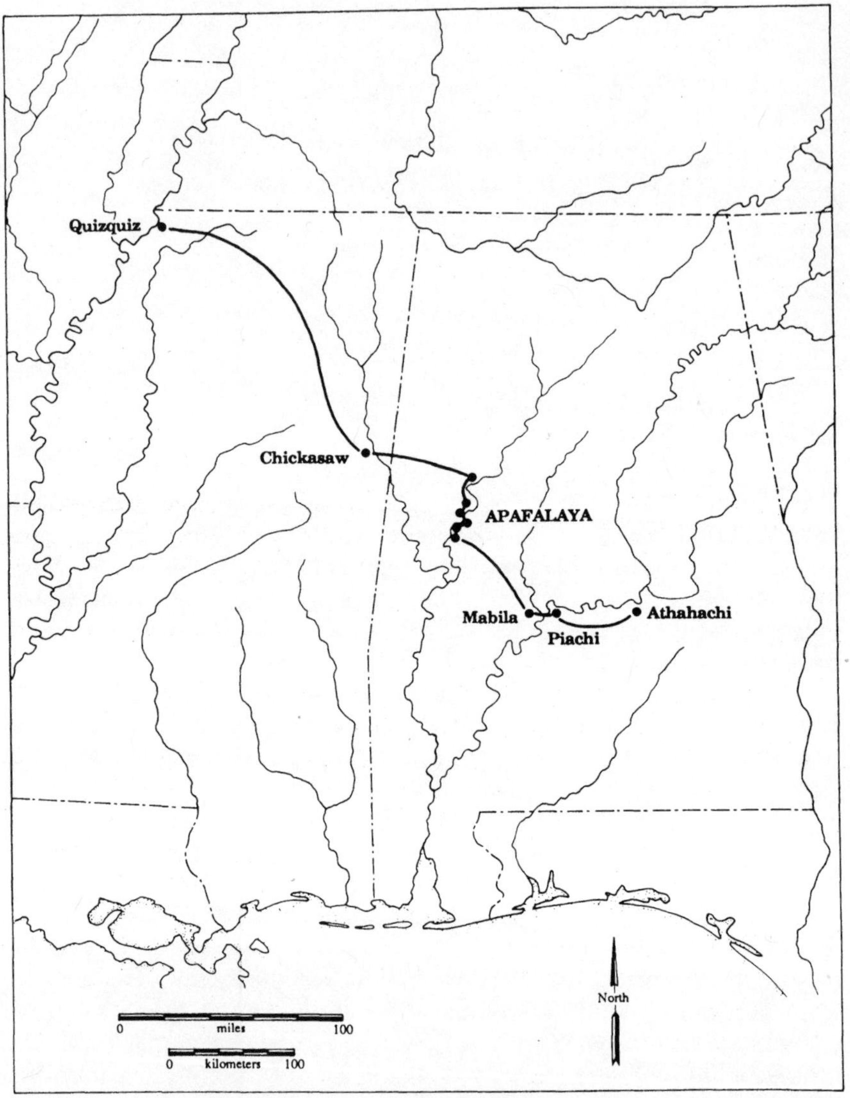

Figure 10-1. The de Soto Route from Athahachi to Quizquiz

incide with polities, the Lower Cahaba River lies near a cultural boundary between Pensacola phase and Moundville III phase to the north as well as a phase farther up the Alabama River that Ned Jenkins has described as a Moundville variant (Jenkins 1982:121; Sheldon and Jenkins 1986:98–99). Moreover, uninhabited buffer zones apparently lay between our proposed location for Mabila and the Moundville complex to the northwest and also between our proposed location for Mabila and a cluster of sites near the junction of the Coosa and Tallapoosa rivers to the east (Knight 1988). Finally, these locations are consistent in a general way with events that occurred during the Luna expedition (Hudson et al. 1985a).

A recent survey of the Lower Cahaba River by Boyce Driskell under the auspices of the Alabama Historical Commission has revealed substantial late prehistoric sites at two locations on the river—Old Cahawba and Haysop Creek near Centreville—and very little in between (Carey Oakley, personal communication). The site at Haysop Creek does not fit our reconstruction of the route of the expedition. It is too far distant from the Alabama River to have been the site of Mabila. Although it is possible that this survey failed to detect one or more sites in between these two, on the basis of the evidence that is now available, Old Cahawba is the most likely site for Mabila on the Cahaba River itself.

Even more recently (February 1987), Vernon J. Knight, Jr. conducted a limited excavation at Old Cahawba to determine whether this location could be eliminated as the site of Mabila. Knight found postmolds of a palisade like that described by the chroniclers; surrounding them was a ditch (not mentioned by the chroniclers); and in the bottom of the ditch were pieces of Late Pensacola phase pottery that date to the sixteenth century. Near the site is a relic channel of the Cahaba River that could be the "pond" mentioned by Fidalgo of Elvas (1866:89). At present we can only conclude that Mabila *could* have been at the Old Cahawba site but that far more evidence will be needed before the case can be said to be persuasive.

Mabila could have been at Old Cahawba, but it also could have been elsewhere in the vicinity. For example, Clarence B. Moore excavated a mound (32 feet [9.6 m] wide and just over 4 feet [1.3 m] high) ca. 1 mile (1.6 km) downriver from Old Cahawba. Moore excavated about three-fourths of this mound; he found few artifacts—only burials and a few projectile points—but the cultural affiliation and time of occupation of the mound are unknown (Moore 1897–1901:302). Other sites are said to exist on the Cahaba River to the west and northwest of present Selma, but they have not been investigated in recent times (Edwards 1921).

On November 14, 1540, de Soto managed to rouse his army and depart from Mabila. Biedma explicitly says that they traveled toward the north (Biedma 1922:21).[1] Because they were carrying only a two-day supply of corn (Fidalgo of Elvas 1866:91), they probably traveled as rapidly as they could. They probably followed a trail that lay close to present Highway 14, skirting the edge of the piedmont hills. Perhaps guided by one or more Indians of Mabila, they were headed for the Moundville archaeological complex. For three days they traveled through an uninhabited area, camping each night. On November 17 they came to a "very fine river"—the Black Warrior, probably just south of present Eutaw, Alabama (Ranjel 1922:128).[2] During these four days of travel they encountered no settlements of people (Fidalgo of Elvas 1866:92). Travel was especially difficult because it was cold and rainy (Biedma 1922:21). The distance from the mouth of the Cahaba River to the vicinity of Eutaw is ca. 60 miles (96 km), requiring them to travel 15 miles (24 km) per day. This schedule would have taxed them, but it was within their capability. If Mabila was located further upstream from Old Cahawba, they would have had a less strenuous rate of travel.

On November 18 they continued their travels, probably following a trail that lay near the present Southern Railroad. On this day they had to cross over bad places and swamps ("*malos pasos y ciénegas*"). Given the rainy weather, it would have been particularly swampy along Big Brush Creek and Five Mile Creek, and they would have had to go to some trouble in crossing these creeks. At the end of the day they came to the village of Talicpacana, where they found corn (Ranjel 1922:128).[3] Presumably, all the people of this town had fled.

Because of recent advances in refining the chronology of the Moundville phase, it is now far easier to identify de Soto period sites than it was just a few years ago (Bozeman 1982; Steponaitis 1983). For reasons to be discussed later, we think it probable that de Soto arrived at the Black Warrior River in the latter part of the Moundville III period. All sites known to have been occupied earlier than the Moundville III period, and later abandoned, can be eliminated from consideration.

Talicpacana was possibly located at the White mound and village site (1HA7/8), which lay ca. 14 miles (22.4 km) to the north of the point at which they came to the Black Warrior River (Figure 10-2). This is the southernmost minor ceremonial center of the Moundville phase, and a strong Moundville III component is present at the site (Bozeman 1982:246–61).

They evidently rested at Talicpacana for two or three days, sending out parties of cavalry to explore the country. They discovered two nearby villages. One, on the opposite side of the Black Warrior River, appeared to be "finely situated" (Ranjel 1922:129). Unfortunately, Ranjel does not name this village, nor does he say that they subsequently reached it, although they probably did. On Sunday, November 21, Vasco Gonçalez found the second village, named Moçulixa, half a league (about 1.7 miles [2.7 km]) distant from "this" (presumably half a league from the village on the opposite side of the river). The people of Moçulixa had fled their town, taking their stores of corn to the other side of the river, where they piled it up and covered it with mats. When the Spaniards appeared, the warriors of Moçulixa on the other side of the river threatened them.

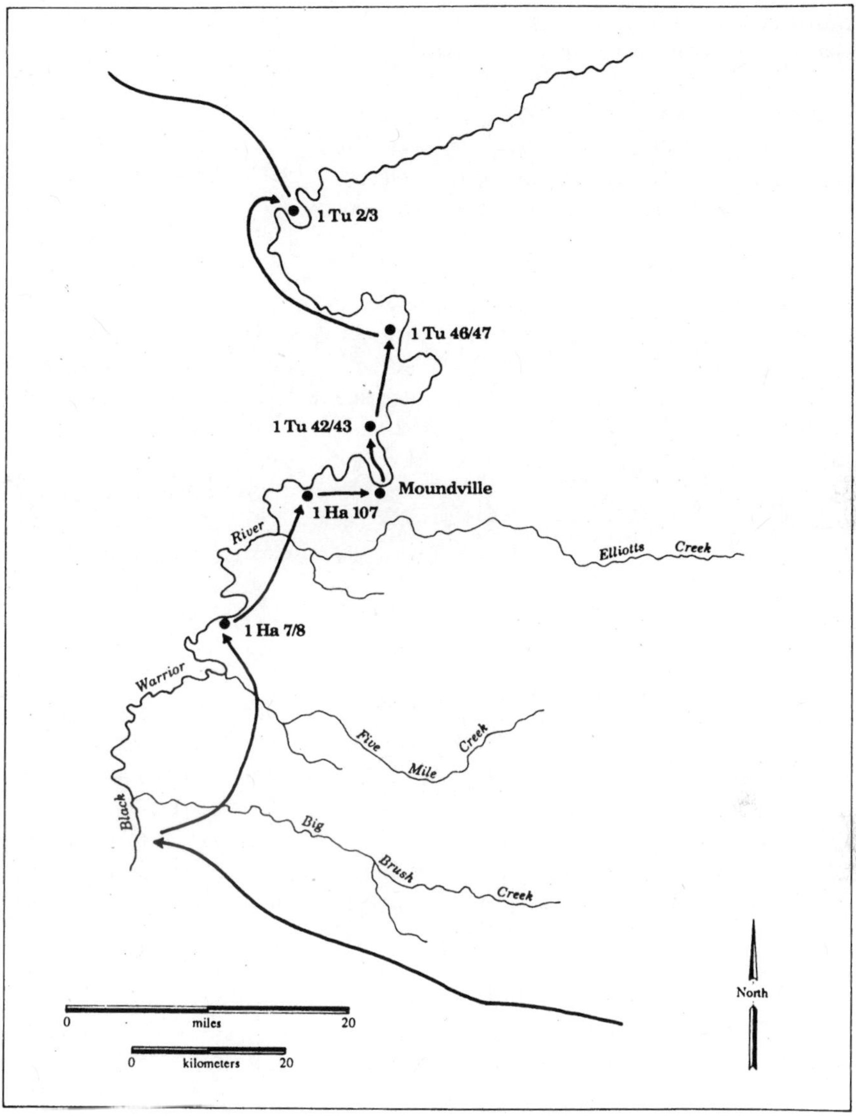

Figure 10-2. De Soto in the Black Warrior River Drainage

The site of the town they saw on the opposite side of the river is unlocated. A possible location of Moçulixa is the 1HA107 site ca. 9 miles (14.4 km) north of 1HA7/8. The mound that once stood at this site has been completely destroyed by agriculture and erosion. It stood on a bluff overlooking the Black Warrior River. None of the chroniclers makes any mention of a mound at Moçulixa. Perhaps because the chroniclers regularly observed mounds at large towns, they ceased mentioning them.

What interested de Soto and his men was not mounds but the corn piled up on the other side of the river. They desperately needed this corn, but the river was too deep to ford. As a way of surprising the Indians on the other side, de Soto set his men to building a flatboat (piragua) in which to convey a small strike force across the river quickly. They built this barge in four days (Fidalgo of Elvas 1866:91), completing it on November 29 (Ranjel 1922:129).

Unfortunately, Ranjel does not specifically say where the flatboat was built. The only thing that is clear is that after they completed it, they had to drag it on a sled half a league upstream to a place where they could launch it into the river (Fidalgo of Elvas 1866:91–92). It is likely that after they discovered the existence of Moçulixa, they moved to that place and built the boat there. This action is implied by Fidalgo of Elvas (1866:91), who says that they moved from Talicpacana to Moçulixa (he erroneously says Zabusta) and built the boat within the town.

De Soto wanted the boat built in secrecy, and site 1HA107 would have been a good place to do it. Because the town was elevated, the Indians could not easily have seen from the opposite side of the river what the Spaniards were doing. The Spaniards had only to defend the perimeter of the area they occupied. The bluff would have made an attack from the river difficult for the Indians.

Garcilaso telescopes and confuses the crossing of the Black Warrior River with the Spaniards' later crossing of the Tombigbee. But at least some of the details of his account fit the action at Moçulixa. In particular, Garcilaso describes the steep banks of the river. And he makes it clear that the Indians all retreated to the northern side of the Black Warrior, where they tried to defend this bank against the Spanish invaders. He also says that at first the Indians tried to come in canoes at night to attack de Soto's men. There were three places where they could make a landing. But the Spaniards dug foxholes at these places, and harquebusiers and crossbowmen hid inside them. When the Indians appeared, the Spaniards shot at them and then attacked them with swords, doing them great injury. After two or three such encounters, the Indians attacked no more (Garcilaso 1951:394).

Garcilaso adds yet another detail. In the interest of secrecy, he says, they built the boat in the woods, at some distance from the camp. However, his distances from the camp are probably incorrect—a league and a half upriver, and a league from the bank—and he is incorrect in saying that they built several flatboats.

By November 29 the flatboat and the sled on which it was to be transported were completed. Early on November 30, the mules and horses (and possibly Indian slaves) pulled the sled to a place where it could be launched into the river. In some places the terrain was so rough the men had to carry it. Garcilaso says that they launched the boat into the river at dawn (Garcilaso 1951:395). Ranjel says the boat was launched at Moçulixa (1922:129), whereas Fidalgo of Elvas says they pulled it a half league upstream (1866:91–92). Ranjel says that 60 soldiers entered the boat, whereas Fidalgo of Elvas puts the number at 30. As the flatboat was launched and boarded, some Indians on the other side of the river heard the commotion, and they ran to the riverbank to resist the landing. They shot many arrows, but, according to Ranjel, only three or four Spaniards were wounded. When the boat reached the opposite shore, the Indians ran away and hid in some canebrakes (Fidalgo of Elvas 1866:92; Ranjel 1922:129).

It is not clear whether the piles of corn were still on the other side of the river. Though none of the chroniclers says so, the boat probably made additional crossings, ferrying over contingents of cavalry and infantry to secure the other side of the river. Two loads of 30 men would have made up the 60 mentioned by Ranjel. Meanwhile, the cavalry on the Moçulixa side of the river went upstream to a town named Zabusta to secure a place where the entire force could cross over. On the following day, Wednesday, December 1, the entire army crossed the river at Zabusta in the flatboat and in several canoes they found at this town.

Zabusta was possibly located at the Moundville site (1TU50), which at its zenith occupied as much as 250 acres (100 ha) and which fronted a steep bluff on the Black Warrior River. On both sides of this bluff were places where boats or canoes could be landed (Kandare 1983:72). Again, none of the chroniclers mentions mounds at Zabusta, and this is surprising in light of the large number of mounds at the site. But there are good indications that the political fortunes of the Moundville site had declined during the late Moundville III period, so that by 1540 parts of the site may have been abandoned and partially overgrown. Also, de Soto and his men would have been there for only the time it took to ferry everyone across the river.

After they crossed the river, they probably went to the Wiggins site (1TU42/43), ca. 4 miles (6.4 km) upstream from Moundville.[4] From Ranjel (1922:120) it appears that they spent the night of December 1 at yet another village. This village was probably at site 1TU46/47, which is ca. 4 miles (6.4 km) upstream from site 1TU42/43. This site, 1TU46/47, has a mound that measures 165.0 by 148.5 feet (50 by 45 m) at the base, 82.5 by 82.5 feet (25 x 25 m) at the summit, and is 6.6 feet (2 m) high. It was occupied during the Moundville II and III phases (Bozeman 1982:112–28).

Presumably they continued on the next day, Thursday, December 2, traveling to yet a third town, where they captured a chief who was named Apafalaya (Ranjel 1922:129–30). It is probable that this town was the Snow's Bend site (1TU2/3), located ca. 14 miles (22.4 km) upstream from site 1TU46/47 (DeJarnette and Peebles 1970). The Snow's Bend site has one of the largest Mississippian mounds in the Warrior Valley outside of the Moundville site itself. It measures 138.6 by 138.6 feet (42 x 42 m) at the base, 85.8 by 89.1 feet (26 by 27 m) at the summit, and it is 11.6 feet (3.5 m) high. It was definitely occupied during the Moundville III phase (Bozeman 1982:94–109).

Unfortunately, none of the chroniclers specifically mentions the name of this third town. It is clear that "Apafalaya" was the name of the chief, and of the province (Fidalgo of Elvas 1866:91), and of the river (Ranjel 1922:130). In most of the places where de Soto traveled, the chief and the main town of the chiefdom had the same name, and this was probably the case at Apafalaya. Long ago John R. Swanton noted that the place names de Soto encountered at Apafalaya appear to have Choctaw etymologies (1985:218).

These identifications of named towns with specific archaeological sites on the Black Warrior River are tentative. Much additional archaeological survey and excavation will be needed before much confidence can be placed in locating Talicpacana at 1HA7/8, Moçulixa at 1HA107, Zabusta at 1TU50, and Apafalaya at 1TU2/3. But it is clear that the distribution of major and minor ceremonial centers along the Black Warrior River is consistent with the overall pattern of activities of the de Soto expedition. That is, the southern centers are on the eastern side of the river and the northern centers are on the western side of the river.

It is too bad that none of the chroniclers had anything to say about whether the people of Apafalaya had been allies of Tascaluza in the battle at Mabila. It is clear that unlike the people who lived within a 4-league radius of Mabila who had been so devastated they could field no warriors after the battle, Apafalaya was able to field many warriors, but despite their numbers they did not stand and fight. They would only shoot arrows at the Spaniards while they were crossing the river. One would expect a more forceful military response from the people of such an impressive archaeological site and area. There are several possible explanations for why the Indians of Apafalaya put up so little resistance: (1) perhaps the Moundville chiefdom was past the apex of its social power and was, in fact, in decline; (2) perhaps these people did sustain losses in the battle at Mabila and were too weakened to mount a significant resistance; or (3) perhaps they were neither in decline nor had participated in the battle at Mabila but had learned of the carnage at Mabila and wished to avoid similarly devastating loss of life.

No diagnostic middle-sixteenth-century Spanish artifacts have yet been found on the Black Warrior River, and for this reason the terminal date of the

Moundville III phase cannot be ascertained with precision. The best authorities place the transition from Moundville III to Alabama River phase in the neighborhood of 1500 to 1550 (Peebles 1986:26; Steponaitis 1983:114–26). Thus, the terminal date of Moundville III falls very near the time of the de Soto expedition, but it matters very much on which side of the terminal date the expedition occurred. Even though it may have been unstable, it is clear that a chiefdom existed on the Black Warrior River during the Moundville III phase, whereas in the Alabama River phase that followed there is no evidence of mound building nor of hierarchical social organization. It is clear as well that the overall health of the population declined sharply during the Alabama River phase (Peebles 1986:31).

We infer from our reading of the de Soto documents that a chiefdom was in place on the Black Warrior River when de Soto arrived, and the existence of a chiefdom implies that de Soto arrived at the Black Warrior during the Moundville III phase. The encounter of de Soto and his men with the people along the Black Warrior is similar to their encounter with chiefdoms elsewhere in the interior of the Southeast. "Apafalaya" was the name of the "province" as well as the name of the only chief in the area mentioned by the chroniclers. Moreover, de Soto made hostage the chief of Apafalaya just as he did the chiefs of Coosa and Tascaluza.

If our provisional scenario for de Soto's encounter with various sites on the Black Warrior River proves to be accurate, it implies that in 1540 the center of the chiefdom was not at Moundville (1TU50) but at Snow's Bend (1TU2/3). This placement is consistent with existing archaeological evidence that the Moundville III phase was a time of "decline," though "instability" may be a more accurate description. After ca. 1400 the Moundville site ceased to grow in size and in internal complexity, and within another century or so the site was abandoned, or nearly so (Peebles 1986:29–30).

It is now abundantly clear that Mississippian chiefdoms were dynamic, competitive societies, so that their developmental trajectories were not onward and upward forever (DePratter 1983:204–11). In some ways the situation de Soto found in the chiefdom of Apafalaya was similar to what he found in the chiefdom of Coosa (DePratter et al. 1985; Hudson et al. 1985b). That is, de Soto apparently visited the Etowah site, the largest Mississippian site in the territory of Coosa, but it was not at that time the center of the chiefdom. The center lay to the north, at the Little Egypt site, a less spectacular mound center. But for all that, in 1540 Coosa was still a vital chiefdom, which survivors of the de Soto expedition remembered as a rich province (cf. Hudson et al. 1985a). In Coosa at least, instability did not *necessarily* entail decline, although such an interpretation must be provisional until reliable population estimates for Coosa can be obtained.

With the time of transition from Moundville III to Alabama River phase falling so near the time of the de Soto expedition, one wonders whether this transition represented a demographic and social transformation similar to that which occurred in the chiefdom of Coosa. Namely, the people of Coosa ceased building mounds within about 50 years after the de Soto expedition, and at about the same time there is evidence of depopulation and abandonment of towns, as well as a diminution of social hierarchy (Smith 1987:97–98). The principal cause of this latter change was Old World pathogens. Hence, we interpret the social changes of the Moundville III phase as "normal" Mississippian social dynamics, whereas the changes entailed in the Alabama River phase were fundamental structural changes.

The foregoing is not the only way in which the social nature of the Moundville III and the Alabama River phases can be interpreted. For example, Christopher S. Peebles argues that the transitions could have occurred either just before or just after the de Soto expedition (1986:25). He argues that the kinds of changes that occurred during the latter Moundville III phase and the Alabama River phase period were all cut from the same cloth: "At the time of the earliest Spanish expeditions, the Alabama River phase communities already had begun the process of realigning themselves internally and externally. A new social order and a new order of alliances were already emerging. This new social formation was affected by de Soto, de Luna [*sic*], and Pardo, but they cannot be counted among its causes." (1986:33). Peebles states that "the population does not seem to have declined markedly between the Moundville phase and the Alabama River phase," but he does not in fact cite actual evidence that the population remained the same (1986:31).

Peebles is at pains to dispel what he claims is a widely held "myth" that "only an external force such as the Europeans and their diseases could have caused [the] destruction [of Mississippian societies]" (1986:24). The decline in the Moundville III phase was not caused by a decline in the health of the people nor by a major change in the subsistence economy, Peebles claims, but rather by a change in the nonsubsistence part of the economy. Namely, after 1450 the number of imported artifacts declined sharply: "One of the effects of this depression in foreign exchange was to weaken and ultimately remove one of the major props that supported the elite. If they were supported through tribute—food and locally manufactured goods—but could not make prestations in the form of socially valuable goods, then their legitimacy was called into question" (1986:30).

Peebles lists several possible causes of this "depression of foreign exchange." But whatever the cause, the effect was that the Moundville elite were first undercut and then they were cast adrift like so many deposed Romanovs. With no one in command, competition for resources increased, and this competition led to malnutrition and disease. And he adds that the malnutrition and disease may have been exacerbated when the fallen elite began practicing agriculture and were exposed to soil fungi (Peebles 1986:30–32).[5]

This is not the place to argue the merits of our conception of Moundville III–Alabama River phase as opposed to Peebles's. But we do observe that he appears to confuse the substance of social power with the symbols of social power. If, for example, BMWs and Mercedes-Benzs were to become unavailable in America, it is unlikely that this shortage would have any effect on American social structure. Moreover, we see no merit in his attempt to explain all of the changes entailed in late Moundville III and Alabama River phase in terms of endogenous causes, whereas exogenous causes are generally acknowledged to have been important elsewhere in the Southeast. By exogenous causes we mean the direct military actions that occurred in the sixteenth-century entradas (as at Mabila), political instability set in motion by de Soto (particularly along the Mississippi River), and Old World pathogens that could have been introduced by the sixteenth-century entradas and most certainly during the century following these entradas. But little real headway can be made in deciding between these two scenarios until more is known about the archaeology of the Black Warrior River.

De Soto and his men rested for a week at the town where they captured the chief of Apafalaya. On December 9 they set out in search of the next chiefdom with the chief of Apafalaya as their guide and interpreter (Ranjel 1922:129–130).[6] On their sixth day of travel, December 14, they arrived at the River of Chickasaw after having traversed several swamps and cold rivers.[7] Biedma makes it clear that after departing from Mabila they traveled northward all the way to Chickasaw (1922:21). Because they encountered high water, it is likely that they traveled north from Apafalaya and then took a western turn somewhere south of present Fayette and from there they crossed the Sipsey River (perhaps near present Moores Bridge) and Luxapallila Creek (perhaps near present Millport). Continuing, they would have had to cross Mud Creek, Yellow Creek, and Cooper Creek. The most difficult of these crossings would have been the Sipsey River.[8] A second, less likely, possibility would have taken them on a more westerly than northerly route. Namely, they could have traveled south of Luxapallila Creek, following a trail that lay on or near the present Gulf, Mobile, and Ohio Railroad. For the entire distance from Apafalaya to the Tombigbee River they encountered no human habitations (Fidalgo of Elvas 1866:92).[9]

At five places old roads and trails crossed fords in the segment of the Tombigbee River west of Apafalaya: the Columbus Crossing, west of Columbus; the Waverly Crossing, above the mouth of Tibbee Creek, just northwest of

Columbus; the Barton Ferry Crossing, just below the mouth of Buttahatchee River; the Morgan's Crossing, near the present town of Aberdeen; and the Cotton Gin Port Crossing, southwest of present Amory. Traveling north of Luxapallila Creek from Snow's Bend it is about ca. 75 miles (120 km) to the Columbus Crossing, the Waverly Crossing, and the Barton Ferry Crossing, which, given six full days of travel, the Spaniards could have easily made by averaging 12.5 miles (20 km) per day. In fact, this pace was a little slower than they usually traveled. From Snow's Bend to the Aberdeen Crossing, it is ca. 84 miles (134.4 km), a distance they could have made by traveling 14 miles (22.4 km) per day. From Snow's Bend to the Cotton Gin Port Crossing, it is ca. 90 miles (144 km), a distance they could have made by averaging 15 miles (24 km) per day. This latter rate of travel was within their capability, but given their lack of food supplies and the cold and wet conditions under which they traveled, it would have required strenuous effort.

When they reached the Tombigbee River, they found it to be flooded and out of its banks. As Rufus Ward (1986:62–68) has pointed out, because of this flooding, two of the five possible crossings mentioned above would have been favored above the others. Their northerly route would have enabled them to circumvent most of the water that would have been backed up in Luxapallila Creek and its tributaries. But backwater along the margins of the Buttahatchee would have made for a very difficult crossing, and the documents indicate no such crossing. If we assume that Chickasaw was north of Tibbee Creek, this would favor the Barton Ferry or Waverly crossings. But when the Tombigbee River is in flood, the floodplain adjacent to the Waverly Crossing is much wider than that which is adjacent to the Barton Ferry Crossing. Hence, the Barton Ferry Crossing would have been an easier place for them to have crossed.

But if we assume that the main town of Chickasaw was south of Tibbee Creek, and if they traveled north of Luxapallila Creek, then the Columbus Crossing was the one they must have taken. This crossing was favored above the others in having high ground on both sides of the river with a relatively narrow space in between. It would have been the best place to make the crossing when the river was high, and there is historical evidence that nineteenth-century Indians in the area were cognizant of the advantages of this crossing (Rufus Ward, Jr., personal communication). If they traveled south of Luxapallila Creek, they would have had a somewhat wider expanse of water to cross. Here their crossing would have been between the mouth of Luxapallila Creek and the mouth of Magowah Creek.

On the opposite side of the flooded river they could see many Indians, armed and threatening, and according to Ranjel, they had many white flags ("*banderas blancas*") (1922:131).[10] De Soto ordered his men to build a piragua, and he ordered Baltasar de Gallegos to take 30 horsemen, all of whom were good swimmers, upstream to find a crossing place and to cross the river and attack the Indians on the other side. When the piragua was completed, de Soto sent an Indian across in it to try to make peace with the Chickasaws, but they killed him at once (Fidalgo of Elvas 1866:92). One wonders whether this unfortunate ambassador was the chief of Apafalaya.

The Indians on the other side of the river learned of Gallegos's attempted crossing, and they retreated from their position. It is not clear whether they learned of Gallegos's maneuver before or after he crossed with his cavalry. In any case, de Soto was able to cross the Tombigbee without incident on December 16 or 17 (Fidalgo of Elvas 1866:92; Ranjel 1922:132).

Once on the other side of the river, de Soto set out in search of the main town of Chickasaw with a force of cavalry. Late that night they arrived at the main town of Chickasaw, and they found that it had been abandoned by its people (Fidalgo of Elvas 1866:92; Ranjel 1922:132).

The site of the main town of Chickasaw has not yet been found. From events that transpired during de Soto's stay, it is clear that the chief of Chickasaw was a paramount chief. Each paramount chief whom de Soto encountered to this point lived at a mound center. A notable cluster of mounds occur on Upper Chuquatonchee Creek and its tributary Houlka Creek. Five mounds occur at the Owl Creek site (22CS502), with the largest being ca. 20 feet (6 m) high (Brookes 1977). They are arranged in a typical Mississippian pattern on a generally level terrace of Chuquatonchee Creek. The Paden site, also originally with several mounds, lies ca. 3 miles (4.8 km) southwest of the Owl Creek site on a small tributary of Chuquatonchee Creek. Yet another cluster of mounds occurs at the Thelma site (22CS501), on Upper Houlka Creek. But all three of these mound sites date to the Middle Mississippian period. All had been abandoned by the time of the de Soto expedition (Atkinson 1986:21; Atkinson and Johnson 1985).

Another concentration of Mississippian mounds occurs between Line-Tibbee Creek and the Noxubee River. Of the seven known mounds in this area, only two have been professionally excavated (Atkinson 1987:63). These are the Lyon's Bluff site on the southern bank of Line Creek and the Lubbub Creek archaeological locality on the western side of the Tombigbee River near the mouth of Lubbub Creek. Both of these sites are known to have had sixteenth-century occupations (Marshall 1986; Peebles 1987). While it would be imprudent to assume that a Mississippian chiefdom could not exist in the absence of a mound center, most of them did have mound sites as their centers, and therefore mound sites should be thoroughly explored before possible nonmound centers are explored.[11]

The only documentary clue is that the chroniclers make it clear that after crossing the river, de Soto led a party of cavalry to the main town and arrived there late at night. On another occasion, when the expedition was in the present state of Georgia, de Soto led a small force of cavalry from the Flint River to the Ocmulgee River. They traveled ca. 12 leagues (ca. 41.5 miles [66.4 km]) in 18 hours, an average speed of ca. 2.3 miles (3.7 km) per hour. If we assume that de Soto set out from Barton Ferry at, say 11:00 A.M. and reached the main town of Chickasaw at 7:00 P.M., well after sunset in this season of the year, he and his men would have had eight hours in which to travel. They could have traveled ca. 18.5 miles (29.6 km), which means that the main town of Chickasaw was somewhere in southwestern Monroe County, central or eastern Clay County, eastern Oktibbeha County, or Lowndes County, or even northern Noxubee County.

According to Garcilaso de la Vega (who is often in error on matters of place and time), the main town of Chickasaw was situated on a flat hill that extended north and south, it was surrounded by ravines that contained little water, and near it were many groves of oaks and "walnuts" (that is, walnuts or hickories or both) (Garcilaso 1951:397). The town appears to have been palisaded, with gates in the palisade, though the precise form of the gates is not described (Fidalgo of Elvas 1866:95).

The land around Chickasaw was fertile, and the settlement pattern reminded Fidalgo of Elvas of Mabila (1866:92). That is, the houses were spread out over the countryside, only short distances apart, with much of the land under cultivation. Though none of the chroniclers says so, Apafalaya may have had a similar settlement pattern.

There is much in favor of James Atkinson's recent proposal that Chickasaw was located on the watersheds of Magowah Creek, Catalpa Creek, and the southern tributaries of Tibbee Creek (1987). This territory is within the broad area specified above, and as we shall presently see, it is consistent with the travel time of the expedition as it traveled northwest to Quizquiz and the Mississippi River. Whether Atkinson's precise delimitation of the central territory of Chickasaw will hold up will depend upon the results of future archaeological research. Certainly this area should receive priority if in the future an archaeological search for Chickasaw is conducted.

De Soto and his men spent a difficult winter at Chickasaw. At Christmas there was a severe snowstorm, which they compared to the snow that falls in the Burgos area of Spain, and this weather was especially hard because they were short of clothing. Also, in the main town of Chickasaw they found only 20 houses that they could use (Fidalgo of Elvas 1866:92), and by Christmas they still had not built enough additional houses to shelter all of the men (Fidalgo of Elvas 1866:92; Ranjel 1922:132). In building these additional houses they apparently got most of their materials from tearing down houses in nearby hamlets (Garcilaso 1951:390).

The Chickasaws kept up continual pressure on the Spaniards, not so much by actually attacking as by creating several alarms each night. When the Spaniards ran out to attack them, the Indians faded into the countryside. The purpose of these maneuvers was possibly to keep the Spaniards from sleeping, or perhaps to dull their vigilance and make them think that the Indians were cowards, thus paving the way for a surprise attack (Garcilaso 1951:398).

In time the Spaniards succeeded in capturing a few Chickasaws, including one who was very close to the chief. With this man in captivity, de Soto sent a messenger to the chief saying that he wanted to talk to him and to be on a friendly footing (Fidalgo of Elvas 1866:92–93). On January 3 the chief of Chickasaw came to see de Soto. As befitted a paramount chief, he was carried upon the shoulders of some of his people. As tribute, he gave de Soto some deerskins and some little dogs. Subsequently, the people of Chickasaw brought the Spaniards many rabbits to eat (Biedma 1922:22).[12]

The chief of Chickasaw also provided de Soto with guides and interpreters to go to Caluça, a province that was said to have had over 90 villages. It was a province that was much feared because the warriors of Caluça were said to be very savage, and the soil of that region was said to be fertile (Ranjel 1922:132). None of the chroniclers tells where the Caluça were located. It is possible that the Caluça were people who lived on the Yalobusha-Yazoo River or the Big Black River. Wherever these people were located, de Soto presumably sent a detachment of men to look them over.

The chief of Chickasaw brought two chiefs who were subjects of his to meet with de Soto. These were Miculasa, chief of the Chakchiumas, and the chief of the Alibamu.[13] These chiefs likewise brought tribute gifts of shawls, skins, and rabbits (Fidalgo of Elvas 1866:93).

The chief of Chickasaw came to visit de Soto many times. Having been evicted from his own town, he lived at a town that was located in a savannah about a league or so from de Soto's camp (Fidalgo of Elvas 1866:94). When de Soto wanted to see the chief of Chickasaw, he would send over a horse on which the chief was to ride to and from de Soto's camp (Fidalgo of Elvas 1866:93). On one of his visits the chief of Chickasaw complained to de Soto that one of his subject chiefs had rebelled and was refusing to pay tribute. The chief asked de Soto to assist him by sending some Spanish soldiers along with his own warriors into this rebellious chief's territory to punish him. De Soto agreed to do it, and he led 30 cavalry and 80 infantry, along with 200 of Chickasaw's warriors. They marched to Chakchiuma to find that the town had been burned and abandoned. But all was not as it seemed. De Soto began to suspect that the chief of the Chickasaws intended to lay a trap for him by contriving to split his force so that the Indians could gain an advantage in a planned attack. Whether or not his was the Indian's intent, the Spaniards were too vigilant for the Chickasaws to achieve the surprise they needed. The Spanish-Chickasaw army returned to de Soto's camp, and de Soto invited several of the chiefs to dine with him; he served them pork, which was very much to their liking. After they returned, Chief Miculasa sued for peace, and messengers came to de Soto from Talapatica, a town or person not mentioned again (Fidalgo of Elvas 1866:93–94; Ranjel 1922:132–33).

The question is, where were Alibamu and Chakchiuma located? As we shall presently see, when de Soto departed from Chickasaw, going in a northwestwardly direction, he came to Alibamu at the end of his first day of travel. This would seem to indicate that Alibamu was located on the upper tributaries of Tibbee Creek. James Atkinson proposes that the territory of Alibamu lay north of Line Creek and west of Chuquatonchee Creek, but including the lower course of Houlka Creek, where there is a known concentration of late prehistoric sites (1987:67–68). One question that might be raised about his delineation of the Alibamu territory is that it does not include a mound site. A very slight extension of this territory to the south would include the Lyon's Bluff site (22OK1), a solution that ought to be given careful consideration.

The Lyon's Bluff site was laid out in the same way in which the main town of Chickasaw was laid out. That is, the Lyon's Bluff site was a small palisaded town situated on the edge of a steep bluff (10 to 11.6 feet [3 to 3.5 m]) above a flood plain. It contained some 14 houses, comparable to the 20 houses in the main town of Chickasaw. The Lyon's Bluff site was occupied as early as A.D. 1130, and its occupation continued into the Protohistoric period, until at least A.D. 1550 (Marshall 1977). The cultural materials at Lyon's Bluff, which are contemporary with de Soto, are from the Sorrels phase. These materials are similar to the Moundville III phase associated with Apafalaya. The Sorrels phase also shows some ceramic resemblances to cultures in eastern Arkansas and northwestern Mississippi, the cultures de Soto visited after departing from Chickasaw (Sheldon and Jenkins 1986). Also consistent with information in the de Soto chronicles is the fact that, in addition to compact villages on the bluffs, many Sorrels phase households existed down in the floodplain of Tibbee Creek, near the fields where the people farmed (Marshall 1986).

Because Chickasaw was a paramount chief, it makes sense to expect that he was centrally located between his two tributaries. Therefore, with Alibamu to the north, one would expect to find Chakchiuma to the west or south. Atkinson argues that Chakchiuma lay to the west of Chickasaw. One piece of evidence he cites is the so-called de Soto map, which shows "Niculas" (probably "Miculasa," chief of the Chakchiumas) almost directly west of the Chickasaws (1987:64). The problem with this map is that it contains many evident inaccuracies, particularly east of the Mississippi River. For example, it shows Apafalaya on the western side of the Tombigbee River, when it was clearly on the eastern side.

Atkinson also cites evidence that some or all of the Chakchiumas were located in this area in the early eighteenth century, as attested by the Marigny map of 1743. A late prehistoric occupation has been established for this same area, as well as the occurrence of European artifacts from the late sixteenth and/or seventeenth centuries—the Rolling Hills complex (Atkinson 1979). The historical evidence for the Chakchiumas having lived in the vicinity of Starkville in the early eighteenth century is well established, and there is historical evidence that some of them lived on the Yalobusha River at about the same time (Atkinson 1987:66–67). But it remains to be definitely established that the people who occupied this area in the sixteenth century were ancestral to those who lived there in the eighteenth century. One of the lessons learned in our recent researches is that eighteenth-century ethnic locations in the Southeast do not necessarily coincide with sixteenth-century locations.

As an alternative hypothesis for the location of the Chickasaws and their tributaries, we propose the following. The core territory of Chickasaw was located west of the Tombigbee River opposite present Columbus, including the watersheds of Magowah Creek and Catalpa Creek.[14] The territory of Alibamu lay on Line Creek and its tributaries, as well as the western tributaries of Chuquatonchee Creek. The territory of Chakchiuma lay between the Noxubee River and the Tombigbee River, including the Lubbub Creek Archaeological Locality (Galloway 1982a:154–55). The location of these territories provides a context for explaining why the Summerville II-III phase of the Lubbub Creek Archaeological Locality so closely resembles the Sorrels phase of the Lyon's Bluff area (Jenkins 1982:138).

One difficulty with this interpretation is that in going with Chickasaw warriors to attack Chakchiuma, de Soto would have had to split his army for several days. It would have taken a day or more for an army of cavalry and infantry to reach Chakchiuma, at least a day may have been spent there, and a day or more to return. Thus de Soto's army would have been split for three or more days. But the very fact that the chroniclers express concern about a conspiracy suggests that de Soto may, in fact, have split his army for this long a period. Also, during the Tristán de Luna expedition of 1559–61, a similar instance occurred. A company of Luna's soldiers who were sent from Luna's camp to the chiefdom of Coosa split their ranks and accompanied a party of Coosa warriors to chastise the Napochies, a tributary people who were refusing to pay tribute. On this particular venture, the Spaniards had to travel just over two days to reach the Napochies (Hudson et al. 1985a).

De Soto was probably conciliatory toward the Chickasaws because he wished to save his forces. On one occasion, he very nearly beheaded two of his own men for stealing skins and shawls from the Indians, who had reacted to the theft by making a great commotion and abandoning their houses (Fidalgo of Elvas 1866:94).

In early March, de Soto prepared to depart from Chickasaw. He asked the chief of Chickasaw to supply him with 200 carriers. But the Chickasaws became openly angry at de Soto's request, and he recognized that they were hostile as he rode about on the evening of March 3, the day before he was to depart (Ranjel 1922:133).[15]

Before dawn on the morning of March 4, the Indians of Chickasaw struck de Soto with a very costly surprise attack. A strong north wind was blowing (Garcilaso 1951:398), and several hundred Indians eluded three Spanish sentinels and crept up to the camp with fire concealed in little pots. Then with a great tumult they set fire to the houses. The fires spread quickly, and the half-clothed, half-armed Spaniards ran out into the night. In this attack, the Chickasaws killed 59 horses and 12 Spaniards (Ranjel 1922:134–35).[16] Many of the other Spaniards were wounded, and they lost much of their clothing and many of their weapons in the fire. Many of the horses burned to death in their stalls, which were presumably inside the town (Fidalgo of Elvas 1866:96). But the Chickasaws did not press their advantage to the fullest. They may have thought that some of the horses that were stampeding about in the dark were cavalry (Fidalgo of Elvas 1866:96). The Chickasaws withdrew. More than 300 of de Soto's hogs burned up in the pigsty that had been built by the Spaniards (Biedma 1922:23; Garcilaso 1951:404).[17]

Three days later, the Chickasaws returned at night to again attack. But it so happened that it began to rain, and when this rain wet the strings of their bows, the Indians turned back without attacking (Garcilaso 1951:407–08).

After the battle, on or about March 8, de Soto and his men moved a league or so away from their burned-out camp, to the little town in the savannah (*sabana*) where the chief had been living (Fidalgo of Elvas 1866:97–98; Garcilaso 1951:405; Ranjel 1922:134). Alonso de Carmona referred to this place as Chicaçilla ("Little Chickasaw") (Garcilaso 1951:405). Here they made camp on a sloping hillside (Ranjel 1922:135). They built a forge, making a bellows out of a bearskin, and with this they retempered their steel weapons, which had been in the fires. They made new lances, saddles, and shields (Biedma 1922:23; Fidalgo of Elvas 1866:98; Garcilaso 1951:406; Ranjel 1922:135).

While de Soto was at Chicaçilla, he kept up a particularly vigilant watch at night. And during the day he sent out cavalry with orders to kill every Indian they saw. Later the cavalry swore that not an Indian was alive within a radius of 4 leagues of their camp (Garcilaso 1951:408).[18]

On March 15, the Chickasaws again attacked in the early morning. But this time the Spaniards were prepared. Their camp was on ground that was more level, and probably more open, where they could use their horses to good advantage. The Indians killed two horses and wounded two, but they killed no Spaniards (Garcilaso 1951:409). On the other hand, the Spaniards succeeded in killing only a few Chickasaws (Ranjel 1922:135).

From some of the prisoners he had captured, de Soto got information on the route he would follow to the Mississippi River. He knew that he would have to cross an uninhabited area that would require 7 to 12 days to cross and in which the only food they would have was that which they carried (Biedma 1922:24; Fidalgo of Elvas 1866:99).

On April 26 they departed from Chicaçilla and traveled through an area that included many small communities of 15 to 20 dwellings each, arriving that night at the town of Alibamu (Fidalgo of Elvas 1866:99; Ranjel 1922:136).[19] They traveled toward the northwest, and according to Garcilaso, they traveled 4 leagues on this day (Biedma 1922:24; Garcilaso 1951:415). They found very little corn in Alibamu; apparently the Indians had hidden it. Traveling northwest, they would have first encountered Alibamu towns on Line Creek and its tributaries.

Because he needed all the food he could get, on April 27 de Soto sent out three parties of cavalry and infantry in different directions to scout the country. Juan de Añasco went in the direction that de Soto would later follow, presumably to the northwest, and nearby he found a very strong fortification containing about 300 Indian warriors (Biedma 1922:24). It was built on a savannah near a small stream with very steep banks. A likely location of this fortification was on a tributary of Line or Houlka Creek. Many creeks in this part of Mississippi have remarkably steep banks.

For several reasons, de Soto determined to attack this fortification. He felt that if he failed to do so, the Indians would be emboldened and would harass the Spaniards mercilessly (Fidalgo of Elvas 1866:100). He may also have assumed that the fortification contained something that was worth obtaining (Biedma 1922:24). According to Garcilaso it was a square, palisaded fort with two transverse interior walls that ran parallel to the first wall and joined the side walls. The fortification had doors in the front and back walls that were so small a horse could not go through them. When the Spaniards breached the exterior wall, the Indians fell back behind the first interior wall. And when the Spaniards penetrated this one, the Indians retreated to the second interior wall, until finally they ran out the back of the fort and across crude, narrow footbridges laid across the stream (Garcilaso 1951:416–19). These bridges were high above the water, suggesting very steep banks (Garcilaso 1951:419). The use of these footbridges implies that the channel of the creek was not very wide. Because of the steep banks, the horses could not cross this stream, and the Alibamu had cleverly deprived de Soto of his most important military advantage.

The Spaniards incurred heavy casualties in taking this fort. Seven or 8 of them were killed and 25 or 26 were wounded, some of whom died later (Biedma 1922:24; Fidalgo of Elvas 1866:100). The Indians had shot many of them through the thighs, which were unprotected by armor. The Spaniards only succeeded in killing a few Indians, and they captured a few.[20] From these they learned that the Indians of Alibamu had built the fort for the purpose of testing themselves against the Spaniards (Biedma 1922:25). The fortification contained no food.

On April 30 they departed from the Alibamu fortification and began traveling toward the Mississippi River. According to Ranjel, they reached the first village of Quizquiz at midday on May 8, after eight and a half days of travel from the Alibamu fort (and nine and a half from Chicaçilla) (Biedma 1922:136). Along the way there were many pondy areas that they had to ford or swim (Fidalgo of Elvas 1866:101). Starting from Chicaçilla they went northwest to the vicinity of present Houston, Mississippi.

Beyond present Houston they could have gone in two directions. One possibility is that they could have veered off to the west, going south of the Yalobusha River to present Grenada, where they could have picked up an Indian trail known as Charley's Trace, which would have taken them to Friar's Point or Sunflower Bend (Atkinson 1987:68–69). Using this route, the total distance from the Alibamu fort to the first town of Quizquiz would have been ca. 128 miles (204 km), which would have required them to travel 15 miles (24 km) per day. They would have had to cross ca. 70 miles (112 km) of the Mississippi River meander belt. This route has been favored both by the United States De Soto Expedition Commission (Swanton 1985) and by Jeffrey P. Brain, Alan Toth, and Antonio Rodriguez-Buckingham (1974).

The second route would have taken them due north of present Houston to cross the small upper tributaries of the Little Tallahatchie River near present New Albany. From here they would have swung northwest, crossing the Tippah River and several small creeks to go through the vicinity of present Holly Springs. Circling around through the small upper tributaries of the Coldwater River, they headed directly west. Then they picked up a trail on the headwaters of Johnson Creek that they followed to encounter towns of the chiefdom of Quizquiz and the Mississippi River. In this approach to the Mississippi River, they would have had to cross fewer than 20 miles (16 km) of the Mississippi River meander belt, and most or all of this travel would have been along the levee ridge of the Walls meander, which then would have been a lake. The total distance of this route from the Alibamu fort would have been ca. 132 miles (211 km), which they could have achieved by averaging the travel time of 15.5 miles (24.8 km) per day.

Either of these routes is possible on the basis of the average daily travel they would have required. But we favor the second of these possible routes for three reasons. First: It is consistent with Biedma's statement that they traveled northwest after departing from Chicaçilla. Biedma is the only chronicler who frequently mentioned the direction in which they traveled, and when he did so it was usually a general direction they followed for several days.

Second: The northwestward route is more consistent with the chroniclers' characterization of the terrain they traversed. On the northwestern route they certainly encountered woods and swamps ("*montes e ciénegas*") when they crossed small streams and creeks. For example, in February 1740, M. Céleron led an army of Frenchmen and Indians from Fort Assumption at present Memphis southward to raid the Chickasaws. The creeks, rivers, and canebrakes they crossed are similar to those described by the de Soto chroniclers. Even if Céleron's army did not follow precisely the same trail that de Soto followed, they surely traversed the same kind of terrain (de Lery 1923).[21] None of the de Soto chroniclers mentions crossing the vast swamps they would have encountered had they crossed 70 miles of the Mississippi River meander belt, and it is instructive that when they were on the western side of the Mississippi River, on the two occasions on which they had to cross swampy terrain in the Mississippi River meander belt and Western Lowlands, the chroniclers complain about the terrible swamps (Hudson 1985:3–6).

Third: A crossing of the Mississippi River in the vicinity of the Walls meander makes it possible to construct a route west of the Mississippi River that has fewer geographical and archaeological contradictions and inconsistencies than a crossing at either Friars Point or Sunflower Landing. For example, Brain and his colleagues argue that Aquixo was at the Old Town site (3PH20) and that Casqui was near the junction of the L'Anguille and St. Francis rivers. When de Soto went from Aquixo to Casqui, he had to travel through a terrible swamp. But travel from the Old Town site to the junction of the L'Anguille and St. Francis rivers could have been done rather easily by following along the natural levee ridges of the Mississippi and St. Francis rivers. It is for this reason, presumably, that Brain and his colleagues argue that de Soto took a far more difficult route by going behind Old Town Lake and through the Big Creek/Lick Creek drainages (Brain et al. 1974:268). But why would de Soto have taken this route when it was always his practice to have Indian guides show him the best trails?

A more serious inconsistency occurs when Brain and his co-workers take de Soto to Pacaha, which they place at the Belle Meade site (CT30). To take this route, de Soto and his men would have had to cross the Lower St. Francis River, yet the only crossing the chroniclers mention was a kind of lake across which the people of Casqui built a footbridge, which they strung together along trees growing in the "lake." Finally, when they went from Casqui to Quiguate, which Brain and his colleagues place at the Dupree site (3PH1), de Soto would necessarily have traveled back through Aquixo territory, but none of the chroniclers mentions doing so. Finally, this reconstruction of the route also directly contradicts Ranjel, who says that Quiguate was on the River of the Casqui. That is, Quiguate was on the same river that Casqui was on, and that river was definitely not the Mississippi.

A further test of the credibility of these two routes from Alibamu to the Mississippi could be done if enough archaeological survey information could be obtained. Namely, it is quite clear that the Spaniards traveled the entire eight and one-half days through an uninhabited area. Hence, if significant Late Mississippian populations existed along either of these routes, then that route could be eliminated. One body of information that favors the northwestern route (that is, does not contradict it) is Janet Rafferty's survey in the vicinity of the Ingomar Mounds site in which no significant amounts of Late Mississippian materials turned up (Rafferty 1983). Also, no Mississippian mounds have been recorded in northern Pontotoc County nor in Union County (Janet Rafferty, personal communication).

Returning now to our reconstruction of the route, let us follow de Soto's approach to the Mississippi River. He and his men came to the first village of Quizquiz on May 8, taking the people there completely by surprise. This first village was possibly at the Irby site (22DS516), a town they would have encountered immediately after having come down from the bluff. All of the men were absent from this village because they were working in their cornfields. De Soto captured more than 300 women, including the mother of the chief (Biedma 1922:25; Fidalgo of Elvas 1866:101; Ranjel 1922;137). De Soto sent word for the chief to come, but the chief replied that he would only do so if de Soto released his captives. Because his men and horses were so weak from hunger, de Soto agreed to the chief's demand.

Even though de Soto released his prisoners, the Indians of Quizquiz came and threatened to fight (Biedma 1922:25). While de Soto waited with his army near a "river bank," the Indians approached to within a crossbow shot and stopped. This encounter probably occurred near the lake formed by the Walls meander (David H. Dye, personal communication). Interestingly, the Spaniards found little corn in this first town (Fidalgo of Elvas 1866:102).

Garcilaso says that the chief of the first town lived on a high mound with two stairways leading up to the summit (Garcilaso 1951:423).[22] Garcilaso is unreliable with respect to time and location, but this description may indicate that at least one of the towns of Quizquiz that they visited had a mound. De Soto and his men left this first town and went to a second town of Quizquiz a league away, probably the Lake Cormorant site (22DS501), where they found plenty of corn. And a league beyond this they came to another similarly well-provisioned town, and near here they saw the Mississippi River for the first time. This may be the same town that Fidalgo of Elvas says was half a league from the river (1866:102) and was possibly the Norfolk site (22DS513). All of this would seem to imply that the first town of Quizquiz was 2.5 or 3 leagues from the river.[23]

The day on which de Soto first saw the river is not reported. It may have been as early as May 9 or 10, but certainly it was before May 21, when the Spaniards moved to a savannah between the village and the river and began building four piraguas in which to cross to the other side. The chroniclers describe the bluffs on both sides of the river as being quite steep and high (Garcilaso 1951:428).

The most frustrating thing about de Soto's route from Mabila to the Mississippi is that along it lay two of the sites that should most be expected to yield significant quantities of sixteenth-century Spanish artifacts in identifiable contexts—Mabila and the main town of Chickasaw. But at this time neither site can be pinpointed. We are confident, however, that in looking for these needles we have located some promising areas in the haystack for archaeologists to examine.

Finally, although we intended only to take our hero to the Mississippi River, we would be remiss if we did not briefly summarize some of his activities on the other side of the river to demonstrate that our reconstruction avoids the inconsistencies and contradictions mentioned earlier (Hudson 1985:3–6). De Soto crossed the Mississippi River and entered the chiefdom of Aquixo, comprising the Walls phase towns north of Horseshoe Lake. Like Quizquiz, Aquixo was tributary to Pacaha, a powerful chief whose center was located up the river. From Aquixo the expedition traveled westward, spending an entire day crossing a very bad swamp, traveling through the vicinity of present Simsboro, Greasy Corner, and Round Pond. On the other side of this swamp they reached the first towns of Casqui, where they picked up the levee ridges along the St. Francis River and followed them upstream to the main town of Casqui, which Dan and Phyllis Morse have located at the Parkin site (Morse and Morse 1983:309).

From Parkin they traveled to Pacaha after the people of Casqui first built a footbridge over a "lake"—that is, across Gibson Bayou or Big Creek. The territory of Pacaha certainly comprised Nodena phase, as the Morses have argued (1983:311), and the main town was somewhere north of Wapanocca Lake. It is interesting that Nodena phase and its tributary Walls phase were once classified into the same phase, suggesting that pottery and polities do sometimes coincide (Phillips 1970:933–34). De Soto remained in Pacaha for about a month, sending out two expeditions to the north, one of which probably reached the Campbell site (23PM5). Had this latter expedition gone any farther north it would no doubt have entered into the wilderness of the Vacant Quarter.

De Soto and his men returned to Casqui and continued on until they came to a river across which the people of Casqui ferried them in canoes—namely, the St. Francis River near present Madison. After traveling down the western side of the St. Francis for three days they came to the chiefdom of Quiguate, which must have comprised Kent phase sites. Many of these were located near the River of the Casqui—the St. Francis (House 1987).

The difficulties in the routes proposed by Swanton, Brain and co-workers, and Atkinson also apply to a segment of the de Soto route recently proposed by Richard A. Weinstein. Weinstein traces the route from Chickasaw to Quizquiz (1985). He places the main town of Chickasaw in the vicinity of Tupelo or else in northern Chickasaw County or southern Pontotoc County. From here he traces the route west-southwest along the Skuna River. He argues that the expedition crossed the Yalobusha near present Grenada, then swung southwest to a few miles north of present Greenwood, and then turned northwestward to pass near Schlater, Minter City, and Tutwiler before turning west to the Oliver site (22CO503) near the Sunflower River, where they encountered the first village of Quizquiz. From here he says they went west and then northwest to Sunflower Landing, where they came to the third village of Quizquiz.

Weinstein's reconstruction directly contradicts information in the documents. Biedma specifically says that they set out in a northwestward direction from Chicaçilla ((1922:24), whereas Weinstein takes them west-southwest from Chicaçilla. Moreover, Weistein asserts that the Oliver site (with 23 mounds) was the first town of Quizquiz and also the dominant town (1985:17), whereas Fidalgo of Elvas specifically says that de Soto found little corn in this first town but that when he moved on to another town he found a sufficiency of corn (1866:102).

As Weinstein himself recognizes, his placement of these towns of Quizquiz is inconsistent with Ranjel's statement that they were spaced about a league apart—a distance of ca. 7 miles (11.2 km) from the first to the last. But the distance from the Oliver site to Sunflower Bend is ca. 17 miles (27 km), more than twice the distance specified by Ranjel (Weinstein 1985:19).

In contrast, the placement of the three towns of Quizquiz at the Irby, Lake Cormorant, and Norfolk sites accords well with Ranjel's description.

One of the advantages Weinstein claims for his route is that it takes the de Soto expedition near Schlater, Mississippi, where a halberd (possibly sixteenth-century Spanish) has been found, and also to the Oliver site, where a large number of probable and possible de Soto artifacts have been found (that is, glass, a large quantity of turquoise beads, two cut quartz beads, a brass point, and at least two and possibly five Clarksdale bells). But all of these small artifacts could easily have been carried from one place to another. The discovery of these artifacts does not prove that de Soto visited these sites (cf. Hudson 1987:7–8). In Hudson's reconstruction of the de Soto route west of the Mississippi River (1985) he located Aminoya in the vicinity of the Friars Point and Sunflower Bend area. Aminoya was the place where the survivors of the expedition spent their last winter and where they built the boats in which they would make their escape down the river. To lighten their load, they left behind everything they could spare, and one can easily explain the artifacts at Schlater and the Oliver site as having come from Aminoya.

Acknowledgments

For information, advice, assistance, and criticism we are grateful to the following individuals: Lawrence Alexander, Jim Atkinson, Sam Brookes, David H. Dye, Caleb Curren, Ned Jenkins, Marvin Jeter, Vernon J. Knight, Richard Marshall, Roger Nance, Gerald P. Smith, Vincas P. Steponaitis, and Rufus Ward, Jr. But we bear full responsibility for our interpretation. We are grateful to Julie Smith for drawing the maps.

Notes

1. Many scholars have plowed this same field. The work of the U.S. De Soto Expedition Commission is the point of departure for all modern research on the route (Swanton 1985). Caleb Curren's recent reconstruction of de Soto's route through Alabama resembles that of the U.S. De Soto Expedition Commission in many respects (Curren 1987). George E. Lankford's reconstruction of this same segment of the route is novel in that he places the chiefdom of Apafalaya on the Black Warrior River in approximately the same location as we do (Lankford 1977), and James Knight has concurred with Lankford (Alexander 1982).

2. The 1909 soil map of Hale County, U.S. Department of Agriculture, shows this area to be free of swamps.

3. Fidalgo of Elvas had this as Taliepataua, obviously a mistranscription or a misprint. Garcilaso says that it took them three days to make this leg of the journey, and he erroneously thought that Apafalaya was a part of the chiefdom of Chickasaw (Garcilaso 1951:393).

4. Subsequently 1TU42/43 became the largest town in the Moundville area (Vernon J. [James] Knight, personal communication). That is, during the Alabama River phase, this is a site where several populations coalesced.

5. In 1987 Peebles published a fuller statement of his theory of Mississippian decline in western Alabama. Constraints of space do not allow us to reply here.

6. One is tempted to infer from this occurrence that the languages of Apafalaya and Chickasaw were mutually intelligible, though of course the chief of Apafalaya could have been bilingual.

7. We have adopted the convention of using the English name of "Chickasaw." The Spanish spelling is "Chiçaca."

8. See James Adair's (1775:325–26) vivid account of crossing this same stretch of country in December 1747.

9. Biedma says they traveled from Mabila to Chickasaw in 10 or 12 days, suffering greatly from cold and rain (1922:21). Actual trail time, counting the crossing of the Tombigbee River was ca. 13 days. Garcilaso, who erroneously thought that Apafalaya was the first town of Chickasaw, says that after crossing the river (the Black Warrior), they marched for 4 days over level ground to the main town of Chickasaw. He gives no account of crossing the Tombigbee River (1951:397).

10. The use of white flags in this situation seems inconsistent with southeastern color symbolism. White was the color of peace.

11. The chiefdom of Guatari, which Juan Pardo encountered in western North Carolina in 1566–68, appears to have lacked mounds (Hudson n.d.).

12. Fidalgo of Elvas says 150 "conies." It is striking that they were given no venison. Could this gift have been motivated by sympathetic magic—that is, rabbits to induce timidity (cf. Mooney 1890:14)? Or did the Chickasaws truly lack venison?

13. We have adopted the convention of using English "Chakchiuma" instead of Spanish "Sacchuma." Fidalgo of Elvas has Miculasa spelled "Niculasa," and he has Sacchuma spelled "Saquechuma" (1866:93). "Miculasa" may contain the root *mico*.

14. In 1985 Richard Marshall constructed a similar hypothesis about the location of Chickasaw. We are grateful to him for sharing his thoughts with us as well as providing us with documentary information.

15. Elvas says it was on March 8, which is probably a mistranscription.

16. Biedma says 57 horses were killed. Fidalgo of Elvas says 11 Spaniards died (1866:97).

17. Fidalgo of Elvas says 400 hogs were lost, but he mistakenly writes that they were burned up at Mabila—an interesting error (1866:97).

18. This distance was the same as that which they burned over the country around Mabila.

19. Fidalgo of Elvas says they departed on April 25.

20. Fidalgo of Elvas says they killed only three Indians (1866:100).

21. We are grateful to James Atkinson for calling this document to our attention.

22. Garcilaso erroneously calls Quizquiz "Chisca."

23. We are grateful to David H. Dye and Gerald Smith—and particularly the latter—for assistance in working out de Soto's point of crossing the Mississippi River. After working this out, we discovered that Gregory Perino came to essentially the same solution 20 years ago (Perino 1967:140–41).

11

The Terminal Mississippian Period in the Arkansas River Valley and Quapaw Ethnogenesis

Michael P. Hoffman

The first scientific study of the Quapaw Indians was accomplished in 1827 when George Izard reported certain facts about them to the American Philosophical Society (Bizzell 1982:72). He was given an account of the Quapaw migration tradition by Paheka, the grandfather of the principal chief.

I imagine that Paheka's first three words to Territorial Governor Izard in Quapaw were, "a'nax q sikkattoka wajini," which means, "Listen up, turkey-Whiteman"—although I have no proof. What follows is an English translation of Paheka's story of Quapaw movement to Arkansas.

> When we abandoned our former lands we set out without knowing whither we were going. Our motive for leaving the country we occupied was the scarcity of game.... On arriving at the mouth of the Ohio River [Ny-Tonka], our chiefs determined on separating the nations.... After our separating, our party followed the course of the Ny-Tonka[1] [Mississippi]. The first red skins [Indians] whom we met with were settled some way below the Ny-Whoutteh-Junka[2] [the Little Muddy River, now the St. Francis]; they were called Tonnika. We attacked and put them to flight. Some time afterwards we entered this river, which we call Ny-Jitteh [Red River, now the Arkansas]. We soon discovered that there were other red skins in the country. Parties were sent out to look for them. They were found encamped in the Great Prairie [between the Post of Arkansas and the town of Little Rock]. We attacked them; they made a valiant resistance, but we beat them and drove them away. This nation called itself Intouka; the whites at that period gave them the name of Illinois. Then we were left entire masters of this country. (Bizzell 1982:72)

This story of their migration to Arkansas was, and is, very important to the Quapaw because it was told repeatedly to Europeans and Americans and is still being told today. Sometimes it has fallen on deaf ears, however.

This paper focuses on the origins of the Quapaw tribe and its protohistory. Various lines of evidence including archaeological evidence are brought to bear on these subjects. As we shall see, contemporary archaeological interpretations and other sources of information often conflict, because archaeological remains attributed to the Protohistoric Quapaw show continuity with Central and Lower Mississippi Valley Mississippian period cultures, while tribal traditions and linguistic, ethnological and ethnohistorical data often do not. This "Quapaw paradox" realization is not new (Hoffman 1986:30) and most sources of information about it are not either. However, this paper does present some new data and options for resolving the conflict between archaeological and other information sources.

Who Are the Quapaw?

The Quapaw are a tribe, an ethnic group of Native Americans, who lived in four or more villages on the Mississippi and Arkansas rivers in the latter part of the seventeenth century when first contacted by the French (Figure 11-1). The term "Quapaw" comes from the word "Ogaxpa," which has been translated variously as "the downstream people," or "downwind people," or "into the south wind people" (Dickinson 1984:201). They were called "Arkansa" (Arkamsea, Arkansea) by the seventeenth-century Illinois Indians. The Quapaw are an enduring people. Most other Central and Lower Mississippi Valley ethnic groups had disappeared by the late eighteenth century. The Quapaw numbered about 1,900 in 1985.

Sources of Information and Perspectives about Quapaw Origins

Tribal Oral Traditions

The Quapaw and the Osage, Omaha, Ponca, and Kansa, the other tribes in the Dhegiha language group, share a common belief, passed on through oral tradition, that they were at one time a single people in the area of the Lower Ohio River (La Flesche 1915:459). With the exception of the Quapaw, the Dhegiha tribes managed to cross the Mississippi River and moved westward. The word "Omaha" means upstream and refers to the idea that they

Figure 11-1. Quapaw Indian ca. 1700 (Charles Banks Wilson, artist)

moved up the Mississippi River to cross it. The Quapaw moved downstream, as Paheka relates, and drove the Tunica out of the Lower St. Francis River country. S. D. Dickinson, Arkansas's foremost ethnohistorian, thinks the "Intouka," whom the Quapaws drove out of the Grand Prairie area, were actually the Tunica instead of the Illinois, based on linguistic evidence (Dickinson 1984:202). Paheka was the grandfather of Principal Chief Heckaton in 1827; thus this oral tradition dates at least well back into the eighteenth century. Quapaw oral tradition does not include when the movement southward took place.

Linguistic Relationships

The Quapaw language is not spoken anymore, although several tribal members know many words. Archaeologists thought until recently that the tribe's linguistic affiliations were poorly known (Griffin 1960:813; Morse and Morse 1983:321). Robert Rankin, a linguist at the University of Kansas (1982, 1985, personal communication), has compiled nineteenth- and twentieth-century Quapaw vocabularies, some of which he collected himself, and demonstrates conclusively that Quapaw is a Dhegiha Siouan dialect or language. His most recent paper, incidentally, was stimulated by the linguistic uncertainties of Griffin and the Morses.

The Quapaw language is very closely related to the others of the Dhegiha group—the Omaha, Osage, Ponca, and Kansa. The language does not show indications of amalgamation with other Siouan or non-Siouan peoples (Rankin 1985:5). If extensive amalgamation did occur, Rankin postulates that there should be linguistic indications of such mixing such as a "Siouan-based pidgin language, a creole" (1985:5). Rankin further says (1985:6): "The Quapaw language is virtually free of traces of language contact in proto or prehistoric times; the only borrowings, calques or transparent coinages result from contact with English, or French (either directly or via the Muskogean-based Mobilian trade jargon). . . . Pidginization and creolization are ruled out by the fidelity with which Quapaw inflection matches that found in the other Dhegiha languages."

He further states (1985:6–12):

> Quapaw material culture may have developed in situ near the mouth of the Arkansas River, but examining the phonology, grammar, and vocabulary of the language, it is impossible to imagine it has descended from anything other than common Dhegiha Siouan, wherever that may have been situated. The language shows none of the usual features of simplifications associated world-wide with pidgins, compromise languages or mixed languages of any sort whatsoever. . . . I think it unlikely that the Quapaw language was somehow transferred from one people to another sometime during the protohistoric period. Such events leave traces in the vocabulary and/or grammar of a language especially over an elapsed period of only a few hundred years. . . . Quapaw linguistically is the quintessence of Dhegiha, whatever that may portend in or for ethnohistory or archaeology.

Two independent glottochronological studies of Dhegiha have yielded very similar results. Grimm (1985:16) estimates that Quapaw speakers separated from the other Dhegihas between A.D. 1070 and 1470. Viers's estimate (1983) for the separation is sometime between A.D. 1289 and 1513. A third study, that of Headley (1971:78), estimates the separation at ca. A.D. 950. These estimates must be evaluated in light of the many criticisms of the glottochronological technique.

Siouan linguists don't agree about what, ethnically, great linguistic similarity indicates. Rankin (1985) and Grimm (1985) imply the traditional view that great similarities indicate relatively recent common culture and identity. John Koontz (1985:57), on the other hand, perceives that the Dhegiha languages represent "a dialect continuum," contiguous tribes residing near each other but not considering themselves a single people. Such as linguistic continuum could have existed in the Lower or Central Mississippi Valley.

Quapaw Ethnology

Quapaw ethnology is another source of information useful for studying origins, affiliations, and protohistory of the tribe. The logic of using ethnological comparison is that Quapaw cultural characteristics should be very similar to other Central and Lower Mississippi Valley tribes, if the Quapaw were long native to the territory. Swanton (1911, 1946) and Hudson (1976) have demonstrated that many cultural practices were widely present in the Lower Mississippi Valley and indeed, most of the Southeast, at least in the seventeenth and eighteenth centuries. Likewise, if the Quapaw were very similar ethnologically to their Dhegiha kinsmen, the time of Dhegiha separation would be recent and presumably the time since the tribe lived near the mouth of the Ohio River would be short. One immediate problem is that no Quapaw tribal ethnographies exist. The late-nineteenth- and early-twentieth-century Siouan scholars, such as Dorsey (1884, 1885, 1897), McGee (1897), and La Flesche (1915), neglected the Quapaw for a variety of reasons. For the most part, cultural characteristics of the Arkansas Quapaw have to be picked out of French and early American accounts that cover the period of 1673 to 1834. Small amounts of late nineteenth- and early-twentieth-century Quapaw ethnographic information are available in Dorsey, McGee, and La Flesche.

Recent research by Hudson and DePratter (1984) seems to demonstrate that the Arkansas River Valley was densely populated in the 1540s. It is probable that the Quapaw and Carden Bottoms phases were represented. However, because at least some doubt exists that there is a one-to-one identification of these phases with Quapaw or Quapaw-related peoples, ethnographic data from the de Soto narratives are not included here.

The Quapaw people of 1673 to 1834 were village farmers, living in long-term settlements composed of long, dome-shaped houses covered with bark. Multiple family bark-, mat-, or skin-covered summer village long-houses were also characteristic of other Dhegiha tribes and most midwestern and northeastern Native Americans (Driver 1961:Map 15). Except for the Shawnee and Algonquian tribes on the northern fringe of the Southeast, this type of house was absent from the Southeast away from the mid-Atlantic coast.

Fletcher and La Flesche (1911:35) refer to the five Dhegiha peoples as "cognate tribes" that "bear a strong resemblance to one another not only in language but in tribal organization and religious rites." Such similarities include patrilineal descent, the Omaha cousin system, patrilineal clans, specific clan names, and moieties. Most other southeastern groups, particularly the Gulf-speaking tribes, were matrilineal with matri clans and Crow cousin terminology. The Quapaw in the late 1600s and early 1700s had independent village chiefs whose power was not autocratic, characteristics different from strongly ranked Lower Mississippi Valley groups like the Natchez. However, the post–de Soto depopulation of much of the Southeast must have strongly affected the ranking system of many groups.

The ancient Quapaw religion is little known, but in common with other Siouan-speaking people, Wah Kon Tah was the sacred dynamic force in the universe that could occupy people and things. Swanton (1911:166) notes the absence of a specialized temple structure among the Quapaw, a characteristic of most of the other Historic tribes of the Lower Mississippi Valley. Guardian spirits were important to individual Quapaw as sources of power and protection (Feiler 1962:63), a concept foreign to most southeastern tribes. Dogs were sometimes eaten at special rituals, an act present in several midwestern and northeastern Indian groups but not common in the Southeast (Feiler 1962:62).

A qualitative characteristic of the Quapaw is the tribe's adaptability, a trait that no doubt has contributed to its survival. The Quapaw have freely intermarried with other tribes and with whites. They learned how to keep their land in the face of U.S. government pressures by increasing their numbers through adoption and then dividing their land voluntarily into individual landholdings in the late nineteenth century (Baird 1980:127). In the same era they enthusiastically adopted a new religion, the Peyote religion, which helped them cope with the pressures of acculturation. The new religion stimulated the abandonment of many of the old ways (Wilson 1978). Much of Quapaw Indian culture today has been borrowed from their Oklahoma Indian neighbors, including the form of the intertribal powwow and gourd, war, and stomp dances.

The two general features of Quapaw ethnology emphasized here—strong similarities with the other Dhegiha peoples and general relationships to the Midwest and Northeast, and Quapaw cultural adaptability—may both be important in understanding their origins and protohistory.

Quapaw Ethnohistory

A basic question in Quapaw ethnohistory is where to start. Early interpretations of the chiefdom of Pacaha or Capaha in eastern Arkansas that was visited by de Soto identify it as Quapaw (Swanto 1911:186; Thomas 1959:333). This identification was based mainly on the linguistic similarity of the words "Capaha" and "Ogaxpa." Swanton (1939:51–52) later changed his mind about the identification because the seventeenth-century Quapaw lacked the specialized temple structure of the Pacaha and because of some Pacaha linguistic similarities with the Tunican tongue. Morse and Morse (1983:320) revived the idea of the Pacaha and Quapaw relationship at least to some extent.

Charles Hudson and Chester DePratter (1984), in their recent study of de Soto's route west of the Mississippi River, identify the River of the Cayas as the Arkansas River. The River of the Cayas was densely populated with farming people. Aside from the chiefdom of Cayas, with its major center of Tanico, other chiefdoms on the Arkansas River included Autiamque, south of Little Rock, and Ayays and Anilco further south on the river. Apparently no sharp linguistic break occurred between the chiefdoms of northeastern Arkansas (Aquixo, Casqui, Pacaha, and Quizquiz) and those of the Arkansas River. A linguistic break did occur, however, between chiefdoms along the Arkansas River and Tula, which is thought to have been a Caddoan-speaking group of the Ouachita Mountains.

The only Protohistoric archaeological phases known in the Lower and Central Arkansas Valley regions are the Quapaw and Carden Bottoms phases. Brain (1985b:xli), Dickinson (1980), and Swanton (1911) identify the Ouachita River as the River of the Cayas. Dickinson (1980:2) presents linguistic evidence that may indicate that Tanico was a Tunican word and thus thinks the tribe was located along the Ouachita River. Coligua, another town, is thought by Brain (1985b:xl) and others to have been on the Arkansas River near Little Rock.

The first certain European descriptions of the Quapaw tribe began in 1673 with that of Father Marquette, who listed 10 possible Quapaw villages on his 1673 map (Tucker 1942), eight of which were on the Arkansas River (one had the unlikely Quapaw name of Akoroa). He visited only Kappa on the Mississippi River. When contacted again four years later by La Salle, the Quapaw lived in three villages on the Mississippi River with only one up the Arkansas River a short distance (Baird 1980:23). La Metairie and Joutel also refer to several Quapaw villages on the Arkansas River in the latter part of the seventeenth century (Thomas 1959:334). The French received information about the recent former haunts of the Quapaw from Illinois and Miami Indians living near the Lower Ohio River area. The Lower Wabash and mouth of the Ohio River regions were called "the River of the Arkansea" because the tribe was said to have told at least three times that the Quapaw were recently living on the Lower Ohio River (Hathcock 1983:15).

The Quapaw population rapidly decreased in the late seventeenth century and throughout the eighteenth century. It may have been 15,000 or more in A.D. 1682 (Baird 1980:37) but had fallen to fewer than 1,000 by 1700 and fewer than 500 in 1834. European diseases caused the decline. The population loss caused or stimulated the abandonment of old villages and the founding of new villages, coalescence of villages, and movement of Quapaw population. Only three pure Quapaw villages existed in 1806, all located on the Arkansas River, although a fourth village, near the mouth of that stream, was composed of Choctaw men who had married Quapaw women (Baird 1980:51).

Under pressure to leave Arkansas, some Quapaw moved to live with the Caddos on Red River in 1826 (Baird 1980:71). Later many Quapaw people lived temporarily with the Osage in Oklahoma. During both of these episodes their ethnicity was preserved.

In summary, it is possible that the Arkansas River valley was densely inhabited during the de Soto era, but whether the Indians were Quapaw- or Tunican-speaking is not known. The late-seventeenth-century Quapaw lived in four major villages, three on the Mississippi River and one on the Arkansas River. There are persistent references to additional villages up the Arkansas River as well as a possible Koroa-Tunica town or towns. There is also reasonably good indirect evidence that the Quapaw were living in the Lower Ohio River area at a time not too distant from the late seventeenth century.

Bioarchaeology

Bioarchaeological investigations should also contribute to an understanding of Quapaw tribal origins. Aware of the question of Quapaw origins, Rose and Marks (1985:94) outline nonmetrical skeletal traits that might be used to measure genetic affinities among Quapaw, other Dhegiha-speaking peoples, and Lower Mississippi Valley populations, and they suggest two relevant hypotheses when adequate samples become available (Rose and Marks 1985:97):

1. Culture change in the Central Arkansas River Valley was the product of indigenous development (i.e., genetic continuity) and not the product of population movement.
2. The Arkansas Valley Quapaw, a Siouan-speaking group, are a product of the enculturation [*sic*] of an indigenous population.

Unfortunately such skeletal samples do not exist. The only skeletal samples of Late Mississippian period peoples from the Central and Lower Arkansas River Valley are those collected by Moore at the Greer and Menard sites that were studied by Aleš Hrdlička (Moore 1908), those found by Ford (1961) at Menard, and a group collected by the University of Arkansas Museum at the Kinkead-Mainard site in 1933 (Hoffman 1977). These samples exhibit high caries and high porotic hyperostosis rates, features that are related to high consumption of maize (Rose and Marks 1984:96).

The Protohistoric Archaeology of the Lower and Central Arkansas River Regions

The Protohistoric archaeological manifestations of the Lower and Central Arkansas River regions are relatively well known, even if the data on which they are based are not recent. Both regions evidenced strong terminal Woodland period Coles Creek–related occupations, scanty and/or poorly known Early-Middle Mississippian indications, and a strong Late Mississippian–Protohistoric presence. Occupations of this latter period are marked by artifactual horizon markers such as Nodena points, small end scrapers, head pots, teapots, pottery types such as Parkin Punctated, Barton Incised, and Wallace Incised, and European trade goods (Hoffman 1986:24).

Quapaw Phase (*Figure 11–2*)

Phillips (1970:943) can be credited, or better, blamed, for the use of the tribal designation in conjunction with the archaeological manifestation. Phillips did so because of Ford's (1961) "conclusive identification" of what had been referred to earlier as the Menard or Wallace phase as the tribal Quapaw at the Menard site. Ford's (1961) careful ethnohistoric study linking the historic Quapaw village of Osotouy with the latest ceramic complex at the Menard site is in fact convincing and widely accepted.

The Late Mississippian Menard site ceramic complex has served as the basis for establishing the distribution of the Quapaw phase. Earlier, C. B. Moore (1908) recognized Lower Arkansas River ceramics as distinctive, and later Dickinson and Dellinger (1940) called the ceramic complex the "Arkansas Valley Culture" and first suggested a linkage with the Quapaw tribe. The ceramic complex and phase have been defined and redefined several times (Hathcock 1983:46–58; Hoffman 1977:32–35, 1983; House and McKelway 1982; Morse and Morse 1983:318; Phillips 1970:943–44). Mortuary

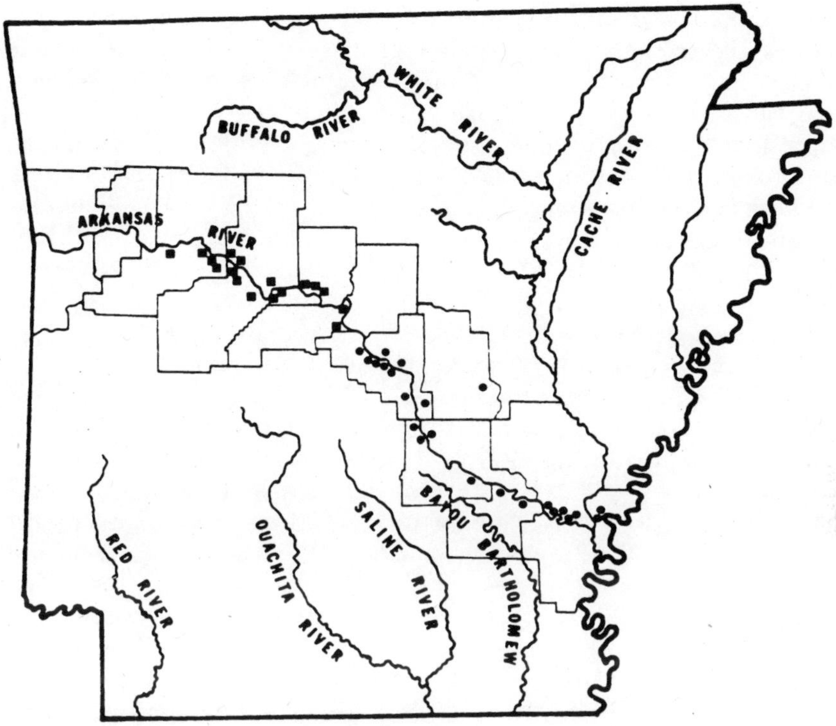

Figure 11-2. Protohistoric Phases Along the Arkansas River

pottery accounts for almost all of the Quapaw phase ceramics; excavated midden pottery is available only from the Menard site. The Quapaw phase combination of types, their varieties and frequencies, when taken together form a distinctive complex unlike any other in the Central or Lower Mississippi Valley.[3] However, the distributions of individual types and varieties and shapes of the ceramic complex extend well beyond the Lower Arkansas Valley. Old Town Red teapots, bottle shapes, Carson Red on Buff, *var. Olmond*, Avenue Polychrome, *var. Avenue*, Nodena Red and White, *var. Dumond*, and other styles similar to those of the Quapaw phase commonly occur in the mortuary pottery of the Kent, Old Town, and Carden Bottoms phases (Hathcock 1983:47–48). Parkin Punctated and Barton Incised styles also overlap to the northeast. Winterville and Leland Incised are found at Quapaw phase sites on the Lower Arkansas River and in much greater quantities to the south and east. The only recent report on a site of the Quapaw phase, that of Hoffman (1977) at Kinkead-Mainard, suffers from a ceramic classification with a Caddoan rather than Lower Mississippi Valley classificatory bias, but certainly there is a continuum in style between Leland Incised and Keno Trailed and between Wallace and Winterville Incised, *var. Belzoni.* Some pots from the Kinkead-Mainard site that had the "Tunica mode" (Brain 1979:224) are similar to Brain's Barton Incised, *var. Trudeau* (Hoffman 1977:Figures 7, 12e, 10, 55a), while others closely resemble Winterville Incised, *var. Tunica* (Hoffman 1977:Figures 4, 2a). Both of these varieties are associated with the historic Tunica (Brain 1979:234–39). Caddoan pottery types similar to those of the Ouachita Mountain area occur in increasing frequency in the upstream Quapaw phase sites. If, as most professional archaeologists think, Wallace Incised in significant quantities is considered a primary marker type for the Quapaw phase, its distribution is mainly from the Kinkead-Mainard site at the eastern end of the Arkansas River trough, southward to the mouth of the Arkansas River, and a short distance north of the Arkansas River mouth on the Mississippi River (Figure 11–2).

The distribution of the Quapaw phase ceramic complex corresponds poorly to the four known late-seventeenth-century Quapaw villages, one of which was on the Arkansas; the others were on both sides of the Mississippi River above its mouth. However, if the villages on Father Marquette's 1673 map of the Arkansas River are assumed to be Quapaw, then the correspondence is better.

Settlement and mortuary characteristics associated with the Quapaw phase are varied. Site plans commonly are compact. They probably were palisaded. The sites range from 2 to 40 acres (0.8 to 16.0 ha) (Menard). Temple mounds, conical mounds containing burials, and plazas are known at several sites (Hoffman 1986). A house structure excavated by Ford (1961), which measured 30 by 20 feet (9 by 6 m), was outlined by postholes. However, evidence of wattle and daub construction in the structure is at odds with Quapaw houses that are described as bark covered (Ford 1961:152–56). The only other published Quapaw phase structure, excavated by Moselage at the Dupree site near the Lower White River, was a small rectangular wattle and daub house (Moselage 1965:1–3).

Euroamerican trade goods occur in at least a half dozen of the approximately two dozen Quapaw phase sites. Although the trade goods have not been studied intensively, it is believed that they predate 1673 and the direct French presence (John House, personal communication). No known Quapaw phase site has yielded great quantities of late-eighteenth- or early-nineteenth-century trade goods.

Carden Bottoms Phase

The Carden Bottoms phase (Figure 11-2), an archaeological manifestation with a ceramic complex closely related to the Quapaw phase, exists in the Arkansas River trough above Little Rock to the vicinity of Paris, Arkansas (Clancy 1985; Hoffman 1986:27–30). It is known only from mortuary pottery in museum collections derived from early pothunter excavations. Recently, the University of Arkansas Museum collection from the major archaeological locality of this phase, Carden Bottoms, was the subject of a master's thesis by Phyllis Clancy; it was the first extended analysis of the ceramic complex. The major ceramic differences from the Quapaw phase are the virtual absence of Wallace Incised, the higher proportion of Barton Incised, primarily *var. Barton*, and the higher amounts of Late Caddoan pottery in the Carden Bottoms phase. Very early European trade goods, including a Clarksdale bell, copper wire bracelets, copper tinkling cones, and a few glass beads, were present at the Carden Bottoms locality.

If Hudson and DePratter (1984) are right in the identification of the River of the Cayas as the Arkansas, then it is probable that the Carden Bottoms phase site, Point Remove, represents Tanico in the de Soto era. The early European trade goods make this identification likely. These upstream people in the de Soto era virtually disappeared by sometime in the late seventeenth or early eighteenth century. It is logical that their survivors joined the villages further downstream on the Arkansas River. Although Wallace Incised, a key ceramic type, is very rare, it is likely that the Carden Bottoms phase people were ethnically similar to those of the Quapaw phase because the other ceramic similarities are so great.

The Quapaw Paradox

Students of the origin of the Quapaw tribe soon discover that the archaeological Quapaw phase data often conflict with tribal oral traditions, linguistic information, and much of the ethnological and historical data. The Quapaw phase is Mississippian in a cultural sense with good evidence of nucleated palisaded villages, possibly the temple mound–plaza complex, maize dependence, and many ceramics that have affinities to Mississippian period phases of the adjoining Central and Lower Mississippi valleys. Based on these resemblances, archaeologists such as Davis (n.d.), House and McKelway (1982), McGimsey (1964), and Morse and Morse (1983) think that the Quapaw phase and, with qualifications, the descendant Historic tribe developed from the Native Mississippian cultures of the area. Scholars, other than professional archaeologists, such as Baird (1980) and Hathcock (1983), accept recent Quapaw migration to the area on the basis of the nonarchaeological sources of information.

Several options for explaining this Quapaw paradox are reviewed here along with suggestions for producing data useful in evaluating the options. These options are not exhaustive.

Both Paheka and Ford Were Right

If Paheka's story of recent tribal separation from their Dhegiha kin and migration southward from the Lower Ohio Valley is basically correct as the linguistic, ethnological, and some ethnohistorical information indicate, and Ford (1961) was right with his identification of the Menard (Quapaw) phase and Quapaw tribe, then the Quapaw must have been very adaptable people. Very adaptable would have meant taking over the ceramics and general lifestyle of the preexisting or contemporary cultures in the area. Great adaptability for survival is in fact a qualitative trait of the Quapaw (Hoffman 1985). However, adopting Mississippian ceramics en masse from neighboring Lower and Central Mississippian Protohistoric groups strains credibility.

The Quapaw people were living on the banks of the Arkansas River in 1541 if Paheka, Ford, and Hudson are right. If the River of the Cayas was the Arkansas River, it was densely populated by people of the Quapaw and related Carden Bottoms phases. There is some ethnohistoric support for the idea of Quapaw tribal sixteenth-century villages up the Arkansas River in Father Marquette's 1673 map. Glottochronological estimates for the time of separation of the Quapaw language from the other Dhegiha tongues (Grimm 1985; Viers 1983) indicate that the separation was complete by the sixteenth century.

There should be some similarity of Quapaw phase archaeological materials with those of the eastern plains Dhegiha groups, if Paheka and Ford were right. Unfortunately, the archaeology of the eastern plains Dhegihas is little known—or little published—in the late seventeenth century and early eighteenth century. Carl Chapman reports that Osage sites in Missouri produced Oneota-related pottery (1974:211). He also believes that the tribe developed in place from the Late Prehistoric Neosho phase in southwestern Missouri. Only the use of wide-trailed incision links the ceramics of the Quapaw phase with the pottery of the historic Osage.

Jerome Rose (personal communication) reports that skeletal samples of eastern plains Dhegiha peoples are available for comparison with those of the Quapaw phase. Unfortunately, only small skeletal samples of the Quapaw phase exist.

Archaeological cultures ancestral to the Quapaw phase should exist in the Lower Ohio Valley if both Paheka and Ford were right. Hathcock (1983:39–42) postulates that the Protohistoric Caborn-Welborn phase of the Lower Ohio and Wabash rivers in Indiana and Kentucky does represent such a culture. There is definitely northeastern Arkansas and southeastern Missouri pottery in the Caborn-Welborn collections, but nothing looks very similar to the Quapaw phase ceramic complex (Green and Munson 1978). Early European trade goods at some Caborn-Welborn sites may indicate that the phase was contemporary with the Quapaw phase rather than ancestral.

If Paheka and Ford were both right, the very adaptable Quapaw had moved from their Lower Ohio River homeland to the Arkansas River before 1541. There are, in fact, archaeological indications of Late Prehistoric (A.D. 1350–1400) population movements from the Cairo Lowlands to the Missouri Bootheel and northeastern Arkansas (Morse and Morse 1983:271–320).

The Historic Quapaw Were Shreds and Patches

The Morses (1983:320) postulate that the historic Quapaw represent an amalgamation of northeastern Arkansas chiefdoms that collapsed after being decimated by post–de Soto European diseases. The survivors formed the Quapaw phase and descendant Quapaw people on the Arkansas River. There is logic in the idea. The general atmosphere of the Mississippi Valley in the era was such that long-distance movements of tribal groups and amalgamations were commonplace (Dobyns 1983:302–12). The heterogeneous nature and origin of the ceramics and burial programs of the Quapaw and Carden Bottoms phases make it likely that such changes were taking place.

An argument against such a process being responsible for the entire Quapaw phase is that the Quapaw and Carden Bottoms phases predate de Soto on the Arkansas River, if the stream is the River of the Cayas.

Larger problems exist if the Quapaw people are thought to have been the result of the amalgamation of several peoples. Rankin (1985) makes it unequivocally clear that the Quapaw language does not reflect amalgamation. Quapaw ethnology, too, seems similar to that of other Dhegiha people. Quapaw oral tradition also does not reflect amalgamation, and within the Quapaw people today there are no signs of remnants of other tribes. Quite often when ethnic groups merged in the Historic Southeast, oral traditions relate the merger and persisting areas of pluralism. For instance, the Yuchi merged with the greater Creeks in the eighteenth century (Robbins 1977:4; Speck 1909:7), yet they have maintained separate ceremonial grounds in Oklahoma. Small groups of Natchez people merged with the Creek and Cherokee in the eighteenth century, yet, like the Yuchi, they are still an identifiable minority in the two tribes and maintain their own ceremonial grounds (Howard 1984:117). The Quapaw people have readily intermarried with other tribes and with white people over the last several hundred years, but a Quapaw cultural system has remained intact to this day.

The Quapaw are Everything

Roy Hathcock (1983) and Kent Westbrook (1982) have expanded the ceramic definition of the Quapaw phase to a Quapaw "culture," which includes Central Mississippi Valley Protohistoric manifestations that used mortuary pottery that was painted and included such pottery shapes as teapots, hourglass-necked bottles, and Conway-style head pots. The Quapaw culture was present in a large area from the Central Arkansas Valley to the St. Francis River and adjacent Mississippi River. This culture includes both the Kent and Old Town phases. Hathcock and Westbrook ignore the prime marker of the Quapaw phase, Wallace Incised, which is virtually absent in the Kent and Old Town phases. Probable sharp differences in nonmortuary ceramics in these phases also are ignored (Phillips 1970:938–40). Another problem is Hathcock's and Westbrook's belief in the post–de Soto migration of the Quapaw tribe into Arkansas and the probable evidence of the contemporaneity of Kent and Old Town phases and the de Soto entrada. Hathcock, in a recent paper (personal communication) considers it a possibility that the people of the Old Town and Kent phases, after suffering depopulation in the post–de Soto era, migrated to the Central Arkansas Valley to establish the Carden Bottoms phase. This idea is based on specific similarities in mortuary pottery of the Lower St. Francis and Central Arkansas Valley regions.

The Mississippian nature of the Quapaw phase, particularly the painted mortuary ceramics, supports the idea of considering together the Lower St. Francis River and adjacent Mississippi River phases with that phase, regardless of the question of Quapaw tribal ancestry. Nonmortuary excavation data are badly needed.

Quapaw tribal identification is another question. The long bark-covered house, if found archaeologically in these phases, would be an important piece of evidence because that feature is one of the most long lasting of Quapaw tribal material traditions (the tribal meeting place is a long concrete structure, and individual camps at the ceremonial ground are usually rectangular ramadas).

It is entirely possible that Protohistoric chiefdoms visited by de Soto in northeastern Arkansas were Dhegiha-speaking, although Swanton (1985:52) thought there was evidence that these people were Tunican-speaking. If they were Dhegiha-speaking, collapse into the historic Quapaw tribe would have been facilitated and less likely to have left indications of heterogeneity in the historic Quapaw language.

The Quapaw Are Yet Unknown Archaeologically: Paheka Was Right, Ford Was Wrong

The cornerstone of Quapaw archaeology for the last 25 years has been Ford's (1961) identification of the historic late-seventeenth-century Quapaw village of Osotouy with the Menard site. The site is located at the southernmost base of the Little Prairie, the first land that does not flood above the mouth of the Arkansas River, a characteristic emphasized by early French visitors to Osotouy. This fact, combined with significant quantities of Wallace Incised, the latest pottery type in the region, European trade goods, and a rectangular structure (but wattle and daub), clinched the tribal location—and the Menard site became the type site for the Quapaw phase.

Archaeological, ethnological, and ethnohistorical incongruities raise questions about the correctness of the Menard site–Quapaw association. The distribution of the Quapaw phase, based on its characteristic ceramic complex, is primarily the Lower Arkansas River to about Little Rock (Hoffman 1986) and is certainly not limited to the area near the mouth of the Arkansas River and along the adjacent Mississippi River where four Quapaw villages were certainly known in the late seventeenth century. At least a half dozen of these Quapaw phase sites have small amounts of European trade goods (Hoffman 1986) of kinds that appear to be pre-1673 (John House, personal communication). It is strange that no Quapaw phase site is yet known that has yielded large amounts of European trade goods. Inasmuch as the Quapaw survived in Arkansas until 1834, several sites should exist with significant quantities of eighteenth- and nineteenth-century European goods, particularly with the Euramerican settlement of Arkansas Post in their midst.

Excavations at the Quapaw phase sites, Menard (Ford 1961), Dupree on the Lower White River (McGimsey 1965; Moselage 1965), and those by Moore (1908) at several sites on the Arkansas River have failed to make findings that can be related without doubt to the historic Quapaw tribe. The houses at Menard and Dupree were not long, rectangular, bark-covered structures but instead were short, rectangular, wattle and daub houses.

Several students of Quapaw archaeology have indicated unease in the one-to-one correlation of the phase with the ethnic group (Hoffman 1977; John House, personal communication). Linguist Robert Rankin (personal communication) and Dan Morse (personal communication) both suggest that perhaps only the terminal parts of the Quapaw phase may represent the tribe.

If it is possible that the Quapaw phase does not simply represent the archaeological manifestation of the Quapaw tribe, what does it represent? It could, as the Morses (1983) suggest, represent the shreds and patches of the collapsing post–de Soto chiefdoms of northeastern Arkansas. But we have seen difficulties with that idea. Paheka declared to Governor Izard that the pre-Quapaw inhabitants of the territory that later became the Quapaw homeland were the Tunica. The Tunica, to be thought of generically as Tunican-speaking people and/or related Koroa, are thought by Swanton (1985:52) to have lived in northeastern and east-central Arkansas during the de Soto era. Recently, Jeter (1986) has reviewed the evidence of Tunica presence in Arkansas. He concludes (1986:58) that the group was the major one in southeastern Arkansas in Protohistoric times. If the River of the Cayas is the Arkansas River in the de Soto era, then the major "Tanico" town of the Carden Bottoms phase, the Point Remove site, would link the Carden Bottoms phase linguistically with the Tunica. The ceramic complex of the Quapaw phase has similarities both with the Protohistoric Mississippian phases of northeastern Arkansas, thought by Swanton (1985) to have been Tunican-speaking people, and with Lower Mississippi Valley phases related to the historic Tunica people (Brain 1979). Wallace Incised and the Belzoni variety of Winterville Incised, a late Mississippian period ceramic in the Lower Yazoo Basin, are closely related in technique (Phillips 1970:173–74). Thus, an explanation for the discordance between the Quapaw phase and the tribe's linguistic, ethnological, and ethnohistoric characteristics is that the phase represents the Protohistoric remains of the Tunica whom the Quapaw later drove away. This possibility then raises several questions. If the Quapaw phase does not represent the historic Quapaw, what archaeological manifestation does? There is no candidate, but there are many other sites at the base of the Little Prairie (Phillips et al. 1951:419). The Quapaw tribe was indeed a recent post–de Soto immigrant group to the Arkansas Valley if the Quapaw phase represents the Tunica. In this scenario we have yet no archaeological remains of the Quapaw in Arkansas, and that makes it difficult to look for Lower Ohio Valley archaeological antecedents of the Quapaw, because we don't know what to look for.

We may at this point wish that Paheka had been a stoic, silent Indian and that Ford had stayed in Mississippi and New York.

Conclusions

Nothing is yet resolved concerning Quapaw tribal origins and protohistory, but that is not to say that archaeology cannot improve our understandings. Almost all of the archaeological data are very old; extensive modern excavations are needed. A basic question is what the Quapaw phase represents—the Quapaw people, the Tunica people, or an amalgamation of remnant Late Mississippian peoples. Large-scale excavation in the habitation area of a Quapaw phase settlement should quickly reveal whether the Quapaw tribe's most distinctive architectural characteristic is present—the dome-shaped, bark-covered long house. Another way to approach the problem of what the Quapaw phase represents is to locate a Quapaw village of the mid-eighteenth to early nineteenth century. Such a site should have many European trade goods of the late date, long houses (the Quapaw were quite conservative in Arkansas), and a ceramic complex that could be compared to that of the Quapaw phase. If Quapaw tribal identification is solidified by such excavations, then the other questions discussed here can be addressed, such as the issue of continuity between the Quapaw phase and the earlier and neighboring Mississippi period manifestations and Quapaw migration traditions. Bioarchaeological studies promise to offer an independent means to judge closeness of skeletal populations, once good samples are obtained.

The use and reconciliation of diverse sources of information such as attempted in this study—oral tradition, linguistic, ethnological, ethnohistorical, bioarchaeological, and archaeological—are areas of scholarship in which anthropologists boasting a holistic approach should excel. However, attempting to make sense out of such diverse sources of information is, as this study certainly demonstrates, like stepping off into a swamp with no firm ground. If even the formerly assumed firm ground of the Menard-Quapaw identification is suspect, the swamp may prevail.

Acknowledgments

Robert Rankin, Sam Dickinson, and John Belmont read a draft of this paper and provided useful comments, suggestions, and criticisms, and I thank them. The final result is my responsibility, however. Finally, I would like to thank the Quapaw tribe for its help and encouragement of my research. Jesse McKibben, the tribal chairman, and tribal historians Edna Wilson, Odestine McWatters, and Carrie Wilson have been particularly helpful, and I thank them.

Notes

1. Ny-Tonka literally means "big river" or "great river," according to Robert Rankin (personal communication).

2. Ny-Whoutteh-Junka literally means "little grey river" according to Rankin (personal communication).

3. Some important characteristics of the Quapaw phase ceramic complex include the types Wallace Incised, Old Town Red, Carson Red on Buff, *var. Olmond*, Avenue Polychrome, *var. Avenue*, and Nodena Red and White, *var. Dumond*. Bell Plain is uncommon. The few Quapaw phase midden deposits excavated include the types Parkin Punctated and Barton Incised in some amounts. Quapaw phase sites of the Lower Arkansas River area have significant amounts of pottery of Tunican and Natchezan types while Quapaw phase sites near Little Rock include more Late Caddoan culture engraved pottery. Significant Quapaw phase vessel shapes include the Conway style heat pot, Old Town Red teapots, helmet-shaped bowls, Old Town Red rim effigy bowls, and globular bottles with hourglass-shaped necks.

12

Historic Indians of the Lower Mississippi Valley: An Archaeologist's View

Ian W. Brown

In the late fall of 1910, under the agency of a steam-powered craft called the *Gopher*, the archaeologist Clarence B. Moore ascended the Mississippi River to a landing in West Feliciana Parish, Louisiana. He spent little time in the area, as there were no impressive mounds or rich cemetries to hold his attention, but he did stay long enough to view a collection and to do some digging at the nearby Trudeau plantation. Having duly recorded an assortment of eighteenth-century objects (which included a kettle and a catlinite pipe) from this site, Moore packed up his gear and headed upriver toward what were, to him, much more rewarding hunting grounds (Moore 1911:376).

The Mississippian and Plaquemine mound centers in the Tensas and Yazoo basins have been a draw to many an archaeologist since Moore, but those of us who are also interested in historic Indians of the Mississippi Valley appreciate that Moore took the time to record some of the sites that exhibited European contact. The Trudeau plantation site, mentioned above and now simply referred to as Trudeau, was the principal village of the Tunica Indians in the mid-eighteenth century (Figure 12-1). Jeffrey P. Brain has done much to reveal the significance of this site in his 1979 volume, *Tunica Treasure*.

Since Moore's travels in the Lower Mississippi Valley, a large number of historic Indian sites have been added to the survey files. In this paper I will examine these historic sites from the perspective of culture change and continuity. Archaeology has contributed much to an understanding of Mississippi Valley Indian lifeways in the historic period, but the greatest strides, I believe, have been in illuminating the process of culture change.

Fifteen years ago, as an undergraduate with considerable moisture behind the ears, I did my first fieldwork in the Lower Mississippi Valley. Ste-

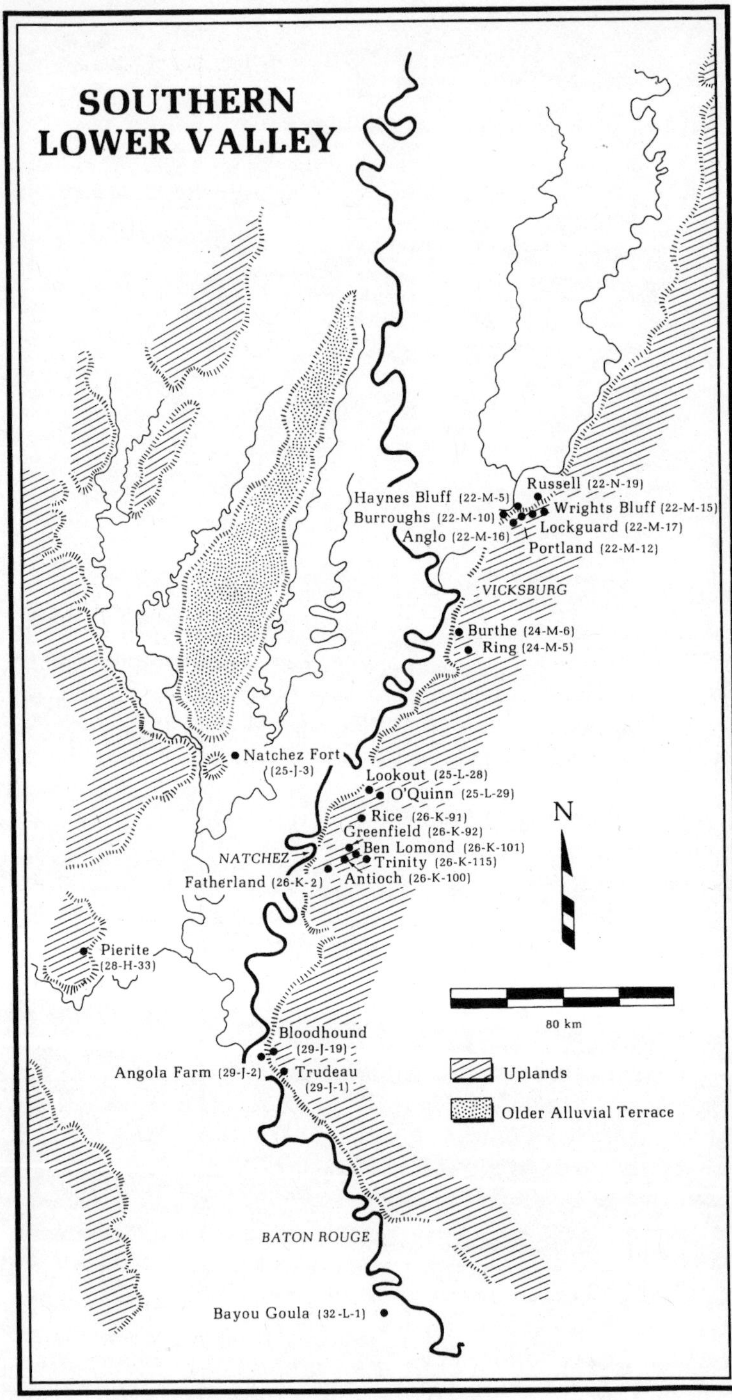

Figure 12-1. Various Historic Indian Sites in the Southern Portion of the Lower Mississippi Valley

phen Williams, director of the LMS, sent Jeff Brain south in 1971 to start a new project in the Natchez Bluffs region of Mississippi (Brain et al. n.d.). Jeff and I made a pilgrimage to Etowah and Moundville on the way, and I was duly impressed with the magnitude of these sites. My initial excitement in the vitality of southeastern lifeways was deflated, however, when we pulled into our fieldquarters at Cannonsburg, a small community located northeast of Natchez. The settlement consisted of us, in our mid-nineteenth-century home, our landlord's house, a broken-down cotton gin, Miss Pearl's grocery, and several shacks of unknown occupancy. Robert "Stu" Neitzel, a grand old man of southeastern archaeology, arrived shortly after us and suggested we drive to a small cafe on Route 61 South to get some plate lunches. As we left the cafe, more than sufficiently stuffed, Stu pointed off toward a grassy patch in the distance along the banks of a muddy creek. "That's Fatherland," he said. "That's where it all began."

The Fatherland site didn't strike me as all that impressive at the time. I, after all, had only recently climbed the mounds at Etowah and Moundville and knew what Indian mounds were supposed to look like. We couldn't even see the puny little tumuli at Fatherland, as a growth of brambles had eaten them up since Stu's excavations in the early 1960s. As the crew assembled in Cannonsburg over the next few days, Stu took every available opportunity to impress upon us the importance of the current project and to regale us with well-oiled stories of the life and time of one R. S. Neitzel. We didn't know whether we could (or should) believe Neitzel's tales, but he did convince us of the importance of the Natchez Indians and their pivotal position in the Mississippi Valley in the dynamic eighteenth century. They and the Tunica deserved much more attention than they had received in the past, and this project, he said, would start us on the path.

We all tend to look romantically at our first field experience, and I am no exception, but when I think back to those sweltering days of June 1971, I still do feel we were onto something very important. Whether or not anything came of it is for others to say, but to have actually walked on the same sites as the Great Sun, the Tattooed Serpent, Le Page du Pratz, and Dumont de Montigny was an experience I was not soon to forget. C. B. Moore, Warren K. Moorhead, John R. Swanton, and James A. Ford must have had the same sensations when they climbed (and dug into) the mounds in the Natchez Bluffs region.

As discussed by Williams (1962), historical archaeology in the Lower Mississippi Valley really began with the work of Jim Ford and Moreau B. Chambers (Brown 1978b:3). As young students, with little professional guidance but lots of enthusiasm, they investigated scores of sites in Mississippi and Louisiana in the late 1920s and early 1930s. A number of these sites were historic and both Ford and Chambers, unlike Moore, realized how significant they were for understanding French-Indian interaction in the eighteenth century. Natchez Fort, Burthe, and Ring were among a number of historic sites investigated by them and reported upon in Ford's classic 1936 volume, *Analysis of Indian Village Site Collections from Louisiana and Mississippi*. The most important sites, however, that aided in the assignment of ethnicity to material culture were Fatherland, Angola Farm, and Haynes Bluff.

Excepting perhaps the Bayou Goula site (which will be discussed shortly), Fatherland is the most intensively excavated historic Indian site in the Lower Mississippi Valley. Chambers and Ford in 1930 concentrated their work on Mound C. Neitzel followed in the early 1960s, around the same time that Stephen Williams (1967; Williams et al. 1966) was investigating the Taensa across the river. Neitzel (1965) intensively excavated Mounds B and C at Fatherland and did some minor work on Mound A. In 1972 he was commissioned by the Mississippi Department of Archives and History to continue work in the flats around the mounds, the prime purpose being to locate structural remains. Neitzel's most recent volume, published posthumously in 1983, has contributed much to our knowledge of settlement patterns at this historic ceremonial center. Although a great deal of continuity existed in the types of buildings constructed on the Fatherland mounds from prehistoric to historic times, there are indications that considerable changes occurred in the form of the surrounding structures (Brown 1985b). Whereas rectangular wall-trench buldings were the norm during the Mississippian period in the Natchez Bluffs region, by historic times the Natchez at Fatherland were also erecting buildings with individually set posts. A large one in the plaza at Fatherland was square with rounded corners. Similar structures had existed to the south in the Baton Rouge and Lower Red River regions throughout most of the Mississippian period, suggesting some influences from that direction in late prehistoric times (Brown 1985b:280–81).

Another aspect of culture change at Fatherland can be seen in the burial patterns. Most of the excavated burials came from Mound C, the location of the historic Natchez temple. The presence of European goods in the burials is an obvious change, but more subtle influences are reflected in the way the bodies were laid out. Most of the 25 burials recorded in Chambers's notes were either primary extended interments or isolated skull burials, but a significant number of individuals received special treatment. The cranial remains of eight mature individuals were contained within two rectangular wooden chests, six skulls in one and two in the other (Neitzel 1965:41–42). The use of wooden chests as coffins is a marked change from the prehistoric pattern and reveals influences from the nearby French population.

Another important historic Indian site in the Lower Mississippi Valley is

Bayou Goula, in Iberville Parish, Louisiana. Although excavated in 1940–41 by the Louisiana State Archaeological Survey (a joint project of the state of Louisiana and the WPA), it was not reported upon until 1957. George I. Quimby summarized the excavations at this site and compared the material to that found at the contemporaneous Fatherland site. Unlike Fatherland, however, a problem existed with Bayou Goula with regard to ethnicity. So many groups had lived there at various times that it was impossible to say just who had deposited the materials. In addition to the Bayogoula, other residents included the Mugulasha (Quinipissa), Acolapissa, Tioux, Taensa, and Houma (Quimby 1957:100). Even a strong French component has resided at the site, as represented by the settlement remains. Although Quimby did not realize it at the time of his writing, it is now reasonably clear that the excavated buildings at Bayou Goula were of French construction, not Indian (Brown 1977). They are identical in shape and size to structures at Fort St. Pierre, a French fort in the Yazoo Bluffs region of Mississippi that dates to 1719–29 (Brown 1975, 1978a, 1979a, 1979b).

A number of remnant tribes occupied the southern Yazoo Basin and adjacent bluffs in the seventeenth and early eighteenth centuries (Galloway 1982b). The most important historic Indian site in the region is Haynes Bluff. Located northeast of Vicksburg along the left bank of the Yazoo River, this mound center was investigated first by C. B. Moore (1908:569–70) and later visited and described by both Calvin Brown (1926:55–56) and Jim Ford (1936:110–11). Ford assigned the surface collections from Haynes Bluff to the Tunican complex, an indication of his belief that the site primarily was occupied by historic Tunica Indians. Of that we can be sure. Numerous early-eighteenth-century accounts by missionaries and explorers pinpoint the Tunica at this location. The only trouble is that, as with the Bayou Goula site, a number of other groups resided at Haynes Bluff in the seventeenth and eighteenth centuries. The Tunica themselves were latecomers to the Yazoo Bluffs region. Having originally resided in the northern Yazoo Basin, it is probable that they were pushed south in the late sixteenth or early seventeenth century by Siouan-related groups, probably the ancestors of the Quapaw. The Yazoo, Koroa, and Tioux had deeper roots in the area, and their ancestors were probably the initial builders of the Haynes Bluff mounds during the Protohistoric Wasp Lake phase. Other groups that lived on or around Haynes Bluff were the Ofo, a Siouan-speaking group, and the Chakchiuma (Brain 1988).

Haynes Bluff was the central concern of another Lower Mississippi Survey project in 1974. This operation was led by Jeff Brain, while Vin Steponaitis served as the field supervisor (Brain 1988). Meanwhile, I teamed up with a Mississippi Department of Archives and History crew led by Bill Wright and conducted excavations at Fort St. Pierre and at a number of small Indian sites in the surrounding area. Whereas Haynes Bluff provided the time depth, the firm chronological sequence, it was the tightly dated sites in the hills that acted as synchronic markers. They have served as snapshots in the rapidly changing demography of the Yazoo Bluffs in the first three decades of the eighteenth century. Some of these important sites that were excavated between 1974 and 1976 were Russell, Burroughs, Lockguard, Portland, Anglo, and Wrights Bluff (Brain 1975a, 1988; Brown 1979a, 1979b). Portland is perhaps the most interesting as it is the best single-component early Tunica site that we have. The trash pits excavated at Portland are thought to date somewhere between 1698 and 1706 (Brown 1976, 1979a:206–14). Lockguard and Wrights Bluff, however, are believed to be contemporary with Fort St. Pierre (1719–29) and were, in all probability, Yazoo Indian sites (Brown 1979a:214–24, 228–31). Russell may be an Ofo Indian component (Brain 1975a, 1988; Williams and Brain 1983:384).

We were drawn initially to Wrights Bluff because of a very interesting burial that had been discovered by a pothunter. This burial, an adult male, had been interred in the flesh in a semiflexed position. Buried with him were parts to as many as seven guns, including sideplates, locks, musket cocks, triggerguards, rampipes, gunflints, musket balls, and so forth. A set of six large open-mouthed bells was also included with this burial, representing a full set of chimes. Other items included a screwdriver, 22 hand-wrought nails, a white glass bead, a couple of Mississippi Plain, *var. Yazoo*, vessels, and a catlinite pipe (Brown 1979a:222). This burial posed many questions. Who was this important individual and how did he fit into the society? What role did he play in French-English-Indian interaction in the area? Was he a gunsmith? The disassembled remains of seven guns might suggest such a specialty. And what of the bells? What were they used for and how did they figure in his life? We do not think of catlinite pipes, presumably a reflection of the calumet ceremony, as being the baggage of a common man, yet this individual was provided with one. To my knowledge, only seven other catlinite pipes of the elbow type have been found in the Lower Mississippi Valley: two at Haynes Bluff (Brain 1975a, 1979:248), and one at Fatherland (Neitzel 1965:Plate 13k), and four at Trudeau (Brain 1976:81, 1979:248; Moore 1911:Figure 1). Fragments of a catlinite object, perhaps a pipe, were also found at the Trinity site in the Natchez Bluffs region (Brown 1985a:172, Figure 112e-g). Such pipes generally are quite rare in the Lower Mississippi Valley and occur only in historic contexts.

Another strange pattern was detected in our excavation of the Wrights Bluff burial. The fill of the burial pit included rims from seven different plain native-made pottery vessels: Mississippi Plain, *var. Yazoo*. The proportion of rim sherds to body sherds was one to four, a rather high ratio. I do not know of another instance in the valley characterized by such an unusual offering of potsherds, but the archaeological literature does testify to the practice of burying large single pieces of pottery with individuals. Such instances have been observed at the Menard site in Arkansas (Thomas 1894:231), at Sorrels Mound in Jackson County, Illinois (Thomas 1894:141), as well as at sites in the Norris and Wheeler basins of Tennessee and northern Alabama (Webb 1938:171, 1939:39, 64–65). Peebles (1983:269–70, 301–02) observed the same at Lubbub Creek in west-central Alabama, and, closer to home, this practice has even been seen at Haynes Bluff. With a burial discovered in the summit of Mound A, Brain (1988) recovered a large sherd of Winterville Incised in addition to a whole pot. Perhaps a sherd by itself, or several sherds, were considered sufficient baggage. Maybe only a symbol of the pot was all that was necessary for the afterlife. Some of the other remains deposited with the individual at Haynes Bluff would seem to suggest as much. Over his head was a patched inverted kettle and he held a very fragmentary musket that was missing a lock. Either the bereaved were poor and hated to give up their precious workable items, or, more likely, it just wasn't necessary to deposit the very best merchandise to ensure a successful trip to eternity.

Inverting kettles over or by the head of a deceased individual is another interesting pattern that has turned up elsewhere. In addition to Haynes Bluff, it has also been seen at the Trudeau site (Brain 1979:277–78). Determining what this practice means, where it came from, and how it relates to the burial urn tradition to the east (Sheldon 1974) would, I feel, reveal much about changing religious beliefs in the historic period.

Having spoken of Haynes Bluff, Portland, and Trudeau, I should discuss, to some extent, the Tunica project of the Lower Mississippi Survey. This project had its beginnings in 1968 when a local Louisiana man contacted Stu Neitzel about selling some curiosities he had dug. Stu called Harvard and Jeff Brain hastened south to examine the materials. In his own words, "It was like Christmas morning" (Brain 1979:4). This vast collection eventually was crated and shipped to the Peabody Museum for analysis in 1970. I cut my archaeological teeth working on the brass trade bells from that magnificent collection of burial furniture (Brown 1979d). And what a collection it is. Only two other southeastern historic collections are comparable: that made by Burke, Brannon, and their associates among Upper Creek sites in central Alabama in the early twentieth century (Alabama Anthropological Society 1920–37; Brannon 1935) and, more recently, the important Chickasaw finds in the Tupelo region of northeastern Mississippi (Stubbs 1984).

The trouble with the Tunica Treasure, however, was that we had a collection without a site, materials without provenience, as the collector kept the site location a secret for over two years. To our chagrin, when he finally led us to it in the summer of 1972, we found it to be the same location as C. B. Moore's Trudeau plantation. It seems logical now that the most important Tunica site should have been in the town of Tunica, situated on Tunica Bayou, not far from Tunica Landing. In fact, we should have just followed the road signs! What may seem obvious now, however, was not all that apparent when we looked for the site in 1971. Of course, as explained in detail in the introduction to the Tunica Treasure volume, the collector was very effective in providing us with red herrings.

Trudeau was but one step in the dynamic history of the very mobile Tunica. In the sixteenth century they were situated far to the north in the Yazoo Basin (Brain 1977:4–6; Brain et al. 1974:262). By the late 1600s they were placed in the vicinity of Vicksburg, and by 1706 they had moved to the region around the mouth of the Red River. Here, at such sites as Angola Farm, Bloodhound, and Trudeau, the Tunica played a major role as entrepreneurs in the relations between the French, the Caddo, and other Indian groups in the Lower Mississippi Valley (Brain 1977, 1981; Ford 1936:129–40). At any one time there usually were a number of secondary villages in addition to the primary Tunica settlement (French 1851:174; Swanton 1911:314). By arranging the Tunica sites chronologically, it has been possible to detect and measure changes in their lifeways through time, the subject of Brain's recent book, *Tunica Archaeology* (1988).

The Tunica story, of course, does not end at Trudeau. In the late eighteenth century, following the American Revolution, the Tunica moved to Marksville, Louisiana, where their descendants still play an active role in the community (Brain 1981:48-49). The Pierite site on the Tunica Reservation provides a view of the material culture and burial practices of this group in the early nineteenth century (Brain 1988; Gregory 1978). The Tunica still bury their dead on the Pierite site. In 1976 I visited the reservation with Jeff Brain and Stu Neitzel, only days after the burial of Chief Joseph Alcide Pierite, the last Tunica chief (Brain 1976:82). Although much had changed in their lives since the Europeans first entered the valley, traces of earlier traditions still persist. Anna Mae Juneau, the daughter of Chief Joe, explained to me that a light bulb was set at her father's grave on the day of the burial and that it had to remain lit at all times. To keep away bad spirits, the late-nineteenth-century Tunica would have placed a fire at the head of the deceased and kept it lit for four successive nights (Swanton 1911:325). Perhaps coincidental, perhaps not, but the persistence of a distinct behavioral pattern certainly left its mark on me.

Following my dissertation work at Fort St. Pierre, I joined the LMS in 1977 and headed off to the bayous of Louisiana. Little in the way of historic Indian archaeology existed in the Petite Anse region surrounding Vermilion Bay, but I did manage to take several side trips to my initial stomping grounds. With the help of Smokye Joe Frank and Robert Prospere, over a dozen new sites were recorded in the Natchez Bluffs between 1978 and 1980, most of which were either protohistoric or historic, and many of which were also in imminent danger of destruction by one agency or another (Frank 1980a, 1980b, 1981).

The LMS conducted a new project in Natchez between 1981 and 1982. Stephen Williams was overall director of the project that was sponsored by the National Endowment for the Humanities and the National Geographic Society, and I conducted the field operations and analysis. The project had as its ultimate goal the investigation of cultural processes (Brown 1982, 1985a). We had a unique situation in Natchez, as compared to the Yazoo Bluffs region, in that the general areas of the Indian villages were known historically. As a result of various French military expeditions against the Natchez, we had a reasonably good idea where the towns such as White Apple, Jenzenaque, and Grigra were located. From the historical accounts we also knew which villages were allied with the French and which leaned toward the English. Moreover, as a result of the earlier 1970s surveys by Jeff Brain (et al. n.d.) and the later ones by Frank, Prospere, and myself, we now knew of a dozen or so sites scattered throughout these village areas. It was presumed, largely on my part, and somewhat naively, that a study of the assemblages from these different sites would reflect the historically recorded sociopolitical differences. It was also believed that we would be able to detect adopted groups among the Natchez (Brown 1982:184–88). I do believe a reasonably good case for status differentiation and for the adoption of northern Mississippian related peoples has been made from a comparative analysis of the Lookout, O'Quinn, and Rice Site investigations, but sociopolitical distinctions unfortunately were not reflected in the site assemblages (Brown 1985a:12–73, 80–97, 188–96).

The factor I neglected to consider was time. I assumed that all of the sites were generally contemporary and that the village areas in the Natchez Bluffs region were stable throughout the late seventeenth and early eighteenth centuries. With the knowledge of hindsight, I should have realized that one cannot call a site Grigra just because it is in the general area where the Grigra village was supposed to be. The Grigra were adopted by the Natchez sometime in the seventeenth century and they settled east of the city in the vicinity of Perkins Creek. By the early 1980s we had recorded 14 protohistoric/historic sites in the area and, rightly or wrongly, for the purposes of the research design I reasoned that they were Grigra. The assemblages from these sites were to be compared to site assemblages from indigenous groups, and it was supposed that the variations would be readily apparent. Unfortunately, they were not. All in all, the protohistoric/historic native-made ceramics look pretty much the same at most sites in the Natchez Bluffs region, despite the heritage or political leanings of the historically recorded groups. In the case of the Grigra, I believe we failed to find the principal settlement of this adopted group. The sites that we did investigate in the "Grigra area," like Antioch, Greenfield, and Ben Lomond, are now thought to predate the Grigra occupation (Brown 1985a:98–161). Here I had forgotten the main tenet of the Direct Historical Approach: to apply the approach one must have absolute proof that the sites in question were occupied by specific groups. I could not designate ethnicity to sites simply because they were in the right general area (Baerreis 1961:51–52; Mason 1976:351–52).

Although it was not possible to determine subgroups for the various historic Indian sites in the Natchez Bluffs region, we have at least been able to say that the occupants all fell under the hegemony of the Great Sun, chief of the Natchez. Even though some of the people were adopted, it seems clear that by the early eighteenth century they had all been integrated into the Natchez sociopolitical structure (Brain 1971:220–21). Despite the fact that the element of time confounded my attempt to pigeonhole the sites into subgroups, an arrangement of the site assemblages according to chronology did reveal some interesting manifestations of change. The earlier sites (late seventeenth century), like Antioch and Ben Lomond, have a much higher incidence of modified artifacts than the later sites (early eighteenth century). Sites like Trinity (Brown 1985a:168–87), a late site in the Grigra area, have greater quantity and variation of historic European objects, but any form of manipulation of the artifacts is rare. For instance, at Ben Lomond and Antioch we found evidence for native materials that had been fashioned into European forms—local chert was chipped into gunflints and pottery was modeled into glass bead forms. Seemingly, the Indians saw things they wanted but that were not yet available in quantities sufficient to fulfill their needs. They did the next best thing in recreating the European form with their own materials. Through time, as European materials became increasingly available, the practice of copying declined and direct substitution occurred (Brown 1985a:195–96).

Some years ago George Quimby and Alexander Spoehr, operating out of the Field Museum in Chicago, wrote an article entitled, "Acculturation and Material Culture—I" (1951). Through their extensive study of museum specimens that exhibited some form of material modification that resulted from contact, they provided a model of change that could be subjected to archaeological testing. They concluded that early contact sites, which represent peoples who had experienced minimal acculturation, tend to have artifacts that reveal stability of form. Unmodified native objects will also be found at these sites, they argued, and so will unmodified European objects, but when modifications do occur, the direction of change will be aimed at preserving the original form. For example, European-derived iron or brass might be highly desired, but odds are that at early sites they would be fashioned into native forms (like iron arrowheads, iron fishhooks, or brass thimbles attached to clothing). At these same early contact sites, native materials will often be used to produce European forms, as in the case of native-made gunflints and pottery beads in the Natchez Bluffs region. At later sites, according to Quimby and Spoehr's model, greater contact and increased transferal of European items is presumed. Less substitution of European materials into native forms and more substitution of the European item itself occurs. The acceptance of the item is reflected by decorating it in a traditional manner. This practice represents the stamp of approval and is an indicator of a higher level of acculturation.

The Quimby-Spoehr model of acculturation is indeed reflected in the archaeological collections retrieved from the Natchez Bluffs. Materials from the earlier contact sites show a greater incidence of modification but, as they predicted, the form of the objects (whether European or native) remained stable. Later sites, like Trinity, do indeed show less modification of objects and considerably less copying.

Although the Quimby-Spoehr model itself is simple, the topic is complex. Other individuals have tried to refine it, but they have often carried it beyond the realm of archaeological data (for more detailed discussions of the Quimby-Spoehr model see Brown 1979c; Fontana 1978; White 1975). You cannot merely count up the objects on a site, determine the degree of modification, and thus assess the amount of acculturation that has occurred. We can shed light on relative degrees of acculturation and find patterns in the processes, but only if the context is controlled. The Tunica Treasure from Trudeau would have been a far greater gold mine for culture change studies had it not been ripped out of the ground by a pothunter who cared little for keeping records. A simple listing of the materials by grave lots would have provided much information on status, sex roles, and specialities (as in the case of the Wrights Bluff possible gunsmith). Moreover, had the collector recorded where the objects were found in relation to the parts of the skeleton, we might be able to say more about the function of the European objects to the mid-eighteenth-century Tunica. An iron ax, for example, is not unifunctional. If found at the hand, it may indeed have served its intended function, but if it was retrieved from the chest where it had been tied to a necklace, clearly it served a function above and beyond its utilitarian design. Until we have compiled an extensive inventory of objects by context of sites, we will have made only minor strides in the task of measuring culture change. However, as plate tectonics have demonstrated, minor strides do have a tendency to cover ground, so the future is bright.

Historic Indian archaeology in the Lower Mississippi Valley has come a long way since C. B. Moore trod upon the Trudeau plantation 75 years ago, and we are asking questions now that he would never have considered applicable to archaeology. I am sure that Moore would not have cared how cultures change or why certain traditions persist, but these are questions that most prehistorians today address in their work. One might think that the task would be much easier for historical archaeology because we have an independent data base from which to draw—that is, the written record. It may be easier, but it certainly is not easy. Examining cultural processes in the historic period is a humbling experience, because one becomes acutely aware of how many factors are at work in producing an outcome. As a result, one is careful not to conclude that a particular situation had to have been brought about by a certain set of factors, or that a combination of events would necessarily result in a well-defined outcome. This is the main lesson of historical archaeology for students of prehistory. Patterns of change are there and, by careful study, they can be illuminated. It is a slow and plodding chore, but a satisfying one when the pieces of the puzzle begin to fit together.

13

Comprehensive Planning for the Protection and Preservation of Mississippian Sites in Tennessee

George F. Fielder, Jr.

Cultural resource management is the term currently used to describe a wide range of activities, decision-making processes, and legal obligations. However, it is not a new concept, inasmuch as the participation of local, state, and federal governments in archaeological research, salvage operations, and site acquisition date back to the 1930s and earlier. The archaeological foundations of Arkansas, Mississippi, Tennessee, and most of the states along the Mississippi River are the result of federal- and state-funded archaeological work combined with that of the universities and amateur archaeological societies (Morse and Morse 1983). This paper does not intend to summarize cultural resource management activities in the Mississippi Valley. Rather, it discusses various techniques that can be used to effect archaeological site preservation.

In historic preservation circles, comprehensive planning means the process by which changes in land use can be anticipated and the impacts on archaeological and historical resources can be assessed. It assumes that existing land use and historic preservation planning tools and techniques can be used to avoid, minimize, or mitigate deleterious impacts on cultural resources. Comprehensive planning also includes resource management processes to determine which properties are important enough to warrant preservation, to allocate time and funds for preservation activities, and to sort large inventories of historic and archaeological properties into meaningful groups.

The distribution of known and predicted archaeological sites and its correlation with areas of future land development is an activity that comes naturally to archaeologists. Whether it is called predictive modeling, "man-land" relationships, or patterns of aboriginal sites selection, the process is basically one of correlating observed patterns with selected variables. Hypotheses are proposed, tested, and refined. The resultant site location distribution models can be expressed in both statistical and nontechnical terms depending on who will be using the results.

The forecasting of where economic growth and consequent construction is going to take place is a less exact science than archaeological predicting. Although we have the advantage of dealing with past human behavior rather than trying to predict future behavior, some of the same techniques can be used. Planners look at observed patterns of growth and correlate them with land and economic variables. Past, current, and projected population density maps are a frequent product of planning studies. When these are superimposed on the archaeological site distribution maps, the composite map indicates where future archaeologist-developer confrontations will occur.

Another future growth prediction technique is the review of requests for federal assistance by local communities and agencies. Most state historic preservation offices review around 1,700 such requests per year. The distribution of requests indicates where future activities may occur.

Another excellent source of information comes from professional planners working for federal, state, and local planning agencies. The state planning offices, regional development districts, and local planning commissions all have the responsibility of trying to understand, predict, and manage future growth for the benefit of society. However, in my experience, very little communication has taken place between planners and archaeologists in the areas of archaeological values, predicted site locations, and land-use planning tools. Increased communication will improve the effectiveness of both fields for the eventual benefit of society and cultural resources.

Tools and Techniques of Archaeological Preservation

Techniques of archaeological site preservation are varied but can be divided into three general categories: public awareness, regulatory processes, and the acquisition of property rights. Each category has its strengths and weaknesses, but the latter two rely primarily on the legal system.

Public awareness techniques can be considered as reactive or proactive. The most common reactive technique is the confrontation news story. The scenario is all too familiar: bulldozer uncovers "ancient remains"—archaeologist stops construction to "dig" in front of television cameras—film at 6 and 11! The resultant public perception is frequently reduced to the idea that archaeology is (1) digging stuff up before it gets destroyed, (2) archaeologists are obstructionists, and (3) if anyone wants to stop a project all you need to do is find an archaeological site in the area.

A better reactive posture for archaeologists is to join forces with other groups, including planning agencies, to form coalitions that are perceived to be in the public interest. The news media will be a critical factor in any attempt to carry out preservation by influencing public opinion because the news editors will decide how the situation is portrayed to the public. It is difficult to predict or control how archaeological values are presented by the media, but the judicious use of written press releases will help.

Proactive public awareness techniques that educate and inform are more effective and productive in the long run than reactive ones. School curricula, public service announcements, "soft" news and feature stories, video tapes for loan, and talks to civic organizations produce both short-term and long-term benefits. For example, Tennessee was recently given an important petroglyph site and adjoining acreage as the direct result of a Boy Scout outing some 30 years before. The young scout had been so impressed with the site that years later he bought the property at auction to prevent adverse development.

Since the passage of the National Historic Preservation Act in 1966, regulatory tools and techniques for archaeological resource preservation have multiplied with the enactment of many laws and regulations, including the Archaeological and Historic Preservation Act of 1974 (16 U.S.C. 469-469c), the Archaeological Resources Protection Act of 1979 (16 U.S.C. 470 aa-470 11), and Executive Order 12372, Intergovernmental Review of Federal Programs (47 Federal Register 30959, July 16, 1982). This latter order overhauled the way in which federal programs are reviewed for archaeological and other environmental impacts by state and local planning agencies.

In addition, most states have laws pertaining to archaeological site destruction and preservation. In Tennessee the archaeological statutes primarily apply to lands owned or controlled by the state. Vandalism on privately owned sites done without owner permission constitutes an act of trespass and is a misdemeanor. However, the penalty can be a felony offense if the vandalism is done on a site listed in the Tennessee Register of Archaeological Sites. Privately owned sites cannot be placed on the register without the permission of the owner.

Site preservation, especially for Mississippian sites that usually contain human burials, can be augmented through the application of the cemetery vandalism laws. In Tennessee it is a felony to disturb, destroy, or otherwise vandalize any grave or burial regardless of temporal or cultural affiliation.

In 1984, legislative debate established that the state cemetery statutes apply to prehistoric as well as historic period burials. Other states have similar human burial laws that are increasingly being used for archaeological preservation.

One remark frequently heard is: "If a site is listed on the National Register of Historic Places then it cannot be destroyed—right?" Wrong; the National Register is a means of establishing the significance of sites, buildings, structures, and districts. The National Register is a planning tool that acts as a threshhold that a property must cross to be considered in the federally mandated planning process. This process, which is codified in the Code of Federal Regulations at 36 CFR 800, requires that federal agencies that impact National Register–eligible sites must consider alternatives that will either avoid, minimize, or mitigate the loss of significant archaeological data.

The amount of avoidance or mitigation is negotiated between three parties—the federal agency, the state historic preservation officer, and the independent federal Advisory Council on Historic Preservation. Who pays for what is determined by the type of federal assistance involved in the project. If federal construction funds are used, then the costs for mitigation come out of the project fund. Mitigation projects funded by federal construction monies have contributed an enormous amount of information on southeastern archaeology since the early 1970s.

However, if there is only a federal permit, license, or loan guarantee but no actual federal funds involved in a project, the Department of Interior (National Park Service) must fund any necessary archaeological mitigation in accord with the provisions of Section 3 (b) of the Archaeological and Historic Preservation Act of 1974. The prescribed process has worked well in the past. The Averbuch site, a Mississippian village and cemetery, is one example. The site, which contained approximately 800 burials, was affected by an FHA-insured subdivision in Nashville and the data recovery was conducted by the University of Tennessee–Knoxville under contract with the National Park Service (Klippel and Bass 1984). Unfortunately, today with cutbacks in the federal budget, the process for mitigation of impacts associated with federal permits and insurance guarantees is ineffectual because the National Park Service has limited funds for this kind of work. It is important to remember that even well-funded archaeological data recovery is not preservation in many ways. Long-term preservation of a site requires the private or institutional desire to preserve and the ability to carry out these desires. An essential element in any preservation scheme is control over the physical site and the right to implement preservation options.

Acquisition of Rights Through Ownership

In this country at present (and I suspect for a long time to come), control over activities on a piece of real property comes from the sanctity of ownership. Legal property rights exist in various forms and degrees. Property ownership may be thought of as a bundle of rights that can be bought, sold, or donated as a whole or in part. Using examples as they may apply to the protection of archaeological sites, an owner may execute a covenant in which he/she agrees to preserve and protect a site as long as he/she owns it. A covenant is a written agreement between two or more parties (one of whom is the owner) that something is done, will be done, or will not be done (Powell and Rohan 1985:60–64).

A stronger preservation tool is the conservation easement. An easement, considered as part of the property rights bundle, is the granting of control or use of real property by the owner to a second party. A commonly known example is the right to cross the property in a designated area such as a power line or highway easement. Generally, an easement must be recorded with the property deed to be enforceable.

A conservation easement is a special type of easement for specific purposes to cause the preservation of conservation values for the benefit of the general public. For example, in Tennessee law a conservation easement is defined as an easement in land or structures that is held for the benefit of the people of Tennessee. It is specifically enforceable by its holder or beneficiary and limits the holder of the land, his/her heirs, and assigns (future owners) with respect to the use and management of the land such that restrictions or obligations are intended to protect, preserve, or enhance the land, water, geological, historical, or archaeological resources of the state of Tennessee. A conservation easement for archaeological sites usually takes the form of granting to the holder the right to control any and all activities that may affect the archaeological integrity and value of the site(s) within the described easement boundary. The conservation easement does run with the property and is recorded with the deed. It can be held only by nonprofit organizations and public bodies or agencies. Easements are more effective preservation tools than covenants because specific rights are described in the legal instrument and the easement is recorded with the property deed. It is binding upon all future owners and, because it is held for public benefit, it is enforceable in court by the public if the terms of the agreement are not being followed by the holder. An easement can be acquired by purchase or donation. In many cases, the value of a donated conservation easement can be used to the tax advantage of the donator.

The Division of Archaeology has successfully negotiated one conservation easement on a Mississippian site in Sumner County and is currently negotiating a conservation easement for a Spanish contact protohistoric site and two prehistoric sites located on Moccasin Bend in Chattanooga.

Fee-simple ownership is having the whole bundle of property rights. This type of ownership, the most common type of preservation control, is also the most expensive. In addition, acquisition through fee-simple purchase can be a difficult and lengthy process if the owner is not inclined to sell or thinks that the property is worth more than appraised value. Local, state, and federal agencies have the right to eminent domain or condemnation. In this process the public agency takes possession of the property and the court decides how much compensation the owner will receive. There is a general reluctance to use such a politically unpopular technique for archaeological preservation purposes where the overall public benefit is much less tangible than a new highway.

Fee-simple acquisition has long been seen as a desirable preservation option. Arkansas, Louisiana, Mississippi, and several other states have sites operated as parks (Kwas 1986). Arkansas is also pursuing the purchase of the Parkin site through the creative use of state, federal, and private funds.

In Tennessee, the state has fee-simple ownership of such important Mississippian sites as Chucalissa in Memphis, Mound Bottom in Cheatham County, Link Farm site on the Duck River in Humphreys County, and the Sellars site in Wilson County. We also own at least 44 Mississippian period sites on other state properties such as park, forest, and prison lands. The Department of Conservation has a fund of four million dollars to acquire significant natural, archaeological, historical, and geological properties, and efforts are underway to identify and purchase several sites. Elsewhere in the region, the Tennessee Valley Authority and the University of Mississippi have begun a long-term study of the techniques to prevent or minimize the impacts of wave erosion on shoreline sites (Thorne 1985). Although ownership does not ensure preservation, it does provide the benefits of extended archaeological research and planning.

Summary

To be effective and successful, comprehensive archaeological resource planning must incorporate a variety of complex and time-consuming factors including scientific knowledge of the resources, communication with professional planners, skillful negotiations, and the use of the wide range of preservation tools and techniques. Multiple avenues exist for site preservation that require initiative, creative thinking, and hard work. The rewards are those tangible and intangible benefits that come from knowing that future archaeologists will have answers to seek and the public will benefit from our future endeavors.

References Cited

Acsadi, Gy, and J. Nemeskeri
1970 *History of Human Life Span and Mortality.* Akademia Kiado, Budapest.

Adair, James
1775 *The History of the American Indians.* London.

Alabama Anthropological Society (AAS)
1920–37 *Arrow Points*, vols. 1–22.

Alexander, Lawrence S.
1982 *Phase I Archaeological Reconnaissance of the Oliver Lock and Dam Project Area, Tuscaloosa County, Alabama.* Report of Investigations 33, Office of Archaeological Research, The University of Alabama, University.

Armelagos, George J., James H. Mielke, Kipling H. Owens, Dennis P. Van Gerven, John R. Dewey, and Paul E. Mahler
1972 Bone Growth and Development in Prehistoric Nubia. *Journal of Human Evolution* 1:89–119.

Atkinson, James R.
1979 A Historic Contact Indian Settlement in Oktibbeha County, Mississippi. *Journal of Alabama Archaeology* 25:61–82.
1986 The Origins of the Chickasaw Indians as Evidenced by Cultural Remains in the Upper Tombigbee Valley in Northeast Mississippi. Ms. on file, Natchez Trace Parkway Headquarters, Tupelo.
1987 The De Soto Expedition through North Mississippi in 1540–41. *Mississippi Archaeology* 22:61–73.

Atkinson, James R., and Jay K. Johnson

1985 New Data on the Thelma Mound Group in Northeast Mississippi. Paper presented at the 6th Annual Meeting of the Mid-South Archaeological Conference, Mississippi State.

Badger, R. Reid, and Lawrence A. Clayton (editors)

1985 *Alabama and the Borderlands: From Prehistory to Statehood*. University of Alabama Press, University.

Baerreis, David A.

1961 The Ethnohistoric Approach and Archaeology. *Ethnohistory* 8:49–77.

Baird, W. David

1980 *The Quapaw Indians: A History of the Downstream People*. University of Oklahoma Press, Norman.

Bass, William M.

1979 *Human Osteology: A Laboratory and Field Manual*. 2d ed. Missouri Archaeological Society, Columbia.

Bass, William M., David R. Evans, and Richard L. Jantz

1971 *The Leavenworth Site Cemetry: Archaeology and Physical Anthropology*. Publications in Anthropology 2. University of Kansas, Lawrence.

Beaudoin, Kenneth L.

1953 *A Report of Excavations Made at the T. O. Fuller Site, Shelby County, Tennessee, Between March 8, 1952 and April 30, 1953*. Conducted by the Memphis Archaeological and Geological Society. Privately printed, Memphis.

Biedma, Luis Hernandez de

1922 Relation. In *Narratives of the Career of Hernando de Soto*, translated by Buckingham Smith, edited by Edward G. Bourne, 2:3–40. Reprinted. Allerton, New York.

Bizzell, David W.

1982 A Report on the Quapaw: The Letters of Governor George Izard to the American Philosophical Society, 1825–1827. *Pulaski County Historical Review* 29:66–79.

Black, Thomas K., III

1978 A New Method for Assessing the Sex of Fragmentary Skeletal Remains: Femoral Shaft Circumference. *American Journal of Physical Anthropology* 48:227–31.

Blake, Leonard W.

1965 Interim Report on Plant Remains from the Chucalissa Site. Letter on file, C. H. Nash Museum, Department of Anthropology, Memphis State University, Memphis.

n.d. Report on Corn from the Chucalissa Site. Ms. on file, C. H. Nash Museum, Department of Anthropology, Memphis State University, Memphis.

Blake, Leonard W., and Hugh C. Cutler
1979 Plant Remains from the Upper Nodena Site (3MS4). *Arkansas Archeologist* 20:53–58.

Bocquet-Appel, Jacques P., and Claude Masset
1982 Farewell to Paleodemography. *Journal of Human Evolution* 11:321–33.

Bogan, Arthur E.
1983 Evidence for Faunal Resource Partitioning in an Eastern North American Chiefdom. In Animals and Archaeology: 1. Hunters and Their Prey, edited by J. Clutton-Brock and C. Grigson. *BAR International Series* 163:305–24.

Bourne, Edward G. (editor)
1904 *Narratives of the Career of Hernando de Soto*. Vols. 1 and 2. A. S. Barnes, New York.

Boutton, T. W., P. D. Klein, Mark J. Lynott, James E. Price, and L. L. Tieszen
1984 Stable Carbon Isotope Ratios as Indicators of Prehistoric Human Diet. In Stable Carbon Isotopes, edited by Judith R. Turnland and Phyllis E. Johnson. *American Chemical Society, Symposium Series* 258:191–204.

Bozeman, Tandy K.
1982 Moundville Phase Communities in the Black Warrior River Valley, Alabama. Ph.D. dissertation, University of California, Santa Barbara.

Brain, Jeffrey P.
1971 The Natchez "Paradox." *Ethnology* 10:215–22.

1975a The Archaeology of the Tunica: Trial on the Yazoo. In *National Geographic Society, Research Reports* 15:67–80.

1975b Artifacts of the Adelantado. *Conference on Historic Site Archaeology Papers* 8:129–38.

1976 From the Words of the Living: The Indian Speaks. In *Clues to America's Past*, pp. 75–103. National Geographic Society, Washington, D.C.

1977 *On the Tunica Trail*. Anthropological Study 1. Louisiana Archaeological Survey and Antiquities Commission, Baton Rouge.

1979 (editor) *Tunica Treasure*. Papers vol. 71. Peabody Museum of Archaeology and Ethnology, Harvard University, Cambridge.

1981 Tunica Archaeology. In Traces of Prehistory: Papers in Honor of William G. Haag, edited by Frederick H. West and Robert W. Neuman. *Louisiana State University, School of Geoscience, Geoscience and Man* 22:43–50.

1985a The Archaeology of the Hernando de Soto Expedition. In *Alabama and the Borderlands: From Prehistory to Statehood*, edited by R. Reid Badger and Lawrence A. Clayton, pp. 96–107. University of Alabama Press, University.

1985b Introduction: Update of De Soto Studies Since the United States De Soto Expedition Commission Report. In *Final Report of the United States De Soto Expedition Commission*, by John R. Swanton, pp. xi–lxxii. Reprinted. Smithsonian Institution Press, Washington D.C. Originally published 1939. U.S. Government Printing Office, Washington, D.C.

1988 *Tunica Archaeology*. Papers vol. 78. Peabody Museum of Archaeology and Ethnology, Harvard University, Cambridge.

Brain, Jeffrey P., Ian W. Brown, and Vincas P. Steponaitis

n.d. Archaeology of the Natchez Bluffs. Ms. on file, Lower Mississippi Survey, Peabody Museum of Archaeology and Ethnology, Harvard University, Cambridge.

Brain, Jeffrey P., Alan Toth, and Antonio Rodgriguez-Buckingham

1974 Ethnohistorical Archaeology and the De Soto Entrada into the Lower Mississippi Valley. *Conference on Historic Site Archaeology Papers* 7:232–89.

Brannon, Peter A.

1935 *The Southern Indian Trade*. Paragon Press, Montgomery.

Bray, Robert T.

1963 Southern Cult Motifs from the Utz Site, Saline County, Missouri. *Missouri Archaeologist* 25:1–39.

Brookes, Samuel O.

1977 The Owl Creek Site. *Mississippi Archaeology* 12:2–28.

Brose, David

1980 How Capt. Riggs Hunts for Mound Builders' Relics: An Historical Investigation of Some Influences on C. B. Moore. *Southeastern Archaeological Conference, Bulletin* 22:145–52.

Brown, Calvin S.

1926 *Archeology of Mississippi*. Mississippi Geological Survey, University.

Brown, Ian W.

1975 Excavations at Fort St. Pierre. *Conference on Historic Site Archaeology Papers* 9:60–85.

1976 The Portland Site (22-M-12), an Early 18th Century Historic Indian Site in Warren County, Mississippi. *Mississippi Archaeology* 11:2–11.

1977 A Reexamination of the Houses at the Bayou Goula Site, Iberville Parish, Louisiana. *Louisiana Archaeology* 3:193–205.

1978a Artifact Patterning and Activity Areas: The Evidence from Fort St. Pierre, Mississippi. *The Conference on Historic Site Archaeology Papers* 12:309–21.

1978b *James Alfred Ford, the Man and His Works*. Special Publication 4. Southeastern Archaeological Conference, Memphis.

1979a Early 18th Century French-Indian Culture Contact in the Yazoo Bluffs

Region of the Lower Mississippi Valley. Ph.D. dissertation, Department of Anthropology, Brown University, Providence.

1979b Functional Group Changes and Acculturation: A Case Study of the French and the Indian in the Lower Mississippi Valley. *Mid-Continental Journal of Archaeology* 4:147–65.

1979c Historic Artifacts and Sociocultural Change: Some Warnings from the Lower Mississippi Valley. *The Conference on Historic Site Archaeology Papers* 13:109–22.

1979d Trade Bells. In *Tunica Treasure*, edited by Jeffrey P. Brain, 71:197–205. Peabody Museum of Archaeology and Ethnology Papers, Harvard University, Cambridge.

1982 An Archaeological Study of Culture Contact and Change in the Natchez Bluffs Region. In *LaSalle and His Legacy: Frenchmen and Indians in the Lower Mississippi Valley*, edited by Patricia K. Galloway, pp. 176–93. University Press of Mississippi, Jackson.

1983 Historic Aboriginal Pottery from the Yazoo Bluffs Region, Mississippi. *Southeastern Archaeological Conference, Bulletin* 21:1–17.

1985a *Natchez Indian Archaeology: Culture Change and Stability in the Lower Mississippi Valley*. Archaeological Report 15. Mississippi Department of Archives and History, Jackson.

1985b Plaquemine Architectural Patterns in the Natchez Bluffs and Surrounding Regions of the Lower Mississippi Valley. *Mid-Continental Journal of Archaeology* 10:250–305.

Buikstra, Jane E.

1984 The Lower Illinois River Region: A Prehistoric Context for the Study of Ancient Diet and Health. In *Paleopathology at the Origins of Agriculture*, edited by Mark N. Cohen and George J. Armelagos, pp. 215–34. Academic Press, Orlando.

Bullen, Adelaide K.

1972 Paleoepidemiology and Distribution of Prehistoric Treponemiasis (Syphilis) in Florida. *The Florida Anthropologist* 25:133–75.

Butler, Brian M.

1983 Kincaid Revisited: The Mississippian Sequence in the Lower Ohio Valley. Paper presented at the Mississippian Symposium, Iowa City, Iowa.

Cassidy, Claire M.

1972 Comparison of Nutrition and Health in Preagricultural and Agricultural Amerindian Skeletal Populations. Ph.D. dissertation, Department of Anthropology, University of Wisconsin, Madison.

1984 Skeletal Evidence for Prehistoric Subsistence Adaptation in the Central Ohio River Valley. In *Paleopathology at the Origins of Agriculture*, edited by Mark N. Cohen and George J. Armelagos, pp. 307–45. Academic Press, Orlando.

Chapman, Carl H.
1974 *The Origin of the Osage Indian Tribe.* Osage Indians vol. 3. American Indian Ethnohistory Series. Garland, New York.

Chapman, Carl H., and Leo O. Anderson
1955 The Campbell Site, a Late Mississippi Town Site and Cemetery in Southeast Missouri. *The Missouri Archaeologist* 17:10–119.

Clancy, Phyllis M.
1985 The Carden's Bottom Puzzle Elucidated. Master's thesis, Department of Anthropology, University of Arkansas, Fayetteville.

Clark, George A., Donald S. Boy, and M. Cassandra Hill-Clark
1982 Size, Shape and Stress Indicators in the Human Tibiae from the Dickson Mound Populations (A.D. 950–1300). Paper presented at the 81st Annual Meeting of the American Anthropological Association, Washington, D.C.

Clarke, David L.
1968 *Analytical Archaeology.* Methuen, London.

Clay, R. Berle
1961 Excavations at the Tinsley Hill Village, 1960. Ms. on file, Office of State Archaeology, University of Kentucky, Lexington.

1963a *The Tinsley Hill Mound.* Department of Anthropology, University of Kentucky, Lexington. Submitted to the National Park Service, Richmond.

1963b *Tinsley Hill Village, 1962.* Department of Anthropology, University of Kentucky, Lexington.

1979 A Mississippian Ceramic Sequence from Western Kentucky. *Tennessee Anthropologist* 4:111–28.

1984 Morris Plain: And Other West Kentucky Ceramic Smoking Guns. *Tennessee Anthropologist* 9:104–13.

Coe, Michael D., and F. William Fischer
1959 Barkley Reservoir—Tennessee Portion Archaeological Excavations, 1959. Ms. on file, Office of State Archaeology, University of Kentucky, Lexington.

Cohen, Mark N., and George J. Armelagos (editors)
1984 *Paleopathology at the Origins of Agriculture.* Academic Press, Orlando.

Cole, Gloria G., Mary C. Hill, and Hale B. Ensor
1982 Bioarchaeological Comparisons of the Late Miller III and Summerville I Phases in the Gainseville Lake Area. In *Archaeology of the Gainesville Lake Area: Synthesis*, edited by Ned J. Jenkins, pp. 185–258. Archaeological Investigations in the Gainesville Lake Area of the Tennessee-Tombigbee Waterway, vol. 5. Office of Archaeological Research, The University of Alabama, University.

Connaway, John M.
1981 *Archaeological Investigations in Mississippi*. Archaeological Report 6. Mississippi Department of Archives and History, Jackson.

Connaway, John M., and Sam McGahey
1970 *Archaeological Survey and Salvage in the Yazoo-Mississippi Delta and in Hinds County, 1968–1969*. Preliminary Report 1. Mississippi Department of Archives and History, Jackson.

Cook, Della C.
1976 Pathologic States and Disease Process in Illinois Woodland Populations: An Epidemiologic Approach. Ph.D. dissertation, Department of Anthropology, University of Chicago, Chicago.
1984 Subsistence and Health in the Lower Illinois Valley: Osteological Evidence. In *Paleopathology at the Origins of Agriculture*, edited by Mark N. Cohen and George J. Armelagos, pp. 235–69. Academic Press, Orlando.

Cook, Della C., and Jane E. Buikstra
1979 Health and Differential Survival in Prehistoric Populations: Prenatal Defects. *American Journal of Physical Anthropology* 51:649–64.

Cordell, Linda S.
1984 *Prehistory of the Southwest*. Academic Press, New York.
1985 Why Did They Leave and Where Did They Go? *Exploration: Annual Bulletin of the School of American Research*, pp. 35–39. Sante Fe.

Crane, Verner W.
1956 *The Southern Frontier, 1670–1732*. University of Michigan Press, Ann Arbor.

Culin, Stewart
1907 *Games of the North American Indians*. Annual Report 24. Smithsonian Institution, Bureau of American Ethnology, Washington, D.C.

Curren, Caleb
1987 *The Route of the Soto Army Through Alabama*. De Soto Working Paper 3. Alabama De Soto Commission, Alabama State Museum of Natural History, University.

Cybulski, Jerome S.
1977 Cribra Orbitalia, A Possible Sign of Anemia in Early Historic Native Populations of the British Columbia Coast. *American Journal of Physical Anthropology* 47:31–39.

Damon, P. E., C. W. Ferguson, A. Long, and E. I. Wallace
1974 Dendrochronologic Calibration of the Radiocarbon Time Scale. *American Antiquity* 3:350–66.

Davis, Darrell H.
1923 *The Geography of the Jackson Purchase*. Series 6, vol. 9. Kentucky Geological Survey, Lexington.

Davis, Hester A.

1966 An Introduction to Parkin Prehistory. *The Arkansas Archeologist* 7:1–40.

1973 The Hazel Site, Historical Background. *The Arkansas Archeologist* 14:21–31.

n.d. Quapaw. Ms. submitted for the Plains volume of the Handbook of North American Indians.

DeJarnette, David L., and Christopher S. Peebles

1970 The Development of Alabama Archaeology—the Snow's Bend Site. *Journal of Alabama Archaeology* 16:77–119.

de Lery, M. Chaussegros

1923 Journal of the Campaign made by a Canadian Detachment Against Chickasaws in February, 1740. Rapport de L'Archiviste de la Province de Quebec pour 1922–1923. Translation on file, Natchez Trace Parkway Headquarters, Tupelo.

DePratter, Chester B.

1983 Late Prehistoric and Early Historic Chiefdoms in the Southeastern United States. Ph.D. dissertation, Department of Anthropology, University of Georgia, Athens.

DePratter, Chester B., Charles M. Hudson, and Marvin T. Smith

1985 The Hernando de Soto Expedition: From Chiaha to Mabila. In *Alabama and the Borderlands: From Prehistory to Statehood*, edited by R. Reid Badger and Lawrence A. Clayton, pp. 108–27. University of Alabama Press, University.

Dickens, Roy S.

1982 *Of Sky and Earth: Art of the Early Southeastern Indians*. The High Museum of Art, Atlanta.

Dickinson, Samuel D.

1980 Historic Tribes of the Ouachita Drainage System in Arkansas. *The Arkansas Archeologist* 21:1–11.

1984 Lake Mitchegamas and the St. Francis. *Arkansas Historical Quarterly* 43:197–207.

Dickinson, Samuel D., and Samuel C. Dellinger

1940 A Survey of the Historic Earthenware of the Lower Arkansas Valley. *Texas Archeological and Paleontological Society, Bulletin* 12:76–96.

Dobyns, Henry F.

1966 Estimating Aboriginal American Population. *Current Anthropology* 7:395–416.

1983 *Their Number Become Thinned: Native American Population Dynamics in Eastern North America*. University of Tennessee Press, Knoxville.

Dorsey, James

1884 Omaha Sociology. *Smithsonian Institution, Bureau of American Ethnology, Annual Report* 3:205–370.

1885 On the Comparative Phonology of the Four Siouan Languages. *Annual Report of the Board of Regents of the Smithsonian Institution for the Year 1883,* pp. 919–29.

1897 Siouan Sociology: A Posthumous Paper. *Smithsonian Institution, Bureau of American Ethnology, Annual Report* 15:205–44.

Driver, Harold E.

1939 Culture Element Distributions X, Northwest California. *University of California, Publications in Anthropological Records* 1:297–433. Los Angeles.

1961 *Indians of North America*. University of Chicago Press, Chicago.

Dye, Linda O. N.

1976 Status and Biology: An Osteological Analysis. Master's thesis, Department of Geography and Anthropology, Louisiana State University, Baton Rouge.

Edging, Richard (editor)

1985 *Archaeological Investigations at the Turk Site (15Ce6), Carlisle County, Kentucky*. Western Kentucky Project Report 3. Department of Anthropology, University of Illinois, Urbana.

Edging, Richard, and Sandra L. Dunavan

1986 [Adams Site] Botanical Remains. In *Mississippian Towns of the Western Kentucky Border*, edited by R. Barry Lewis, pp. 87–99. Kentucky Heritage Council, Frankfort.

Edwards, Thomas H.

1921 Map of East Central Alabama Showing Aboriginal Mounds, Town Sites & Villages of Upper and Lower Creeks and Alibamos prior to 1832. *Arrowpoints* 2(5), supplement.

El-Najjar, M. Y., D. J. Ryan, Christy G. Turner II, and B. Lozoff

1976 The Etiology of Porotic Hyperpotosis Among the Prehistoric and Historic Anasazi Indians of the Southwestern United States. *American Journal of Physical Anthropology* 44:477–88.

Feiler, Seymour

1962 *Jean-Bernard Bossu's Travels in the Interior of North America, 1751–1762*. University of Oklahoma Press, Norman.

Ferguson, Dick V., and James L. Gray

1971 *Soil Survey of Mississippi County, Arkansas*. U.S. Government Printing Office, Washington, D.C.

Ferguson, John L., and J. H. Atkinson

1966 *Historic Arkansas*. Arkansas History Commission, Little Rock.

Fidalgo of Elvas

1866 *Narratives of De Soto*, translated by Buckingham Smith. Bradford Club, New York.

1904 True Relation. In *Narratives of the Career of Hernando de Soto*, edited by Edward G. Bourne, 1:1–223. Reprinted. A. S. Barnes, New York.

Fiske, Harold N.

1944 *Geological Investigations of the Alluvial Valley of the Lower Mississippi River*. U.S. Army Corps of Engineers, Mississippi River Commission, Vicksburg.

Flannery, Kent V.

1976 Linear Stream Patterns and Riverside Settlement Rules. In *The Early Mesoamerican Village*, edited by Kent V. Flannery, pp. 173–80. Academic Press, New York.

Fletcher, Alice C., and Francis La Flesche

1911 The Omaha Tribe. *Smithsonian Institution, Bureau of American Ethnology, Annual Report* 27:17–672.

Fontana, Bernard L.

1978 Artifacts of the Indians of the Southwest. In *Material Culture and the Study of American Life*, edited by Ian M. G. Quimby, pp. 75–108. W. W. Norton, New York.

Ford, James A.

1936 *Analysis of Indian Village Site Collections from Louisiana and Mississippi*. Anthropological Study 2. Louisiana Department of Conservation, Baton Rouge.

1961 Menard Site: The Quapaw Village of Osotouy on the Arkansas River. *Anthropological Papers of the American Museum of Natural History* 48:133–91.

Ford, James A., and Gordon R. Willey

1941 An Interpretation of the Prehistory of the Eastern United States. *American Anthropologist* 43:325–63.

Fowler, Melvin L.

1978 Cahokia in the American Bottoms: Settlement Archeology. In *Mississippian Settlement Patterns*, edited by Bruce D. Smith, pp. 455–77. Academic Press, New York,

Fowler, Melvin L., and Robert S. Hall

1972 *Archaeological Phases At Cahokia*. Papers in Anthropology 7. Illinois State Museum, Springfield.

Frank, Joseph V.

1980a Another Historic Natchez Village Site: Ben Lomond 22-Ad-519. *Mississippi Archaeological Association, Newsletter* 15(5):6–7.

1980b The Rice Site: A Natchez Indian Cemetery. *Mississippi Archaeology* 15:32–41.

1981 Antioch Site (22-Ad-631): A Natchez Village Site. *Mississippi Archaeological Association, Newsletter* 16(2):3–6.

French, Benjamin F.

1851 *Historical Collections of Louisiana*, Part III. D. Appleton, New York.

Galloway, Patricia K.

1982a Henri de Tonti du Village des Chacta, 1702: The Beginnings of the French Alliance. In *La Salle and His Legacy: Frenchmen and Indians in the Lower Mississippi Valley* edited by Patricia K. Galloway, pp. 146–75. University Press of Mississippi, Jackson.

1982b Historic Tribes in the Yazoo Basin during the French Colonial Period. Appendix to Cultural Resources Along the Yazoo River from Greenwood to Vicksburg, Mississippi. Report submitted in cooperation with the U.S. Army Corps of Engineers, Vicksburg District, in fulfillment of Contract DACW38-81-P-2166.

Garcilaso de la Vega

1951 *The Florida of the Inca.* Translated by John Varner and Jeannette Varner. University of Texas Press, Austin.

Garn, Stanley M., Fredric N. Silverman, Keith P. Hertzog, and Cristabel G. Rohman

1968 Lines and Bands of Increased Density; Their Implications for Growth and Development. *Medical Radiography and Photography* 44(3):58–89.

Gibbon, Guy

1986 Oneota Culture. *Wisconsin Archaeologist* 67:314–38.

Goodman, Alan H., and George J. Armelagos

1985a The Chronological Distribution of Enamel Hypoplasias in Human Permanent Incisors and Canine Teeth. *Archives of Oral Biology* 30:503–07.

1985b Disease and Death at Dr. Dickson's Mounds. *Natural History* 94(9):12–19.

Goodman, Alan H., George J. Armelagos, and Jerome C. Rose

1980 Enamel Hypoplasias as Indicators of Stress in Three Prehistoric Populations from Illinois. *Human Biology* 52:515–28.

1984a The Chronological Distribution of Enamel Hypoplasias from Prehistoric Dickson Mounds. *American Journal of Physical Anthropology* 65:259–66.

Goodman, Alan H., John Lallo, George J. Armelagos, and Jerome C. Rose

1984b Health Changes at Dickson Mounds, Illinois (A.D. 950–1300). In *Paleopathology at the Origins of Agriculture*, edited by Mark N. Cohen and George J. Armelagos, pp. 271–305. Academic Press, New York.

Goodman, Alan H., Debra L. Martin, George J. Armelagos, and George Clark

1984c Indications of Stress from Bone and Teeth. In *Paleopathology at the Origins of Agriculture*, edited by Mark N. Cohen and George J. Armelagos, pp. 13–49. Academic Press, New York.

Gravier, Jacques

1900 Relation or Journal of the Voyage of Father Gravier, of the Society of Jesus, in 1700, from the Country of the Illinois to the Mouth of the Mississippi River. *Jesuit Relations and Allied Documents* 65:100–179.

Green, Dee

1974 Cluster Analysis of 549 Whole Mississippi Ceramic Vessels. Ph.D. dissertation, Arizona State University, University Microfilms, Ann Arbor.

Green, Thomas, and Cheryl A. Munson

1978 Mississippian Settlement Patterns in Southwestern Indiana. In *Mississippian Settlement Patterns*, edited by Bruce D. Smith, pp. 293–330. Academic Press, New York.

Gregory, Hiram F.

1978 A Historic Tunica Burial at the Coulee des Grues Site in Avoyelles Parish, Louisiana. In *Texas Archeology*, edited by Kurt D. House, pp. 146–64. Southern Methodist University Press, Dallas.

Griffin, James B.

1943 *The Fort Ancient Aspect: Its Cultural and Chronological Position in Mississippi Valley*. Anthropological Papers 28. Museum of Anthropology, University of Michigan, Ann Arbor.

1946 Culture Change and Continuity in Eastern United States Archaeology. In Man in Northeastern North America, edited by Frederick Johnson. *Robert S. Peabody Foundation Papers* 3:307–48.

1952a Culture Periods in Eastern United States Archeology. In *Archaeology of Eastern United States*, edited by James B. Griffin, pp. 226–38.

1952b (editor) *Archaeology of Eastern United States*. University of Chicago Press, Chicago.

1961 Some Correlations of Climatic and Cultural Change in Eastern North America Prehistory. *New York Academy of Science Annals* 95:710–17.

1967 Eastern North American Archaeology: A Summary. *Science* 156:175–91.

1981 The Acquisition of a Little-Known Pottery Haul from the Lower Mississippi Valley. *Louisiana State University, School of Geoscience, Geoscience and Man* 22:51–56.

1985a Changing Concepts of the Prehistoric Mississippian Cultures of the Eastern United States. In *Alabama and the Borderlands: From Prehistory to Statehood*, edited by R. Reid Badger and Lawrence A. Clayton, pp. 40–63. University of Alabama Press, University.

1985b An Individual's Participation in American Archaeology, 1928–1985. *Annual Review of Anthropology* 14:1–23.

Grimm, Thomas C.

1985 Time-Depth Analysis of Fifteen Siouan Languages. *Siouan and Caddoan Linguistics* 5:12–26.

Hackett, C. J.

1951 *Bone Lesions of Yaws in Uganda*. Blackwell Scientific Publications, Oxford.

1976 *Diagnostic Criteria of Syphilis, Yaws and Treponarid (Treponematoses) and of Some Other Diseases in Dry Bones* Springer-Verlag, Berlin.

Haltom, W. L., and Alan R. Shands

1938 Evidences of Syphilis in Mound Builder's Bones. *Archives of Pathology* 25:228–42.

Hamilton, Henry W.

1967 *Tobacco Pipes of the Missouri Indians*. Memoir 5. Missouri Archaeological Society, Columbia.

Hampson, James K.

1973 The Nodena Site. In Nodena: An Account of 75 Years of Archeological Investigation in Southeast Mississippi County, Arkansas. *Arkansas Archeological Survey, Research Series* 4:9–21.

Hanke, Lewis

1937 Pope Paul III and the American Indian. *The Harvard Theological Review* 30(2):65–102

Hariot, Thomas

1955 A Briefe and True Report of the New Found Land of Virginia. In the Roanoke Voyages, 1584–1590, edited by David B. Quinn. *Hakluyk Society, Second Series* 104:317–89. London.

Harn, Alan D.

1978 Mississippian Settlement Patterns in the Central Illinois River Valley. In *Mississippian Settlement Patterns*, edited by Bruce D. Smith, pp. 233–68. Academic Press, New York.

1980 *The Prehistory of Dickson Mounds: The Dickson Excavation*. Report of Investigation 36. Illinois State Museum, Springfield.

Harrington, Jean C.

1952 Historic Site Archaeology in the United States. In *Archaeology of Eastern United States* edited by James B. Griffin, pp. 335–44. University of Chicago Press, Chicago.

Harris, H. A.

1933 *Bone Growth in Health and Disease.* Oxford University Press, London.

Hathcock, Roy

1983 *The Quapaw and Their Pottery*. Hurley, Camden, Arkansas.

Haven, Samuel F.

1856 *The Archaeology of the United States*. Smithsonian Contributions to Knowledge 8. Smithsonian Institution, Washington, D.C.

Headley, Robert K., Jr.

1971 The Origin and Distribution of the Siouan Speaking Tribes. Ph.D. dissertation, Department of Anthropology, Catholic University of America, Washington, D.C.

Hill, M. Cassandra

1979 The Alabama River Phase: A Biological Synthesis and Interpretation. Master's thesis, Department of Anthropology, University of Tennessee, Knoxville.

1981a Analysis, Synthesis, and Interpretation of the Skeletal Material Excavated for the Gainesville Section of the Tennessee-Tombigbee Waterway. In *Biocultural Studies in the Gainesville Lake Area*, by Gloria M. Caddell, Anne Woodrick, and Mary C. [M. Cassandra] Hill, pp. 211–334. Archaeological Investigations in the Gainesville Lake Area of the Tennessee-Tombigbee Waterway, vol. 4. Report of Investigations 14. Office of Archaeological Research, The University of Alabama, University.

1981b The Mississippian Decline in Alabama: A Biological Analysis. Paper presented at the 50th Annual Meeting of the American Association of Physical Anthropology, Detroit.

1983 The Woodland/Mississippian Transition in Alabama: A Study in Adaptation. Paper presented at the 52d Annual Meeting of the American Association of Physical Anthropology, Indianapolis.

1984 Site data sections. In *The Protohistoric Period in Central Alabama*, by Caleb Curren. Alabama Tombigbee Regional Commission, Camden.

1985 A Biocultural Assessment of Protohistoric Populations in Alabama. Paper presented at the 84th annual meeting of the American Anthropological Association, Washington, D.C.

1987 Emergent Mississippian Cultures. In *The Emergent Mississippian*, edited by Richard A. Marshall. Cobb Institute of Archaeology, Mississippi State University, Mississippi State.

Hillson, S. W.

1979 Diet and Dental Disease. *World Archaeology* 11:147–62.

Hoffman, Michael P.

1977 The Kinkead-Mainard Site, 3PU2: A Late Prehistoric Quapaw Phase Site Near Little Rock, Arkansas. *The Arkansas Archeologist* 16–18:1–41.

1981 The Father of Us All: S. C. Dellinger and the Beginning of Arkansas Archaeology and Anthropology. Paper presented at the 38th Southeastern Archaeological Conference, Asheville, North Carolina.

1985 The Quapaw, An Enduring People. Paper presented at "The Quapaw, a Living Tradition" Conference, Pine Bluff.

1986 The Protohistoric Period in the Lower and Central Arkansas River Valley in Arkansas. In the Protohistoric Period in the Mid-South: 1500–1700, Proceedings of the 1893 Mid-South Archaeological Conference, edited by David H. Dye and Ronald C. Brister. *Mississippi Department of Archives and History, Archaeological Report* 18:24–37.

Houck, Louis

1908 *A History of Missouri*. 3 vols. R. R. Donnelley and Sons, Chicago.

House, John H.

1987 Kent Phase Investigations in Eastern Arkansas. *Mississippi Archaeology* 22:46–60.

House, John H., and Henry McKelway

1982 Mississippian and Quapaw on the Lower Arkansas. In A State Plan for the Conservation of Archeological Resources in Arkansas, edited by Hester A. Davis. *Arkansas Archeological Survey, Research Series* 21:41–57.

Howard, James H.

1984 *Oklahoma Seminoles: Medicines, Magic, and Religion*. University of Oklahoma Press, Norman.

Howell, Nancy

1982 Village Composition Implied by a Paleodemographic Life Table: The Libben Site. *American Journal of Physical Anthropology* 59:263–69.

Hudson, Charles

1976 *The Southeastern Indians*. University of Tennessee Press, Knoxville.

1985 DeSoto in Arkansas: A Brief Synopsis. *Arkansas Archeological Society, Field Notes*, No. 205:3–12.

1987 *The Uses of Evidence in Reconstructing the Route of the Hernando de Soto Expedition*. De Soto Working Paper 1. Alabama De Soto Commission, Alabama State Museum of Natural History, University.

n.d. The Juan Pardo Expeditions: Exploration of the Carolinas and Tennessee, 1566–1568. Ms. on file, Department of Anthropology, University of Georgia, Athens.

Hudson, Charles, and Chester D. DePratter

1984 The Hernando de Soto Expedition: From the Mississippi River to Guachoya. Ms. on file, University of Arkansas Museum, Fayetteville.

Hudson, Charles, Marvin T. Smith, Chester B. DePratter, and Emilia Kelley

1985a The Tristan de Luna Expedition, 1559–1561. Paper presented at the 42d Annual Meeting of the Southeastern Archaeological Conference, Birmingham.

Hudson, Charles, Marvin T. Smith, David Hally, Richard Polhemus, and Chester DePratter

1985b Coosa: A Chiefdom in the Sixteenth-Century Southeastern United States. *American Antiquity* 50:723–37.

Hudson, Ellis H.

1958 *Non-venereal Syphilis, A Sociological and Medical Study of Bejel.* E. and S. Livingstone, Edinburgh.

Hummert, James R., and Dennis P. Van Gerven

1985 Observations on the Formation and Persistence of Radiopaque Transverse Lines. *American Journal of Physical Anthropology* 66:297–306.

Huss-Ashmore, Rebecca

1978 Nutritional Determination in Skeletal Populations. *American Journal of Physical Anthropology* 48:407.

Huss-Ashmore, Rebecca, Alan H. Goodman, and George J. Armelagos

1982 Nutritional Inferences from Paleopathology. In *Advances in Archaeological Method and Theory*, edited by Michael Schiffer, 5:395–473. Academic Press, New York.

Jenkins, Ned J.

1982 *Archaeology of the Gainsville Lake Area: Synthesis.* Report of Investigations 23, Office of Archaeological Research, The University of Alabama, University.

Jennings, Jesse D.

1941 Chickasaw and Earlier Indian Cultures of Northeast Mississippi. *Journal of Mississippi History* 3:115–226.

Jeter, Marvin D.

1986 Tunicans West of the Mississippi: A Summary of Early Historic and Archaeological Evidence. In The Protohistoric Period in the Mid-South: 1500–1700, Proceedings of the 1983 Mid-South Archaeological Conference, edited by David H. Dye and Ronald C. Brister. *Mississippi Department of Archives and History, Archaeological Report* 18:38–65.

Johnston, Francis E.

1962 Growth of the Long Bones of Infants and Young Children at Indian Knoll. *American Journal of Physical Anthropology* 20:249–54.

Jones, Joseph

1876 *Explorations of the Aboriginal Remains of Tennessee.* Smithsonian Contributions to Knowledge 259. Smithsonian Institution, Washington, D.C.

Kandare, Richard P.

1983 A Contextual Study of Mississippian Dugout Canoes: A Research Design for the Moundville Phase. Master's thesis, Department of Anthropology, University of Arkansas, Fayetteville.

Kellogg, Louise P. (editor)
1917 *Early Narratives of the Northwest, 1634–1699*. Scribner's, New York.

Klinger, Timothy
1977 Parkin Archeology: A Report on the 1966 Field School Test Excavations at the Parkin Site. *Arkansas Archeologist* 16–18:45–80.

Klippel, Walter E., and William M. Bass (editors)
1984 *Averbuch: A Late Mississippian Manifestation in the Nashville Basin*. 2 vols. Submitted to National Park Service, Contract CX5000-9-5943. Copies available from U.S. Department of Commerce, National Technical Information Service, Springfield, Illinois.

Knight, Vernon James, Jr.
1988 *A Summary of Alabama's de Soto Mapping Project and Project Bibliography*. De Soto Working Paper 9. Alabama De Soto Commission, Alabama State Museum of Natural History, University.

Koontz, John E.
1985 On the Existence of the Mississippi Valley Subgroup in the Siouan Language Family. Ms. on file, University of Arkansas Museum, Fayetteville.

Kroeber, Alfred L.
1953 *Cultural and Natural Areas of Native North America*. University of California Press, Berkeley and Los Angeles.

Krogman, Wilton M.
1973 *The Human Skeleton in Forensic Medicine*. C. C. Thomas, Springfield, Illinois.

Kwas, Mary L. (editor)
1986 *Archaeological Parks, Integrating Preservation, and Recreation*. Tennessee Department of Conservation, Division of Parks and Recreation, Nashville.

La Flesche, Francis
1915 Omaha and Osage Traditions of Separation. *Proceedings of the International Congress of Americanists* 19:459–62.

Lahren, Craig H., and Hugh E. Berryman
1984 Fracture Patterns and Status at Chucalissa (40SY1): A Biocultural Approach. *Tennessee Anthropologist* 9:15–21.

Lallo, John W.
1973 The Skeletal Biology of Three Prehistoric American Indian Societies from Dickson Mounds. Ph.D. dissertation, Department of Anthropology, University of Massachusetts, Amherst.

Lankford, George E., III
1977 A New Look at the De Soto's Route through Alabama. *Journal of Alabama Archaeology* 23:10–36.

Larsen, Clark S.

1982 The Anthropology of St. Catherine's Island. Vol. 3. Prehistoric Human Biological Adaption. *Anthropological Papers of the American Museum of Natural History* 57:159–207.

1984 Health and Disease in Prehistoric Georgia: The Transition to Agriculture. In *Paleopathology at the Origins of Agriculture*, edited by Mark N. Cohen and George J. Armelagos, pp. 367–92. Academic Press, Orlando.

Larson, Lewis H.

1972 Functional Considerations of Warfare in the Southeast During the Mississippi Period. *American Antiquity* 37:383–92.

Lawson, John

1709 *A New Voyage to Carolina, Containing the Exact Description and Natural History of that Country*. London.

Lehmer, Donald J., and Warren W. Caldwell

1966 Horizon and Tradition in the Northern Plains. *American Antiquity* 31:511–16.

Lewis, R. Barry

1974 *Mississippian Exploitative Strategies: A Southeast Missouri Example*. Research Series 11. Missouri Archaeological Society, Columbia.

1982 *Excavations at Two Mississippian Hamlets in the Cairo Lowland of Southeast Missouri*. Special Publication 2. Illinois Archaeological Survey, Urbana.

1983 The Mississippi Period in the Cairo Lowland, Missouri. Paper presented at the 48th Annual Meeting of the Society for American Archaeology, Pittsburg.

1986 (editor) *Mississippian Towns of the Western Kentucky Border: The Adams, Wickliffe, and Sassafras Ridge Sites*. The Kentucky Heritage Council, Frankfort.

1987 (editor) The Mississippi Period in Kentucky, University of Illinois at Urbana-Champaign. Submitted to the Kentucky Heritage Council, Frankfort.

1988 Old World Dice in the Protohistoric Southern United States. *Current Anthropology* 29:759–68.

Lewis, R. Barry, and Lynne M. Mackin

1984 The Adams Site Ceramic Assemblage in Regional Perspective. In *Late Prehistoric Research in Kentucky*, edited by David Pollack, Charles Hockensmith, and Thomas Sanders, pp. 187–204. The Kentucky Heritage Council, Frankfort.

Loughridge, Robert H.

1888 *Report on the Geological and Economic Features of the Jackson Purchase Region*. Kentucky Geological Survey, Lexington.

Lovejoy, C. Owen, Richard S. Meindl, Robert P. Mensforth, and Thomas J. Barton

1985 Multifactorial Determination of Skeletal Age at Death: A Method and Blind Tests of its Accuracy. *American Journal of Physical Anthropology* 68:1–14.

Lovejoy, C. Owen, Richard S. Meindl, Thomas R. Prysbeck, Thomas S. Barton, King G. Heiple, and David Kotting

1977 Paleodemography of the Libben Site, Ottawa County, Ohio. *Science* 198:291–93.

McGee, William J.

1897 The Siouan Indians: A Preliminary Sketch. *Smithsonian Institution, Bureau of American Ethnology, Annual Report* 15:157–212.

McGimsey, Charles R., III

1964 A Report on the Quapaw in Arkansas Immediately Prior to the Treaty of 1818. Report submitted to the Quapaw Business Committee, Miami, Oklahoma.

1965 The Dupree Site in Retrospect. *The Arkansas Archeologist* 6:3–8.

McNutt, Charles H., and Guy G. Weaver

1985 *An Above-Pond Survey of the Cultural Resources of the Little Bear Creek Reservoir Area, Franklin County, Alabama*. Occasional Papers 13. Anthropological Research Center, Memphis State University, Memphis.

Magennis, Ann L., and Debra L. Martin

1985 The Black Mesa Anasazi: Patterns of Disease and Activity in a Marginal Environment. Paper presented at the Annual Meeting of the Northeast Anthropological Association, Lake Placid.

Margry, Pierre

1879 *Mémoires et Documents Pour Servir à l'Histoire des Origines Françaises des pay d'Outremer*. 3 vols. Paris.

Marquette, Jacques

1900 Of the First Voyage made by Father Marquette Toward New Mexico, and How the Idea Thereof was Conceived. *The Jesuit Relations and Allied Documents* 59:87–163.

Marshall, Richard A.

1977 Lyon's Bluff Site (22OK1) Radiocarbon Dated. *Journal of Alabama Archaeology* 23:53–57.

1986 The Protohistoric Component at the Lyon's Bluff Site Complex, Oktibbeha County, Mississippi. In The Protohistoric Period in the Mid-South: 1500–1700, Proceedings of the 1983 Mid-South Archaeological Conference, edited by David H. Dye and Ronald C. Brister. *Mississippi Department of Archives and History, Archaeological Report* 18:82–88.

Mason, Carol

1986 The Historic Period in Wisconsin Archaeology. *Wisconsin Archaeologist* 67:370–92.

Mason, Ronald J.

1976 Ethnicity and Archaeology in the Upper Great Lakes. In *Cultural Change and Continuity: Essays in Honor of James Bennett Griffin*, edited by Charles E. Cleland, pp. 349–61. Academic Press, New York.

Medford, Larry D.

1972 Agricultural Destruction of Archeological Sites in Northeast Arkansas. *Arkansas Archeological Survey, Research Series* 3:241–53.

Menaker, Lewis (editor)

1980 *The Biological Basis of Dental Caries, An Oral Biology Textbook*. Harper and Row, New York.

Mensforth, Robert P., C. Owen Lovejoy, John W. Lallo, and George J. Armelagos

1978 The Role of Constitutional Factors, Diet, and Infectious Disease in the Etiology of Porotic Hyperostosis and Periosteal Reaction in Prehistoric Infants and Children. *Medical Anthropology* 2:1–59.

Merbs, Charles F.

1983 *Patterns of Activity-Induced Pathology in an Inuit Population*. Paper 19. Mercury Series, Archeological Survey of Canada, National Museum of Man, Ottawa.

Merchant, Virginia L., and Douglas H. Ubelaker

1977 Skeletal Growth of the Prehistoric Arikara. *American Journal of Physical Anthropology* 46:61–72.

Milner, George R.

1980 Epidemic Disease in the Postcontact Southeast: A Reappraisal. *Mid-Continental Journal of Archaeology* 5:39–56.

1982 Measuring Prehistoric Levels of Health: A Study of Mississippian Period Skeletal Remains from the American Bottom. Ph.D. dissertation, Department of Anthropology, Northwestern University, Evanston.

Mooney, James

1890 The Cherokee Ball Play. *American Anthropologist*, o.s., 2:105–32.

Moore, Clarence B.

1897–1901 Certain Aboriginal Remains of the Alabama River. *Journal of the Academy of Natural Sciences of Philadelphia* 11:289–348.

1908 Certain Mounds of Arkansas and Mississippi. *Journal of the Academy of Natural Sciences of Philadelphia* 13:480–605.

1910 Antiquities of the St. Francis, White and Black Rivers, Arkansas. *Journal of the Academy of Natural Sciences of Philadelphia* 14:255–364.

1911 Some Aboriginal Sites on the Mississippi River. *Journal of the Academy of Natural Sciences of Philadelphia* 14:366–480.

Moore, James, Alan C. Swedlund, and George J. Armelagos

1975 The Use of the Life Table in Paleodemography. *Society for American Archaeology, Memoir* 30:57–70.

Morse, Dan F.

1971 An Historic Indian Grave Near Blytheville, Arkansas. *Arkansas Archeologist* 12:56–60.

1973a The 1973 Field School Excavations at Upper Nodena. *Arkansas Archaeological Society, Field Notes*, No. 106:3–8.

1973b *Nodena, An Account of 75 Years of Archeological Investigation in Southeast Mississippi County, Arkansas.* Research Series 4. Arkansas Archeological Survey, Fayetteville.

1977 The Penetration of Northeastern Arkansas by Mississippian Culture. In For the Director: Research Essays in Honor of James B. Griffin, edited by Charles E. Cleland. *University of Michigan, Museum of Anthropology, Anthropological Papers* 61:186–211.

1980 Chronological Problems. In *Zebree Archeological Project*, edited by Dan F. Morse and Phyllis A. Morse. Submitted to U.S. Army Corps of Engineers, Memphis District. Copies available from U.S. Army Corps of Engineers, Memphis.

1986a The Early Colonial Quapaw. Paper presented at the Arkansas Tricentennial Symposium, Arkansas Before the Americans, Little Rock.

1986b Protohistoric Hunting Sites in Northeastern Arkansas. In The Protohistoric Period in the Mid-South: 1500–1700, Proceedings of the 1983 Mid-South Archaeological Conference, edited by David H. Dye and Ronald C. Brister. *Mississippi Department of Archives and History, Archaeological Report* 18:89–94.

Morse, Dan F., and Phyllis A. Morse (editors)

1980 *Zebree Archeological Project*. Submitted to U.S. Army Corps of Engineers, Memphis District, Contract DACW 66-76-C-0006. Copies available from U.S. Army Corps of Engineers, Memphis.

1983 *Archaeology of the Central Mississippi Valley*. Academic Press, New York.

Morse, Dan F. and Samuel Smith

1973 Archeological Salvage During the Construction of Route 308. *The Arkansas Archeologist* 14:36–78.

Morse, G. Daniel

1967 Tuberculosis. In *Diseases in Antiquity*, edited by Don Brothwell and Ed Sandison, pp. 251–79. C. C. Thomas, Springfield, Illinois.

Morse, Phyllis A.

1978 The First Excavation at Rose Mound. *Arkansas Archeological Society, Field Notes*, No. 163:7–9.

1981 *Parkin: The 1978–1979 Archeological Investigations of a Cross County, Arkansas Site*. Research Series 13. Arkansas Archeological Survey, Fayetteville.

Morse, Phyllis A., Dan F Morse, Michael G. Millon, and Arthur E. Bogan

1974 Armorel Site Report. *Arkansas Archeological Society, Field Notes*, Nos. 117–18:1–19.

Moselage, John H.
1965 A House Pattern from the Dupree Site. *Arkansas Archeologist* 6:1–3.

Muller, Jon
1978 The Kincaid System: Mississippian Settlement in the Environs of a Large Site. In *Mississippian Settlement Patterns* edited by Bruce D. Smith, pp. 269–92. Academic Press, New York.

Munson, Cheryl A., and Thomas J. Green
1973 The Caborn-Welborn Complex of Southwestern Indiana and Oneota Relationships with Lower Ohio Valley. Paper presented at the 50th Annual Meeting of the Central States Anthropological Society, St. Louis.

Murchison, Mark A., Douglas W. Owsley, and Arthur J. Riopelle
1983 Transverse Line Formation in Protein Deprived Rhesus Monkeys. Paper presented at the 10th Annual Meeting of the Paleopathology Association, Indianapolis.

Murray, Katherine
1989 Bioarcheology of the Parkin Site, Cross County, Arkansas. *The Arkansas Archeologist* 27/28:49–61.

Nash, Charles H.
1963 Total Data. *Arkansas Archaeologist* 4:12–15.
1968 *Residence Mounds: An Intermediate Middle-Mississippian Settlement Pattern*. Occasional Papers 2. Anthropological Research Center, Memphis State University, Memphis.
1972 *Chucalissa: Excavations Through 1963*. Occasional Papers 6. Anthropological Research Center, Memphis State University, Memphis.

Neitzel, Robert S.
1965 Archeology of the Fatherland Site: The Grand Village of the Natchez. *Anthropological Papers of the American Museum of Natural History* 51:1–108.
1983 *The Grand Village of the Natchez Revisited: Excavations at the Fatherland Site, Adams County, Mississippi, 1972.* Archaeological Report 2. Mississippi Department of Archives and History, Jackson.

Nie, N. H., C. H. Hull, J. G. Jenkins, K. Steinbrenner, and D. H. Brent
1975 *Statistical Package for the Social Sciences*. 2d ed. McGraw-Hill, New York.

Olson, Alan P.
1962 A History of the Phase Concept in the Southwest. *American Antiquity* 27:457–72.

Ortner, D. J., and Walter G. J. Putschar
1981 *Identification of Pathological Conditions in Human Skeletal Re-*

mains. Smithsonian Contributions to Anthropology, vol. 28. Smithsonian Institution, Washington, D.C.

Parham, Kenneth R., and Gary T. Scott

1980 Porotic Hyperostosis: A Study of Disease and Culture at Toqua (40MR6), A Late Mississippian Site in Eastern Tennessee. In The Skeletal Biology of Aboriginal Populations in the Southeastern United States, edited by Pat Willey and Fred H. Smith. *Tennessee Anthropological Society, Miscellaneous Papers* 5:39–51.

Parmalee, Paul W.

1960 Vertebrate Remains from the Chucalissa Site. *Tennessee Archaeologist* 16:84–90.

1966 Animal Remains From the Banks Site. In The Banks Village Site, Crittenden County, Arkansas, by Gregory Perino. *Missouri Archeological Society, Memoir* 4:142–45.

Parry, John H.

1985 Early European Penetration of Eastern North America. In *Alabama and the Borderlands: From Prehistory to Statehood*, edited by R. Reid Badger and Lawrence A. Clayton, pp. 83–95. University of Alabama Press, University.

Peebles, Christopher S.

1979 *Excavations at Moundville: 1905–1951*. University of Michigan Press, Ann Arbor.

1981 Archaeological Research at Moundville: 1840–1980. *Southeastern Archaeological Conference, Bulletin* 24:77–81.

1983 (editor) Excavations in the Lubbub Creek Archaeological Locality. In Prehistoric Agricultural Communities in West Central Alabama vol. 1. Submitted to U.S. Army Corps of Engineers, Mobile District, Contracts C5861(79) and C5970(79). Copies available from U.S. Army Corps of Engineers, Mobile.

1986 Paradise Lost, Strayed, and Stolen: Prehistoric Social Devolution in the Southeast. In The Burden of Being Civilized: An Anthropological Perspective on the Discontents of Civilization, edited by Miles B. Richardson and Malcolm C. Webb. *Southern Anthropological Society, Proceedings* 18:24–40.

1987 The Rise and Fall of the Mississippian in Western Alabama: The Moundville and Summerville Phases, A.D. 1000 to 1600. *Mississippi Archaeology* 22:1–31.

Perino, Gregory

1966 *The Banks Village Site, Crittendon County, Arkansas*. Memoir 4. Missouri Archaeological Society, Columbia.

1967 The Cherry Valley Mounds and Banks Mound 3. Memoir 1. Central States Archaeological Society, Quincy, Illinois.

Perzigian, Anthony J., Patricia A. Tench, and Donna J. Braun

1984 Prehistoric Health in the Ohio River Valley. In *Paleopathology at the Origins of Agriculture*, edited by Mark N. Cohen and George J. Armelagos, pp. 347–66. Academic Press, Orlando.

Phenice, Terrell W.

1967a A Newly Developed Visual Method of Sexing the Os Pubis. *American Journal of Physical Anthropology* 30:297–302.

Phillips, Philip

1970 *Archaeological Survey in the Lower Yazoo Basin, Mississippi, 1949–1955*. Papers vol. 68. Peabody Museum of Archaeology and Ethnology, Harvard University, Cambridge.

Phillips, Philip, and James A. Brown

1978 *Pre-Columbian Shell Engravings from the Craig Mound at Spiro Oklahoma*, Part 1. Peabody Museum Press, Cambridge.

Phillips, Philip, James A. Ford, and James B. Griffin

1951 *Archaeological Survey in the Lower Mississippi Alluvial Valley, 1940–1947*. Papers vol. 25. Peabody Museum of American Archaeology and Ethnology, Harvard University, Cambridge.

Powell, Mary L.

1983 Biocultural Analysis of Human Skeletal Remains from the Lubbub Creek Archaeological Locality. In *Prehistoric Agricultural Communities in West Central Alabama*, edited by Christopher S. Peebles, 2 (AD-A155 048/2/6AR): 430–77. National Technical Information Service, Washington, D.C.

1985a The Analysis of Dental Wear and Caries for Dietary Reconstruction. In *The Analysis of Prehistoric Diets*, edited by Robert I. Gilbert and J. H. Mielke, pp. 307–38. Academic Press, New York.

1985b Health, Disease and Social Organization in the Mississippian Community at Moundville. Ph.D. dissertation, Department of Anthropology, Northwestern University, Evanston.

1988 *Status and Health in Prehistory, A Case Study of the Moundville Chiefdom.* Smithsonian Institution Press, Washington, D.C.

1989a Endemic Treponematosis and Tuberculosis in the Prehistoric Southeastern United States: The Biological Costs of Chronic Endemic Disease. In *Human Paleopathology: Current Synthesis and Future Options*, edited by Donald J. Ortner and Arthur C. Aufderheide. Ms. in preparation for Smithsonian Institution Press, Washington, D.C.

1989b The People of Nodena, a Biocultural Analysis of a Late Mississippian Community in Northeast Arkansas. Part 2, *Nodena, An Account of 75 Years of Archaeological Investigations in Southeast Mississippi County, Arkansas*. Arkansas Archaeological Survey Research Series 28, Fayetteville.

Powell, Mary L., and J. Daniel Rogers
1980 *Bioarchaeology of the McCutchan-McLaughlin Site*. Studies in Oklahoma's Past 5. Oklahoma Archaeological Survey, Norman.

Powell, Richard R., and Patrick J. Rohan
1985 *The Law of Real Property*. Vol 5. Matthew Bender, New York.

Price, James E.
1974 Mississippian Settlement Systems of the Central Mississippi Valley. Paper prepared for an advanced seminar on Mississippian development, School of American Research, Santa Fe.

1976 Prehistory of the Fourche Creek Watershed. In *An Assessment of the Cultural Resources of the Fourche Creek Watershed*, by James E. Price, Cynthia R. Price, and Suzanne E. Harris, pp. 33–51. American Archaeology Division, University of Missouri. Submitted to USDA Soil Conservation Service, Springfield. Contract USDA-AG-29SCS-00527. Copies available from the University of Missouri, Columbia.

1978 The Settlement Pattern of the Powers Phase. In *Mississippian Settlement Patterns*, edited by Bruce D. Smith, pp. 201–31. Academic Press, New York.

Price, James E., and James B. Griffin
1979 *The Snodgrass Site of the Powers Phase of Southeast Missouri*. Anthropological Papers 66. Museum of Anthropology, University of Michigan, Ann Arbor.

Price, James E., and James J. Krakker
1975 *Dalton Occupation of the Ozark Border*. Museum Briefs 20. Museum of Anthropology, University of Missouri, Columbia.

Price, James E., and Cynthia R. Price
1979 *An Inventory and Assessment of the Leo Anderson Collection of Archaeological and Historical Specimens.* Center for Archaeological Research Report 256. Southwest Missouri State University, Springfield.

1983 Study of Private Artifact Collections. In *Archaeological Investigations in the Ozark National Scenic Riverways, 1981–1982,* by James E. Price, Cynthia R. Price, Roger T. Saucier, and Timothy K. Perttula. Report 447:227–93. Center for Archaeological Research, Southwest Missouri State University, Springfield.

Putnam, Frederick Ward
1881 *Report of the Curator*. Fourteenth Annual Report of the Trustees of the Peabody Museum of American Archaeology and Ethnology, Harvard University, Cambridge.

1882 *Notes on the Copper Objects from North and South America, Contained in the Collection of the Peabody Museum.* Fifteenth Annual Report of the Trustees of the Peabody Museum of American Archaeology and Ethnology, Harvard University, Cambridge.

Quimby, George I.
1957 The Bayou Goula Site, Iberville Parish, Louisiana. *Chicago Natural*

History Museum, Fieldiana: Anthropology 47:91–170.

1960 *Indian Culture and European Trade Goods*. University of Wisconsin Press, Madison.

Quimby, George I., and Alexander Spoehr

1951 Acculturation and Material Culture—I. *Chicago Natural History Museum, Fieldiana: Anthropology* 36:107–47.

Rafferty, Janet

1983 A New Map of the Ingomar Mounds Site, 22UN500. *Mississippi Archaeology* 18:18–30.

Ramenofsky, Ann F.

1982 The Archaeology of Population Collapse: Native American Response to the Introduction of Infectious Disease. Ph.D. dissertation, University of Washington, University Microfilms, Ann Arbor.

Ranjel, Rodrigo

1904 Relation. In *Narratives of the Career of Hernando de Soto*, translated by Buckingham Smith, edited by Edward G. Bourne, 2:47–149. A. S. Barnes, New York.

1922 Relation. In *Narratives of the Career of Hernando de Soto*, translated by Buckingham Smith, edited by Edward G. Bourne, 2:47–149. Reprinted. Allerton, New York.

Rankin, Robert L.

1982 A Quapaw Vocabulary. *Kansas Working Papers in Linguistics* 7:125–152.

1985 Quapaw as a Historically Dhegiha Language: Grammar. Paper presented at the Siouan Language Conference, Tulsa.

Riggs, C. W.

1893 *How We Find Relics: Pleasures and Perils on the Mississippi River in a Floating Home*. W. B. Conkey, Chicago.

Riley, Carrol L.

1982 *The Frontier People: The Greater Southwest in the Protohistoric Period*. Occasional Paper 1. Centers for Archaeological Investigation, Southern Illinois University, Carbondale.

Robbins, Lester E.

1977 The Persistence of Traditional Religious Practices Among Creek Indians. Ph.D. Dissertation, Department of Anthropology, Southern Methodist University, Dallas.

Robbins, Louise

1978 Yawslike Disease Process in a Louisiana Shell Mound Population. *Medical College of Virginia Quarterly* 14:24–31.

Robbins, Stanley L., and Rawzi S. Cotran

1980 *Pathological Basis of Disease*. 2d ed. W. B. Saunders, Philadelphia.

Robinson, Ronald K.

1977 The Skeletal Remains from Chucalissa (40SY1), Shelby County, Tennessee. Ms. on file, C. H. Nash Museum, Department of Anthropology, Memphis State University, Memphis.

Rolingson, Martha P.
1982 Public Archeology: Research and Development of the Toltec Mounds Site. In Arkansas Archeology in Review, edited by Neal L. Trubowitz and Marvin D. Jeter. *Arkansas Archaeological Survey, Research Series* 15:48–75.

Rose, Jerome C.
1981 Bioarchaeology of Two Late Woodland Sites (22IT537) and (22L0530) from the Tombigbee River Multi-Resource District. In *Quarterly Report* January–March, U.S. Department of Interior, Washington, D.C.

Rose, Jerome C., Barbara A. Burnett, Michael S. Nassaney, and Mark W. Blaeuer
1984 Paleopathology and the Origins of Maize Agriculture in the Lower Mississippi Valley and Caddoan Culture Areas. In *Paleopathology at the Origins of Agriculture*, edited by Mark N. Cohen and George J. Armelagos, pp. 393–424. Academic Press, Orlando.

Rose, Jerome C., and Murray K. Marks
1985 Bioarcheology of the Alexander Site. In The Alexander Site, Conway County, Arkansas, edited by E. Thomas Hemmings and John H. House. *Arkansas Archeological Survey, Research Series* 24:76–98.

Sahlins, Marshall D.
1968 *Tribesmen*. Prentice-Hall, Englewood Cliffs.

Sanders, William T., and David Webster
1978 Unilinealism, Multilinealism, and the Evolution of Complex Societies. In *Social Archaeology: Beyond Subsistence and Dating,* edited by Charles L. Redman, Mary J. Berman, Edward V. Curtin, William T. Langhorne, Jr., Nina J. Versaggi, and Jeffery C. Wanser, pp. 249–302. Academic Press, New York.

Saucier, Roger T.
1970 Origin of the St. Francis Sunk Lands, Arkansas and Missouri. *Geological Society of America, Bulletin* 81:2847–54.
1974 *Quaternary Geology of the Lower Mississippi Valley*. Research Series 6. Arkansas Archeological Survey, Fayetteville.

Schmorl, G., and H. Junghanns
1971 *The Human Spine in Health and Disease.* 2d ed. Grune and Stratton, New York.

Schwartz, Douglas W.
1961 *The Tinsley Hill Site*. Studies in Anthropology 1. University of Kentucky Press, Lexington.

Scott, Eugenie C.
1979a Dental Wear Scoring Technique. *American Journal of Physical Anthropology* 51:213–17.
1979b Principal Axis Analysis of Dental Attrition Data. *American Journal of Physical Anthropology* 51:203–11.

Scrimshaw, Nevin S., Carl E. Taylor, and John E. Gordon
1968 *Interactions of Nutrition and Infection.* Monograph Series 57. World Health Organization, Geneva, Switzerland.

Sheldon, Craig T., Jr.
1974 The Mississippian Historic-Transition in Central Alabama. Ph.D. dissertation, Department of Anthropology, University of Oregon, Eugene.

Sheldon, Craig T., Jr., and Ned J. Jenkins
1986 Protohistoric Development in Central Alabama. In The Protohistoric Period in the Mid-South: 1500–1700, Proceedings of the 1983 Mid-South Archaeological Conference, edited by David H. Dye and Ronald C. Brister. *Mississippi Department of Archives and History, Archaeological Report* 18:95–102.

Smith, Bruce D.
1975 *Middle Mississippi Exploitation of Animal Populations.* Anthropological Papers 57. Museum of Anthropology, University of Michigan, Ann Arbor.
1978a Variation in Mississippian Settlement Patterns. In *Mississippian Settlement Patterns*, edited by Bruce D. Smith, pp. 479–503. Academic Press, New York.
1978b (editor) *Mississippian Settlement Patterns.* Academic Press, New York.
1984 Mississippian Expansion: Tracing the Historical Development of an Explanatory Model. *Southeastern Archaeology* 3:13–32.
1986 The Archaeology of the Southeastern United States: From Dalton to de Soto, 10,500–500 B.P. *Advances in World Archaeology* 5:1–92.

Smith, Gerald P.
1969 *Ceramic Handle Styles and Cultural Variation in the Northern Sector of the Mississippi Alluvial Valley.* Occasional Papers 3. Anthropological Research Center, Memphis State University, Memphis.

Smith, Harold, and Donald Willis
1951 A Report of Findings Made by Harold Smith and Donald Willis at an Aboriginal Village Site Near the Loosahatchie River, on Property Formerly Owned by Mrs. John Smith in Woodstock Township, Shelby County, Tennessee. Ms. on file, C. H. Nash Museum, Department of Anthropology, Memphis State University, Memphis.

Smith, Marvin T.
1977 The Chevron Trade Bead in North America. *The Bead Journal* 3:15–17.
1984 Depopulation and Culture Change in the Early Historic Period Interior Southeast. Ph.D. dissertation, University of Florida, University Microfilms, Ann Arbor.
1987 *Archaeology of Aboriginal Culture Change in the Interior Southeast.* University of Florida Press, Gainesville.

Southard, Michael D., and John W. Cottier
1973 Beckwith's Fort Site, 23Mi2, Renamed Towosahgy State Archaeological Site by the Missouri State Park Board: 1967–1972 Excavations. Report submitted to the Missouri State Park Board.

Speck, Frank G.
1909 Ethnology of the Yuchi Indians. *University of Pennsylvania, Anthropological Publications of the University Museum* 1:1–154.

Steinbock, R. Ted
1976 *Paleopathological Diagnosis and Interpretation*. C. C. Thomas, Springfield, Illinois.

Steponaitis, Vincas P.
1983 *Ceramics, Chronology, and Community Patterns: An Archaeological Study of Moundville*. Academic Press, New York.

Stokes, Marvin A., and Terah L. Smiley
1968 *An Introduction to Tree Ring Dating*. University of Chicago Press, Chicago.

Stout, Charles B.
1984a Gross Spatial Patterning at a Large Mississippian Town and Ceremonial Center. Paper presented at the 41st Annual Meeting of the Southeastern Archaeological Conference, Pensacola.
1984b Mississippian Sites in Western Kentucky: Variations on A General Mississippian Theme? In *Late Prehistoric Research in Kentucky*, edited by David Pollack, Charles Hockensmith, and Thomas Sanders, pp. 167–79. Kentucky Heritage Council, Frankfort.
1985 *The Adams Site: A Spatial Analysis—Preliminary Report*. Western Kentucky Project 2. Department of Anthropology, University of Illinois, Urbana.
1987 *Surface Distribution Patterns at the Adams Site, a Mississippian Town in Fulton County, Kentucky*. Western Kentucky Project Report 6. Department of Anthropology, University of Illinois, Urbana.

Stuart-Macadam, Patricia
1985 Porotic Hyperostosis: Representative of a Childhood Condition. *American Journal of Physical Anthropology* 66:391–98.

Stubbs, John D., Jr.
1984 A Report Presenting the Results of Archaeological Survey in Lee County, Mississippi, June 1981 to June 1983. Report submitted to the Chickasaw Indian Cultural Center Foundation, Tupelo.

Stuiver, Minze, and Gordon W. Pearson
1986 High-Precision Calibration of the Radiocarbon Time Scale, AD 1950–500 BC. *Radiocarbon* 28:805–38.

Stuiver, Minze, and P. J. Reimer
1986 A Computer Program for Radiocarbon Age Calibration. *Radiocarbon* 28:1022–30.

Sussenbach, Tom, and R. Barry Lewis

1987 *Archaeological Investigations in Carlisle, Hickman, and Fulton Counties, Kentucky*. Western Kentucky Project Report 4. Department of Anthropology, University of Illinois, Urbana.

Swanton, John R.

1911 *Indian Tribes of the Lower Mississippi Valley and Adjacent Coast of the Gulf of Mexico*. Bulletin 43. Smithsonian Institution, Bureau of American Ethnology, Washington, D.C.

1939 *Final Report of the United States De Soto Expedition Commission*. House Document 71. 76th Congress. U.S. Government Printing Office. Washington, D.C.

1946 *The Indians of the Southeastern United States*. Bulletin 137. Smithsonian Institution, Bureau of American Ethnology, Washington, D.C.

1985 *Final Report of the United States De Soto Expedition Commission*. Reprinted. Smithsonian Institution Press, Washington, D.C. Originally published 1939. U.S. Government Printing Office, Washington, D.C.

Swedlund, Alan C., and George J. Armelagos

1976 *Demographic Anthropology*. W. C. Brown, Dubuque.

Teitelbaum, Michael S.

1972 Factors Associated with the Sex Ratio in Human Populations. In *The Structure of Human Populations*, edited by G. A. Harrison and A. J. Boyce, pp. 90–109. Clarendon Press, London.

Thomas, Cyrus

1894 *Report on the Mound Explorations of the Bureau of Ethnology*. Annual Report 12. Smithsonian Institution, Bureau of American Ethnology, Washington, D.C.

1959 Quapaw. *Smithsonian Institution, Bureau of American Ethnology, Bulletin* 30:333–36.

Thorne, Robert M.

1985 *Preservation is a Use: Archaeological Site Stabilization, an Experimental Program in the Tennessee River Valley*. Archaeological Papers of the Center for Archaeological Research 5. University of Mississippi, University.

Thurston, Gates P.

1897 *The Antiquities of Tennessee and the Adjacent States*. 2d ed. Robert Clarke, Cincinnati.

Trimble, Michael K.

1986 *An Ethnohistorical Interpretation of the Spread of Smallpox in the Northern Plains Utilizing Concepts of Disease Ecology*. J. and L. Reprint, Lincoln, Nebraska.

Trotter, Mildred, and Goldine C. Gleser
1952 Estimation of Stature from Long Bones of American Whites and Negroes. *American Journal of Physical Anthropology* 10:463–514.

Tucker, Sara Jones
1942 *Indian Villages of the Illinois Country*. Scientific Papers vol. 2. Illinois State Museum, Springfield.

Turner, Christy G., II
1979 Dental Anthropological Indications of Agriculture Among the Jomon People of Central Japan. *American Journal of Physical Anthropology* 51:619–35.

Turner, Kenneth R.
1985 Epidemic Disease in the Early Historic Southeast. Paper presented at the 42n Annual Meeting of the Southeastern Archaeological Conference, Birmingham, Alabama.

Ubelaker, Douglas H.
1978 *Human Skeletal Remains, Excavation, Analysis, Interpretation*. Aldine, Chicago.

Viers, Margaret
1983 A Linguistic Comparison of Quapaw and Osage. Paper written for Anthropology 3263, Indians of Arkansas and the South. University of Arkansas, Fayetteville.

Walthall, John A.
1980 *Prehistoric Indians of the Southeast: Archaeology of Alabama and the Middle South.* University of Alabama Press, University.

Ward, Rufus A., Jr.
1986 The Tombigbee Crossing of the De Soto Expedition. *Mississippi Archaeology* 21:62–68.

Weaver, Guy G.
1982 De Soto Park Site: Preliminary Investigation. Ms. on file, C. H. Nash Museum, Department of Anthropology, Memphis State University, Memphis.

Webb, William S.
1938 *An Archaeological Survey of the Norris Basin in Eastern Tennessee*. Bulletin 118. Smithsonian Institution, Bureau of American Ethnology, Washington, D.C.

1939 *An Archaeological Survey of Wheeler Basin on the Tennessee River in Northern Alabama*. Bulletin 122. Smithsonian Institution, Bureau of American Ethnology, Washington D.C.

Webb, William S., and William D. Funkhouser
1933 The McLeod Bluff Site. *University of Kentucky, Reports in Archaeology and Anthropology* 3(1):1–33.

Weinstein, Richard A.
1985 Some New Thoughts on the De Soto Expedition through Western Mississippi. *Mississippi Archaeology* 20:2–24.

Weiss, Kenneth M.
1973 *Demographic Models for Anthropology*. Memoir 27. Society for American Archaeology, Washington, D.C.

Wells, Calvin
1964 *Bones, Bodies, and Disease.* Praeger, New York.

Wesler, Kit W.
1985 *Archaeological Excavations at Wickliffe Mounds, 15BA4: Mound A, 1984*. Report 1. Wickliffe Mounds Research Center, Murray State University, Wickliffe, Kentucky.

Westbrook, Kent
1982 *Legacy in Clay: Prehistoric Ceramic Art of Arkansas*. Rose, Little Rock.

White, John R.
1975 Historic Contact Sites as Laboratories for the Study of Culture Change. *The Conference on Historic Site Archaeology Papers* 9:153–63.

Wilcox, David R.
1984 Multi-Ethnic Division of Labor in the Protohistoric Southwest. *Papers of the Archaeological Society of New Mexico* 9:141–54.

Wilcox, David R., and W. Bruce Masse (editors)
1981 *The Protohistoric Period in the North American Southwest, A.D. 1450–1700*. Anthropological Research Papers 24. Arizona State University, Tempe.

Willey, Gordon R., and Philip Phillips
1958 *Method and Theory in American Archaeology*. University of Chicago Press, Chicago.

Willey, Gordon R., and Jeremy Sabloff
1970 *A History of American Archaeology*. W. H. Freeman, San Francisco.

Williams, J. Raymond
1967 *Land Leveling Salvage Archaeological Work in Southeast Missouri, 1966*. University of Missouri–Columbia. Submitted to National Park Service, Lincoln, Nebraska.
1968 *Southeast Missouri Land Leveling Salvage Archaeology: 1967*. University of Missouri–Columbia. Submitted to National Park Service, Lincoln, Nebraska.

Williams, Stephen
1954 An Archaeological Study of the Mississippian Culture in Southeast Missouri. Ph.D. dissertation, Department of Anthropology, Yale University, New Haven.
1956 Settlement Patterns in the Lower Mississippi Valley, In Prehistoric Settlement Patterns in the New World, edited by Gordon R. Willey. *Viking Fund Publications in Anthropology* 23:52–62.
1957 James Kelly Hampson. *American Antiquity* 22:398–400.
1962 Historic Archaeology in the Lower Mississippi Valley. *Southeastern*

Archaeological Conference, Newsletter 9:53–63.

1963 The Eastern United States. In *Early Farmers and Village Communities,* edited by William G. Haag, pp. 359–72. National Park Service, Washington D.C.

1964 The Aboriginal Location of the Kadohadacho and Related Tribes. In *Explorations in Cultural Anthropology*, edited by Ward H. Goodenough, pp. 545–70. McGraw-Hill, New York.

1966 Historic Archaeology: Past and Present. *Annual Report of the School of American Research for 1966*, pp. 23–25.

1967 On the Location of the Historic Taensa Villages. *The Conference on Historic Site Archaeology Papers* 1:3–13.

1972 Ripping off the Past. *Saturday Review, Science*, October 1972, pp. 44–53. New York.

1980 The Armorel Phase: A Very Late Complex in the Lower Mississippi Valley. *Southeastern Archaeological Conference, Bulletin* 22:105–10.

1981 Some Historical Perspectives on Southeastern Ceramic Traditions. *Louisiana State University, School of Geoscience, Geoscience and Man* 22:115–22.

1982 The Vacant Quarter Hypothesis: A Discussion Symposium. Symposium presented at the 39th Annual Meeting of the Southeastern Archaeological Conference, Memphis.

1983 Some Ruminations on the Current Strategy of Research in the Southeast. *Southeastern Archaeological Conference, Bulletin* 21:72–81.

1985 The Vacant Quarter and Other Late Events in the Lower Valley. Paper presented at the 42d Annual Meeting of the Southeastern Archaeological Conference, Birmingham.

Williams, Stephen, and Jeffrey P. Brain

1983 *Excavations at the Lake George Site, Yazoo County, Mississippi, 1958–1960*. Papers vol. 74. Peabody Museum of Archaeology and Ethnology, Harvard University, Cambridge.

Williams, Stephen, and James B. Stoltman

1965 An Outline of Southeastern United States Prehistory with Particular Emphasis on the Paleo-Indian Era. In *The Quarternary of the United States*, edited by Herbert E. Wright, Jr., and David G. Frey, pp. 669–83. Princeton University Press, Princeton.

Wilson, Carrie B.

1978 The Peyote Religion Among the Quapaw. Honors Thesis, Department of Anthropology, University of Arkansas, Fayetteville.

Wolf, David J.

1976 A Population Model for the Analysis of Osteological Materials. Ph.D. dissertation, Department of Anthropology, University of Arizona, Tucson.

Wolfman, Daniel

1982 Archeomagnetic Dating in Arkansas and the Border Areas of Adjacent States. In Arkansas Archeology in Review, edited by Neal L. Trubowitz and Marvin D. Jeter. *Arkansas Archaeological Survey, Research Series* 15:277–300.

Zinke, Marjorie L.

1975 An Analysis of Mississippian Burial Components From the Hazel Site, Poinsett County, Arkansas. Master's thesis, Department of Anthropology, University of Arkansas, Fayetteville.

Contributors

GEORGE J. ARMELAGOS, professor of anthropology at the University of Massachusetts, Amherst, is the coauthor of *Demographic Anthropology* (with Alan Swedlund) and *Consuming Passions: The Anthropology of Eating* (with Peter Farb). He is also coeditor of *Paleopathology at the Origins of Agriculture*. He has published extensively on the impact of diet and disease on adaptation of prehistoric populations. Armelagos has published articles in *Natural History*, the *New York Times*, and *Cosmopolitan*. He is an avid outdoorsperson and is an expert kayaker.

IAN W. BROWN, associate curator of North American collections at the Peabody Museum of Archaeology and Ethnology, Harvard University, is the author of *Salt and the Eastern North American Indian: An Archaeological Study* and *Natchez Indian Archaeology: Culture Change and Stability in the Lower Mississippi Valley*. Dr. Brown has done considerable work in Mississippi and Louisiana, notably on and around Avery Island, Louisiana.

CHESTER B. DEPRATTER, research associate at the University of South Carolina Institute of Archaeology and Anthropology, has written or co-written more than a dozen publications on southeastern Indians and sixteenth-century Spanish explorations in "La Florida." His articles have appeared in *American Antiquity, Southeastern Archaeology, Florida Historical Quarterly*, and several edited volumes.

GEORGE FIELDER, state archaeologist and director of the Tennessee Division of Archaeology, has been working in the field of cultural resource management sine 1976, when he joined the staff of the Tennessee State Historic Preservation Office. He was appointed state archaeologist in 1983. Fielder's areas of specialty include historical archaeology; legal aspects of archaeological preservation, especially cemetery and burial law; and public archaeology. He is the current secretary-treasurer of the National Association of State Archaeologists.

JAMES B. GRIFFIN, professor emeritus at the University of Michigan and one of the most influential American prehistorians of the twentieth century, is a recipient of the Viking Fund Medal in Anthropology and is a past president of the Society for American Archaeology and the Southeastern Archaeological Conference. He is the author of more than 100 articles and monographs. Two of his works, *The Prehistory of Eastern North America* and *The Fort Ancient Aspect*, are considered classics in American archaeology.

M. CASSANDRA HILL, a Ph.D. candidate in biological anthropology at the University of Massachusetts, Amherst, is the author of several papers and publications on iron-deficiency anemia and biocultural transitions of prehistoric Southeastern populations. She is also one of the editors for volume 4 of *Archaeological Investigations in the Gainesville Lake Area of the Tennessee-Tombigbee Waterway*.

MICHAEL P. HOFFMAN, professor in the Department of Anthropology, University of Arkansas-Fayetteville, and curator of anthropology for the University of Arkansas Museum, is the author of 30 published monographs and papers. His research interests include Caddoan and Arkansas River Valley archaeology; Quapaw, Seneca-Cayuga, Caddo, and Cherokee tribes; and the history of southeastern archaeology.

CHARLES HUDSON, professor in the Department of Anthropology at the University of Georgia, is author of *The Catawba Nation*, *The Southeastern Indians*, and *Elements of Southeastern Indian Religion*. He is editor of *Black Drink: A Native American Tea* and *Ethnology of the Southeastern Indians*. His most recent research has been on the sixteenth-century Spanish explorers of the interior southeastern United States.

R. BARRY LEWIS, associate professor of anthropology at the University of Illinois at Urbana-Champaign, is the author of several monographs and numerous articles on the archaeology of Mississippi period communities in western Kentucky and southeastern Missouri. His contributions include the monographs *Mississippian Explorative Strategies* and *Excavations at Two Mississippian Hamlets in the Cairo Lowland of Southeast Missouri* and the edited book, *Mississippian Towns of the Western Kentucky Border*. He directs the UIUC Western Kentucky Research Project.

DAN F. MORSE, professor of anthropology at the University of Arkansas and Arkansas State University and archaeologist with the Arkansas Archeological Survey, is the author of five books, including *Archaeology of the Central Mississippi Valley* (co-written with Phyllis A. Morse), and over 125 papers. Morse serves on the Arkansas De Soto Trail Commission and is currently researching both the Protohistoric and the Colonial periods of Arkansas. He is also currently co-writing a book on the Dalton period Sloan cemetery. Major research interests include primitive technology and development of complex chiefdoms.

PHYLLIS A. MORSE, research associate with the Arkansas Archeological Survey, is the author of *Parkin: The 1978–1979 Archeological Investigations of a Cross County, Arkansas Site* and coauthor (with Dan F. Morse) of *Archaeology of the Central Mississippi Valley*. Her major research interests include the route of the de Soto expedition in Arkansas and the nineteenth-century historic settlement of Arkansas. She has been a consultant of prehistoric and historic exhibits for the Mud Island Museum, Pink Palace Museum, and the Old State House Museum in Little Rock, Arkansas. Morse lives and works in Jonesboro, Arkansas.

MARY LUCAS POWELL, director/curator of the University of Kentucky Museum of Anthropology in Lexington, is the author of the forthcoming volume on the Nodena Skeletal series to be published by the Arkansas Archeological Survey Research Report series and is senior editor of *What Mean These Bones?: Studies in Southeastern Bioarcheology* to be published by The University of Alabama Press in 1991. Her specializations in bioarchaeology include paleopathology, mortuary patterns, and associations of social status and health status in stratified populations.

CYNTHIA R. PRICE received a B.A. in anthropology from the University of Illinois in 1969 and an M.A. in anthropology from the University of Michigan in 1970. She has conducted fieldwork in the southwestern United States as well as in Missouri. Her research interests include historical archaeology, cultural ecology, and cultural resources management. She has been conducting research in southeastern Missouri since 1973 with particular emphasis on historic frontier settlement in the eastern Ozarks and the development of the Ozark folk culture.

JAMES E. PRICE, research archaeologist with the Center for Archaeological Research, Southwest Missouri State University, and manager of the Southeast Field Station of that institution in Naylor, Missouri, has conducted research in the southeastern Ozarks and the Missouri Bootheel for 23 years. He is currently serving as principal investigator of a five-year multidisciplinary archaeological research project in Ozark National Scenic Riverways for the National Park Service.

GERALD P. SMITH has conducted research in the southeastern United States for over 25 years. He holds the M.A. degree in anthropology from the University of North Carolina and the Ph.D. degree from the University of Missouri. He has been director of Chucalissa since 1968. During that time he has concentrated research on Chucalissa itself and on the Mississippi Valley and its tributaries from the mouth of the Arkansas River to the northern end of the Alluvial Valley. His particular interests include the sociopolitical organization and cultural ecology of Woodland and Mississippian cultures in the region.

MARVIN T. SMITH, assistant professor of anthropology at the University of Georgia, holds a doctoral degree in anthropology from the University of Florida, where he has concentrated on early Spanish exploration of the New World. He is a recognized authority on early Spanish material culture for the Indian trade. Dr. Smith is the author of several articles and a recent book, *The Archaeology of Aboriginal Culture Change: Depopulation during the Early Historic Period.*

STEPHEN WILLIAMS, Peabody Professor of American archaeology and ethnology at Harvard University and curator of North American archaeology in the Peabody Museum, is the author of numerous articles on southeastern archaeology and is coauthor (with Jeffery P. Brain) of *Excavations at the Lake George Site, Yazoo County, Mississippi, 1958–1960.* Areas of new concern include East Africa and the conservation of archaeological sites on a nationwide basis.

Index

Acts: Archaeological and Historic Preservation Act, 241, 242; Archaeological Resources Protection Act, 241; National Historic Preservation Act, 241

Adaptation, 16, 19, 69, 220–21

Advisory Council on Historic Preservation, 242

African, 4 (settlement)

Age determination, 21

Agriculture, 18, 23, 25, 29, 31, 34, 35, 86, 114, 134, 146; destructive practices of, 63, 70, 74, 78, 185; plant diseases, 178. *See also* Diet

Alabama Historical Commission, 183

American Historical School, 54

American influence, 10

Anasazi, 106

Anderson, Leo and Mary Ellen, 61–63, 67, 81

Anderson Collection, 61–63, 65, 68

Archaeological site preservation, 4

Archaeology in the East, 54

Archeological Conservancy, 133

Arkansas Archeological Survey, 77–78, 159

Art, 2, 10; human figures, 94. *See also* Pottery

Artifacts: abraders, 86, 93, 94; adzes, 66, 93, 164; Astralagus dice, 55, 57, 94; awls, 67, 94, 119, 164; beads, 10, 67 (shell), 67–68, 77, 81, 82 (glass and sheet iron), 68, 81 (chevron), 81 (amber), 94, 119, 125, 164, 165, 205, 219, 232; bell, 81 (brass hawk), 125, 172, 205, 232; bifaces, 119; bracelets, 219; calques, 211; celts, 66–67, 93, 153, 164; chisels, 66, 92–93, 164; copper, 6, 10 (plates), 12, 28 (ear-spools), 85 (raw metal), 87, cones, 219; debitage, 119; discoidals, 164; ear pins, 67, 94, 164; fans, 94; fish hooks, 94, 119; flakers, 64; flint (chipped), 12, 85; gorgets, 10, 11, 12, 67, 178; guns, 232; hammerstones, 86, 94, 119; hoes, 10 (flint), 94 (spades), 119, 123 (and hoe chips); knife, 77 (knife and knife scabbard), 81, 82, 93, 94; maces, 10; nails, 77, 82; needles, 119; palettes, 94; pendants, 10, 81; pins, 164; pipes, 10 (human effigy), catlinite, 55, 77 (disk), 67, 74, 77, 232 (catlinite disk pipe); projectile points, 10, 11, 55, 61, 66, 90–93, 95, 106, 121, 153, 164, 183, 205; plates, 48 (Bell paste); punches, 119, 164; pottery anvil, 90; scrapers, 66, 94, 95, 119, 216; screwdrivers, 232; shells, 8, 11, 12, 28, 67, 178; sickles, 94; spoons, 67, 94

Barkley Reservoir, 57

Bat Cave, 61

Biases, 101

Bible, 13
Biedma, 87, 95, 126, 183, 187, 191, 204, 206 (nn. 9, 16)
Big Springs Village, Inc., 62
Boeuf basin, 177
Bones: human, 11, 16, 178. *See also* Skeletons
Bones: animal, 29 (faunal remains), 32, 45 (Table 3–1); 77 (deer)
Buffering system, 17 (Fig. 2–1), 18
Burial procedures, 6, 12, 75–76, 82, 119, 230, 232–33
Burials, 60, 61–62, 72–73, 75, 76, 99, 104, 120, 125, 144, 145, 159, 164, 183
Busycon whelks, 10

Caddoan ecological/cultural zone, 34–35
Catholic Church, 13
Ceremonies, 8, 9, 10, 12, 95, 125, 213
Chapman, Carl and Eleanor, 62
Chiefs, 7, 8, 9, 11, 75; Apafalaya, 188, 189, 191; Casqui, 130–31; Chickasaw, 193–98; Great Sun (Natchez), 236; Hekaton, 210; Miculasa, 195, 196, 197; Pacaha, 87, 97, 130–31; Paheka, 208, 210, 220–21, 224; Tascaluza, 181, 182
Cities: Biloxi, 177; Blytheville, 14, 78; Cairo, 38; Caruthersville, 60; Centerville, 183; Clarksdale, 169; Davenport, 120; Dell, 81; Evansville, 173; Fayette, 191; Forrest City, 155; Greenock, 78; Helena, 131; Holly Springs, 200; Houston, 200; Lafayette, 173; Lepanto, 14; Little Rock, 214, 223, 226 (n. 3); Marksville, 234; Nashville, 240; Natchez, 229; New Madrid, 173; Osceola, 40, 72; Pocahontas, 82; Rolla, 10, 61; Round Pond, 203; St. Genevieve, 136; St. Louis, 7, 10, 62, 170; Schlater, 204–5; Shreveport, 180; Tuscaloosa, 7, 99; Van Buren, 61–67; Vicksburg, 170, 231, 234; Wilford, 143; Wilson, 14, 98
Climate, 98, 175–76
Cohesiveness (of Lower Mississippi Valley core), 56–57
Columbian Exposition (Chicago, 1893), 120
Comprehensive planning, 239–45
Conservation easement, 244–45
Cortez, 13
Counties: Carlisle, 43; Cheatham, 244; Chickasaw, 204; Clay, 194; Coahoma, 138, 147; Crittenden, 78, 126, 138; Cross, 120; DeSoto, 70, 138; Fulton, 27, 46; Humphreys, 244; Iberville Parish, 231; Lauderdale, 140; Lee, 15, 155; Lowndes, 194; Mississippi, 70, 78, 138; Monroe, 15, 194; Noxubee, 194; Oktibbeha, 194; Pemiscot, 78; Pontotoc, 15, 202, 204; St. Francis, 66, 155; Shannon, 61; Shelby, 138, 140; Sumner, 244; Tipton, 140; Tunica, 138, 147; Union, 10, 202; West Feliciana Parish, 227; Wilson, 244
Creeks: Big, 201; Big Bush, 184; Catalpa, 194, 197; Cooper, 191; Chuquatonchee, 193, 196, 197; Eddy, 57; Five Mile, 184; Fourche, 173; Haysop, 183; Houlka, 196, 199; Johnson, 167, 200; Lick, 203; Line, 193, 196, 197, 199; Little Bear, 175; Luxapallila, 192; Magowah, 192, 194, 197; Mud, 191; Nonconnah, 147; Tibbee, 191, 192, 193, 194, 196; Yellow, 191
Crossings, 191–92
Cuba, 13
Cultural resources management, 239
Culture Element Distribution studies, 137
Curtis, Edward, 119

Dating: absolute, 42; cross-dating, 40–42, 56 (errors in), 57; radiocarbon, 42, 45 (and Table 3–1), 54, 57, 76, 95, 125, 147
Death, 19, 23, 25 (Fig. 2–1), 28, 30 (death and death rates), 31, 32, 77, 101, 116
Degenerative pathology, 25 (Fig. 2–1), 31 (increase in)
Demography, 3, 99, 101–4
Dental pathology, 3, 30, 32, 33, 101, 102, 105, 110–15 (and Tables 6–4, 6–5, 6–6); enamel hypoplasia, 21, 22, 25 (Fig. 2–1), 27, 29, 30, 110, 114, 115
de Soto, Hernando, 13, 132–33, 214
De Soto entrada, 3, 13–14, 73, 76, 77, 81, 87, 96–97, 116, 132, 171, 177–78, 179, 181–207, 219, 221, 222
de Soto (Narratives), 7, 82, 87, 95, 131–32, 167–69, 181–207, 212
de Soto, routes, 79, 126–34, 165–68, 171, 181–207
Diet, 2, 5–6, 8, 16, 17, 18, 25, 28, 31, 44, 45 (Table 3–1), 48, 60, 75, 85, 86, 94, 98, 105, 111 (food preparation); 113 (lye water), 116, 123, 145, 146 (Table 8–1), 151, 178, 206 (n. 12); maize, 6–7, 8, 9, 27, 32, 35, 43,

(Diet, *continued*)
45 (Table 3–1), 60, 69, 86, 98, 123, 145, 176, 178, 186, 216, 219
Diseases: human, 2–3, 13, 14, 16, 17, 18, 19–31, 53, 73, 76, 77, 83, 98–117, 133, 167, 168, 178–79, 215, 221; blood poisoning, 107; caries, 109, 216; cribra orbitalia, 116; osteophytosis, 22, 106, 116; osteomyelitis, 106, 110, 116; periostitis, 107, 108 (Table 6–3), 110; syphilis, 108, 109, 110, 116
Diversity, 19
Division of Archaeology, State of Tennessee, 244

Earthquakes, 178
Elvas, Fidalgo of, 87, 126, 167–68, 183, 186, 187, 194, 203, 204, 206 (nn. 3, 12, 13, 15, 16, 17, 19), 207 (nn. 19, 20)
Environment: ecology, 17–19, 29 (ecological zones), 69, 83
European: influence and contact, 1, 2, 10, 12, 13–14, 43, 53, 54, 55, 58, 67–68, 73, 106, 110, 121, 125, 133, 165–68, 177–78, 197, 210, 218, 221, 225, 227, 231, 232, 236–37

Fire basins, 48
Fishing, 7, 8
Ford, James A., 118, 121, 123, 165, 172, 220–21
Fortifications, 8, 9, 24 (Fig. 2–2), 45 (and Table 3–1), 57, 60, 73, 86, 118, 123, 161, 177, 181, 199–200, 201, 219
Fort Pillow Ridge, 151
French accounts, 82, 214
French influence, 4, 76, 81, 96–97, 219, 223, 230, 234, 235

Gadrooning, 65
Games, 2, 8; chunkey, 11, 73, 74, 75
Garcilaso de la Vega, 80, 82, 87, 126, 186, 187, 194, 199, 203, 207 (n. 22)
Genetic inheritance, 16
Gregorian calendar, 43
Griffin, James B., 118, 120, 123, 165
Growth: 17, 19 (disruption), 21 (curves, disruption), 22 (recovery)

Hall, Wilfred, 119–20
Hampson, James K., 69, 70, 72, 73 (map), 75, 77, 94, 99, 101
Harris lines, 21, 22, 25 (Table 2–1), 29, 30, 34
Host resistance factors, 17 (Fig. 2–1), 18, 19
Houses, 7, 48, 73, 74–75, 76, 78, 82, 94, 96, 99, 118, 144, 145, 147, 149, 161, 171, 194–95, 222, 223, 225; council, 7, 8, 95; charnel, 8
Hunting, 7, 86
Hunting-gathering, 28, 31, 43

Indians: Acolapissa, 231; Algonquian, 167; Alibamu, 195, 200; Aquixo, 128; Bayogoula, 231; Caddo, 215, 234; Chakchiumas, 195, 196, 197, 198, 231; Cherokee, 10, 12, 221, 222; Chicasaw, 15, 171, 180, 191–99; Chitimacha, 180; Choctaw, 15, 180, 215; Creeks, 12, 221; "Dog Nomads," 176; Ethnogenesis, 3, 208–26, 231; Grigra, 235–36; Houma, 180, 231; Illinois, 209, 214; Kadohadacho, 180; Kansa, 209, 211; Koroa, 231; Lamar, 177; Mandan, 133; Miami, 214; Michigamea, 81, 97; Mogulasha, 231; Natchez, 4, 75, 171, 172, 221, 229, 235–36; Ofo, 231, 232; Oklahoma, 213; Omaha, 209, 211, 213; Osage, 209, 211, 214, 220; Paleoindians, 5, 30; Peyote, 213; Pueblo, 7, 176; Ponca, 209, 211; Quapaw, 4, 14, 15, 69, 76, 77 (New Kappa), 80, 82, 96–97, 133, 172, 180; Santee, 110; Taensa, 230, 231; Tioux, 231; Tula, 214; Tunica, 180, 210, 224, 225, 227; Yazoo, 231; Yuchi, 221–22
Infants, 18, 32, 75, 102 (Table 6–2)
Izard, George, 97, 208, 224

John III, Pope, 13
Joliet and Marquette, 133, 168

Lakes and waterways: Beaverdam Lake, 151; Great Lakes, 82; Old Town Lake, 201; oxbow lakes, 2, 131; Pensacola Bay, 181; Tampa Bay, 13; Vermillion Bay, 235; Wapanocca Lake, 74, 86, 204. *See also* Rivers
Languages: Apafalaya, 206 (n. 9); Caddoan, 15, 214; Choctaw, 188; Dhegiha, 209, 211, 220, 223; Muskogean, 15; Siouan, 15, 96, 213, 215, 231; Tunican, 214, 215, 223
Legal property rights, 243
Lenses, 120
Looting, 62, 63, 69, 82, 134, 241
Loughridge, Robert H., 46

Louisiana State Archaeological Survey, 231
Lower Mississippi Survey, 231, 233, 234
Lubbub Creek Archaeological Locality, 197

Marquette map, 82, 214, 218, 220
Marquette journal, 96
Meander belts and channels, 83, 84, 98, 142 (Fig. 8–6), 150 (Fig. 8–9), 152 (Fig. 8–10), 153, 154 (Fig. 8–11), 156 (Fig. 8–12), 158 (Fig. 8–13), 160 (Fig. 8–14), 162 (Fig. 8–15), 163 (Fig. 8–16), 166, 200, 201, 202
Mediterranean, 9
Mexico, 5, 6, 7, 12, 13, 176
Mississippi Archaeological Survey, 149
Mississippi Department of Archives and History, 230, 231
Models: ecological, 17–19; "Contact Era," 55; Quimby-Spoehr, 236–38; Site Location Distribution, 240
Moore, C. B., 78, 80, 119, 120, 183, 216, 227, 229, 234, 238
Mounds, 8, 9, 46, 47, 57, 60, 75 (pyramidal), 96, 118 (temple), 121, 143, 144, 151, 159, 164 (temple), 171, 179, 183, 185, 188, 189, 190, 229, 231
Mounds and plaza arrangements, 8, 44, 46, 47–48, 60, 72, 73, 75, 78, 86, 118, 120 (plaza alone), 140, 141, 142, 146 (Fig. 8–1), 150 (Fig. 8–9), 151, 152 (Fig, 8–10), 154 (Fig. 8–11), 156 (and Fig. 8–12), 157, 158 (Fig. 8–13), 160 (Fig. 8–14), 162 (Fig. 8–15), 163 (Fig. 8–16), 164, 218, 219
Museums: Alabama State Museum of Natural History, 70, 73, 88, 99, 101; Buffalo Museum of Science, 120; Chicago Field Museum, 120; Hampson Museum, 99, 104; Peabody Museum (Harvard), 119, 171, 233; Putnam Museum, 81, 120; Smithsonian Institution, 81, 120, 171

National Park Service, 62, 242
National Register of Historic Places, 242
Near Eastern, 9
Neitzel, Robert, 229
Nubian (studies), 21

Paleoepidemiology, 17 (Fig. 2–1)
Paleodemography, 23
Paleopathology, 16–31, 36–37
Parkin archaeological project, 77, 121
Parks: De Soto, 140, 141; Toltec, Mounds State, 133
Pathogens, 20, 107, 111, 190, 191
Periods: Archaic, 5, 6, 8, 26 (Fig. 2–3), 31, 121; Dalton, 66; Early Colonial, 77; Early Historic, 1, 2, 5, 10, 11, 59, 63; Historic, 178, 227–38, 242; Mississippian, 6, 7–9 (culture), 24 (Fig. 2–2), 26 (Fig. 2–3), 27, 29, 32, 33, 34, 38, 40, 41 (Fig. 3–2), 54, 55, 66, 68, 69, 106, 111, 115, 136, 137, 138, 149, 166, 167, 170, 186, 189, 206 (n. 5), 209, 219, 222, 227, 230; Early Mississippian, 29, 94, 96, 216; Middle Mississippian, 25, 28, 29, 30, 121, 125, 130; Late Mississippian, 38–58, 106, 114, 125, 143, 172, 173, 200, 208–26; Prehistoric, 178, 213, 230, 242; Late Prehistoric, 1, 2, 3, 5, 12, 41 (Fig. 3–2), 63, 66, 67, 183, 197, 220, 221; Protohistoric, 1, 2, 3, 32, 59, 62, 63, 64, 65, 66, 76, 95, 106, 119, 122, 128, 131, 133, 134, 171, 176, 178, 196, 211, 214, 216, 217 (Fig. 11–12), 220, 221, 222, 224; Woodland, 6, 8, 26 (Fig. 2–3), 27, 29, 30 (Woodland/Mississippian), 32, 33, 34, 41 (Fig. 3–2), 44 (late), 46, 121, 210
Phase construction, 40–43, 53 (biases)
Phases: Alabama River, 82, 206 (n. 4); Armorel, 55, 56, 59, 64, 68, 173, 175; Beckwith, 40; Belle Meade, 80, 95; Bottle Creek, 177; Boxtown, 147; Caborn-Welborn, 221; Cairo Lowland, 40, 44, 46; Campbell, 163, 164–65; Cane Hills, 41 (Fig. 3–2); Carden Bottoms, 212, 214, 217–26; Commerce, 138, 147–51, 166; Cramor, 159; Dorena, 41 (Fig. 3–2), 46, 56; Fort Walton, 177; Greenbrier, 95; Horseshoe Lake, 138, 149, 157–59, 165; Jackson, 41 (Fig. 3–2), 43, 48, 53, 54, 55, 56, 57; James Bayou, 41 (Fig. 3–2); Jonathon Creek, 57; Jones Bayou, 153–54; Kent, 14, 128, 155–57, 222; Lower Kent, 95; Little River, 95; Markala horizon, 172, 177, 178; Medley, 41 (Fig. 3–2), 44, 46, 48, 55, 56, 57; Menard, 216, 225; Moundville (I-II), 182, 183, 187 (1TV46//47), 188, 189–91, 196; Neosho, 220; Nodena (Pacaha), 2, 3, 14, 69–117, 129 (Table 7–1), 132, 133, 138, 159–61, 162, 169, 203, 204; Old Town, 15, 220; Parkin, 3, 14, 56, 80, 82, 90 118–34 (and 123 Table 7–1); Pemiscot Bayou, 161, 163; Pensacola, 182; Powers, 173; Pueblo IV, 179; Quapaw, 208–26; Sorrels, 196, 197;

(Phases, *continued*)
Summerville, 197; Tillar Complex, 172; Tinsley Hill, 57; Tipton, 138, 151–53, 166; Wallace, 216; Walls, 14, 56, 70, 132 (Walls/Pecon Point Complex), 80, 88, 95, 128, 129 (Table 7–1), 132, 133, 135–69, 203, 204; Wasp Lake, 231

Phillips, Philip, 118, 120, 123, 126, 165

Pits, 48, 60, 121, 159, 232

Pizarro, 13

Population, 3, 8, 18, 27, 53–55, 59, 77, 83, 95–96, 104, 167, 168, 170–80, 213, 215, 240

Porotic hyperostosis, 20 (anemia), 25 (Fig. 2–1), 28, 29–30, 31, 32, 33, 34, 105, 106, 210

Portugal, 13

Pottery, 8, 10, 12, 40, 47, 48, 63–66, 127–29, 131, 168, 216–17

Pottery forms, 149, 153, 155, 157, 165; bowls, 65–66, 136, 226 (n. 3); bottles, 65–66, 87–91, 135, 216–17, 222; disks, 127–30; effigy vessels, 65–66, 87–90, 94–95, 135, 136 (and Fig. 8–2), 226 (n. 3); fin, 61; head pots, 90, 119, 222; plate vessel form, 56

Pottery types: Addis Plain, 135; Alligator Incised, 136; Angel-Sikeston-Nashville Negative Painted Group, 136; Anna-L'Eau-Evangeline Engraved/Incised, 135; Avenue, 129 (Table 7–1), 135, 141 (Fig. 8–5), 217, 226 (n. 3); Barton Incised, 64, 120, 135, 140, 141 (Fig. 8–5), 147, 148, 151, 155, 157, 159, 160, 161, 164, 216, 217, 226 (n. 3); *vars. Barton and Kent*, 48, 50 (Fig. 3–6), 51 (Fig. 3–7); *var. Kent*, 119, 125; *var. Togo*, 119; Bayou Goula Incised, 135; Beaverdam, 129 (Table 7–1); Beckwith Incised, 136; Blanchard, 123; Bell Plain, 48, 57, 63, 64 (and Table 4–1), 88, 119, 129 (Table 7–1), 135, 147, 148, 153, 155, 156, 159, 161, 226 (n. 3); Boxtown, 147; Caddoan, 172, 219; Late Pensacola Phase, 183; Cahokia Cord Marked, 136; Cahokia Red, 136; Cairo Lowland Tradition, 136; Campbell Applique, 64 (Table 4–1), 65; Campbell Punctate, 164; Carson Red-on-Buff, 64 (Table 4–1), 65, 119, 129 (Table 7–1), 164, 217, 226 (n. 3); Chevalier Stamped, 135; Chickasawba, 129 (Table 7–1); Coles Creek Hardy, 135; Coles Creek Plain, 135; Dupree Incised, 135; Evansville Punctate, 136; Fatherland Incised, 135; Fortune Noded, 64 (Table 4–1), 65, 119, 129 (Table 7–1), 135, 141 (Fig. 8–5), 155, 157, 159, 164; French Fork Incised, 135; Greenhouse Incised, 135; Harris, 129 (Table 7–1); Hollywood White Filmed, 64 (Table 4–1), 164; Horn Lake, 129 (Table 7–1); Hull Engraved, 135, 140, 141 (Fig. 8–5); Jones Bayou Brushed, 153; Keno Trailed, 218; Kimmswick Fabric Impressed, 136; Kent Incised, 64 (Table 4–1), 129 (Table 7–1), 138 (Fig. 8–3), 140, 141 (Fig. 8–5), 151, 157, 159, 160, 161, 164, 217; Larto Red, 136; Late Pensacola Phase, 183; Manchac Incised, 135; Leland Incised, 217–19; Manly Punctated, 64 (Table 4–1), 161; Matthews Incised, 64 (Table 4–1), 141 (Fig. 8–5), 161, 163, 164; *vars. Beckwith, Matthews and Manly*, 48, 50 (Fig. 3–6), 51 (Fig. 3–7); Matthews/Moundville/Carthage Incised, 136; Mazique Incised, 136; Mississippi Plain, 48, 57, 63, 88, 119, 232; Monks Mound Red, 136; Mound Place Incised, 64 (Table 4–1); Neeley's Ferry Plain, 63, 64 (Table 4–1), 119, 135, 136, 140, 147, 149, 153, 155, 157, 159, 161; Northern Delta Tradition, 135–36, 137 (Fig. 8–2); Nodena Red and White, 64 (Table 4–1), 65, 119, 129 (Table 7–1), 135, 164, 216 (*var. Dumond*), 226 (n. 3); Owens Punctated, 141 (Fig. 8–5), 147; Parkin Punctated, 64 (Table 4–1), 65, 119, 122, 129 (Table 7–1), 135, 136 (Fig. 8–2), 140, 141 (Fig. 8–5), 147, 148, 151, 155, 157, 159, 160, 161, 164, 216, 217, 226 (n. 3); O'Byam Incised (*var. O'Byam*) 48, 50 (Fig. 3–6), 51 (Fig. 3–7), 56, 57, 136; Old Town Red, 64 (Table 4–1), 119, 135, 136, 140, 141 (Fig. 8–5), 147, 149, 151, 155, 157, 159, 160, 161, 164, 217, 226 (n. 3); Plaquemine Brushed, 135; Powell Plain, 136; Ramey Incised, 136; Ranch Incised, 64 (Table 4–1), 129 (Table 7–1), 138 (Fig. 8–3), 140, 141 (Fig. 8–5), 155, 159, 161, 164; Rhodes Incised, 64 (Table 4–1), 129 (Table 7–1), 135, 138 (Fig. 8–3), 140, 141 (Fig. 8–5), 155, 159, 164; Southern Delta Tradition, 135; St. Clair Plain, 136; Tibbets Bean Pot, 136; Varney Red Filmed, 64 (Table 4–1); Vernon Paul Applique, 64 (Table 4–1), 119, 125, 129 (Table 7–1), 141 (Fig. 8–5); Wallace Incised, 216, 218 (*var. Belzoni*), 219, 222, 223, 224, 226 (n. 3); Walls Engraved,

64 (Table 4–1), 65, 129 (Table 7–1), 135, 137 (Fig. 8–3), 140, 141 (Fig. 8–5), 155, 159, 161, 164; Winterville Incised, 217, 218 (*vars. Belzoni, Tunica*), 224 (*var. Belzoni*), 233
Productivity, 17
Provinces: Alibamu, 196–97, 202; Anilco, 214; Apafalaya, 188, 189, 194, 196, 205; Aquixo, 14, 80, 95, 131, 132, 165, 167, 201, 203, 214; Autiamque, 214; Ayays, 214; Calpista, 95; Caluca, 95, 195; Capaha (Kappa), 80, 96, 214; Casqui, province of, 15, 97, 98, 126–34, 201, 204, 214; Cayas, 214; Chackchiuma, 196–97, 206 (n. 13); Chickasaw, 191–207; Coosa, 178, 189, 197; Guatari, 206 (n. 11); Oneota Homelands, 177; Pacaha (society and province), 14, 80, 87, 96, 98, 126, 128, 130–31, 132–33, 202, 203, 204, 214; Palisema, 95; Quiguate, 14, 127, 202; Quixila, 95; Quizquiz, 80, 95, 128, 132, 202, 204, 207 (n. 22), 214; Tascaluza (chiefdom), 3
Public awareness techniques, 240–41

Quarries: 52, 85; Crescent, 10, 94; Mill Creek flint, 10

Ranjel, Rodrigo, 87, 98, 126, 184, 186, 187, 192, 204
Regions and states: Alabama, 25 (Table 2–1), 31, 82, 109, 110, 114, 136, 171, 175, 177, 206 (n. 5), 233; American Bottom, Illinois, 40; Appalachians, 10; Arkansas, 114, 132, 169, 171, 175, 208–26, 222, 239; Lower Arkansas Valley, 12, 14; Arkansas Valley, 208–26; Eastern Arkansas, 3, 196; Northeastern Arkansas, 12, 55, 56, 69–117, 131, 166, 172, 177; Southeastern Arkansas, 68; Atlantic Coast, 6, 10, 13; Baton Rouge area, 230; Big Bottoms, 38; Black Bottom, 173; Cairo Lowland, 38; Cane Hills, 38; Chicasaw Bluffs, 60; Crowley's Ridge, 85, 131; Cumberland Plateau, 109; Florida, 109, 130, 172, 177; Georgia, 25 (Table 2–1), 31 (coastal area), 109, 110, 194; Gulf of Mexico, 6, 10, 13, 119; Hatchie Valley, 151; Hatchie River, 166; Illinois Valley, 6, 7, 25 (Table 2–1; lower and middle), 29, 30, 110 (lower); Illinois (state of), 32, 109; Indiana, 136, 175, 221; Southwestern Indiana, 8; Kentucky, 30, 38–58, 109, 111, 221; Western Kentucky, 2; Lake Superior Basin, 10; Little Prairie, 223, 224; Little Black River Area, 8; Loosahatchie River and Valley, 143, 147; Louisiana, 55 (coast), 109, 177, 229, 244; Lower Bayou de Chien Valley, 46; Malden Plain, 59; Memphis area, 3, 4, 5, 7, 14, 95, 136, 140, 143, 169, 170, 201; Mesa Verde, 176, 177; Mid-Continent, 42; Mid-South, 2, 4, 55, 58; Midwest, 213; Mississippi Valley, 1, 2, (Central) (Lower), 6, 38, 42, 55, 56, 63 (lower), 25 (Table 2–1, lower and central), 44, 52, 54 (central and northern lower), 56 (northern lower), 120, 166, 170, 171 (lower), 173, 209, 212, 213, 215, 217, 219, 224, 227–39; Central, 69, 73, 87, 92, 95, 97, 121, 128, 133, 176, 209, 212, 217, 219, 221, 222, 239–45; Mississippi, 82, 182, 239; Western Mississippi, 15; Southwestern Mississippi, 55; Eastern Mississippi, 55; Northwestern Mississippi, 196; Missouri, 38–58, 131, 173; Eastern Missouri, 2; Southeastern Missouri, 1, 2, 9, 12, 14, 40, 59–68, 81, 95, 171; Southwestern Missouri, 220; Nashville Basin, 12, 173; Natchez and its area, 171, 177; Natchez Bluffs, 1, 179, 229, 230, 232, 235, 236, 237; Norris Basin, 233; Northeast, 208; North Carolina, 109, 206 (n. 11); Ohio-Mississippi Confluence, 2, 38–58; Ohio, 6 (southern); 29 (southeastern and southwestern); 30 (southwestern and general); Ohio Valley, 7, 25 (Fig. 2–1, central); 30, 31 (central), 54 (lower), 221 (lower), 224 (lower); Oklahoma, 111; Ouachita Mountains, 214, 218; Ozark Highlands: 61, 62, 92–93; Petite Anse region, 235; Prairies, 6 (eastern); St. Francis River area, 3 (Arkansas), 66 (Missouri); St. Francis Mountains, 83, 86; Sikeston Ridge, 59; South Carolina, 109, 110; Tensas Basin, 227; Tennessee, 29, 109–36 (eastern), 30, 55 (western), 56 (central), 135, 136 (Tennessee/Kentucky), 166, 171, 233, 239–45; Tennessee-Cumberland Valley, 56, 57 (Jackson Purchase); Warrior Valley, 188; Western Lowlands, 59 (Missouri), 165, 201; Wheeler Basin, 233; Yazoo Basin, 227, 231, 234; Yazoo Bluffs, 171, 233; Yucatan, 177
Religion, 2, 12, 75 (temple), 213
Riggs, C. W., 119–20
Risk, 17

Rivers: Alabama, 32, 33, 182, 183, 206 (n. 4); Apafalaya, 188; Arkansas, 14, 15, 81, 135, 169, 173, 208, 209, 211, 214, 215, 218, 219, 221, 222, 223; Bayou Bartholomew, 177; Cumberland, 173; Current, 61; Flint, 194; Guadalquivir, 130, 131; Jacks Fork, 61; L'Anguille, 201; Little (Pemiscot Bayou), 2, 14, 59, 60, 68, 78, 81, 83, 86; Big Black, 195; Black Warrior, 32, 33, 184, 185 (Fig. 10–2), 186, 187, 188, 189, 205, 206 (n. 9); Buttahatchee, 192; (Lower) Cahaba, 181, 182, 183, 184; Coldwater, 200; Coosa, 181, 183; Little Tallahatchie, 200; Mississippi, 2, 3, 9, 14, 15, 16, 38, 67, 78, 82, 83, 131, 132, 133, 135, 136–37, 139 (Fig. 8–4), 140, 141, 148, 149, 151, 157, 165, 166, 167, 169, 171, 179, 191, 199, 200, 202, 203, 207 (n. 23), 209, 210, 212, 218, 221, 223, 227; Noxubee, 193, 197; Ocmulgee, 294; Ohio, 14, 57, 59, 133, 136, 177, 208, 209, 210, 215, 221; Ouachita, 177, 179, 209, 214; Rio Grande, 167, 175, 179; Chickasaw, 191; St. Francis (Quiguate), 14, 60, 95, 118, 120, 122, 123, 131, 132, 165, 166, 169, 172, 177, 178, 201, 203, 204, 208, 209, 221; St. Lawrence, 82; Sipsey, 191; Skuna, 204; Sunflower (Bend and Landing), 204, 205; Tennessee, 6, 136 (and Cumberland); Tallapoosa, 181, 183; Tombigbee, 32, 33, 186, 191, 192, 193, 197, 206 (n. 9); Tunica Bayou, 234; Tyronza, 14, 119, 123, 132; Wabash, 67, 175, 214, 221; White (Coligua), 15, 95, 126, 223; Wolf, 147; Yalobusha, 197, 200, 204; Yalobusha-Yazoo, 195; Yazoo, 231, 234 (basin)
Roman bridge, 131

Salt, 10, 85 (and copper), 87
Scratching, 65
Settlement patterns, 86, 95, 118, 121, 131, 132, 142 (Fig. 8–6), 143–44, 148 (Fig. 8–7), 150 (Fig. 8–9), 152 (Fig. 8–10), 154 (Fig. 8–11), 156 (Fig. 8–12), 158 (Fig. 8–13), 160 (Fig. 8–14), 161, 162 (Fig. 8–15), 163 (Fig. 8–16), 178–79, 193, 194, 212–13, 218
Settlements, 5, 6, 7, 8 (towns), 9, 11, 14, 27, 59, 60, 61, 70, 82, 122, 123; settlement hierarchy, 43, 54; abandoned, 43–44, 59
Sex, assessment of, 104
Sexual activity, 12
Sexual dimorphism, 29, 31, 33, 34, 35
Site preservation, 239–45
Sites: Adams, 44, 45 (Table 3–1), 46 (and Fig. 3–3), 48, 49, 54; Adena/Hopewell-Fort Ancient, 30; Angel, 8, 173; Angola, 234; Arkansas Post, 223; Averbach, 242; Banks (*see* Wapanocca Lake Cluster); Barrett, 156; Barton Ranch, 123, 132; Bayou Goula, 231; Beaverdam, 149; Beck, 129 (Table 7–1), 131, 157; Beckwith's Fort, 45 (and Table 3–1); Belle Meade, 126, 131, 202; Berry, 60, 62, 63, 67; Big Eddy, 155; Bloodhound, 234; Blytheville, 77, 82; Burroughs, 232; Caborn-Wabash complex, 67; Burthe, 230; Caddo, 113 (Tables 6–5, 6–6); Cagle Lake, 60, 164; Cahokia, 7, 8, 10, 12, 27, 28, 29, 173; Cairo Lowland, 52, 136; Callahan-Thompson, 44, 45 (Table 3–1), 48, 53, 54; Campbell (Cooter), 60, 61, 62, 63, 65, 67, 68, 78, 81 (Pemiscot-Bayou Cluster), 82, 83, 111, 164, 172; Canon, 149; Cheatham, 140, 143, 166; Castile Landing, 155; Cherry Valley, 56; Chucalissa, 14, 96, 140, 143, 144, 145, 146, 168, 170, 244; Clay Hill, 155, 157; Commerce, 149; Commerce Landing, 126, 127, 165; Cummings, 128, 129 (Table 7–1); Davis, 155; DeSoto Park Mounds, 143, 170; Dickson Mounds, 27, 28, 29, 106; Duck River, 244; Dogwood Ridge, 166; Dundee, 149, 151; Dupree, 223; Edmondson, 157; Etowah, 189, 223; Eutaw, 184; Evansville, 149; Fatherland, 75, 171, 229, 230, 231; Flowers #3, 146; Fort Ancient Complex, 24 (Fig. 2–2), 30, 177; Fort Pickering, 140; Fort St. Pierre, 231, 232, 234; Fortune, 119, 120; Fourche Maline, 111, 113 (Tables 6–5, 6–6); Friar's Point, 126, 128, 200, 201, 205; Glover, 123; Greer, 216; Halcomb, 119; Hardin Village, 111; Haynes Bluff, 177, 231, 232, 233; Hazel, 56, 76, 81, 120, 121, 125; Horseshoe Lake, 165; Hess, 44, 45 (Table 3–1), 48, 53, 54; Hickman, 54; Ingomar Mounds, 202; Irby, 131, 167, 205; James Bayou, 41, 43; Jerusalem Church, 156, 157; Jonathan Creek, 57; Jones and Borum Place, 128; Kent, 156; Kinfolk Ridge, 161, 163; Kersey II (Cagle Lake), 60, 61, 63; Kincaid, 173; Kinkead-Mainard, 218; Lake Cormorant, 128, 131, 140, 167, 203, 205; Little Egypt, 189; Link Farm, 244; Lockguard, 26;

Lookout, 235; Lubbub, 106, 111, 113 (Table 6–5, 6–6), 193, 233; Lyons Bluff, 196, 197; Mabila (Old Cahawba), 181, 183, 184, 188, 191, 203, 206 (n. 9); McCoy (Chute), 60, 61, 62, 63, 67, 164; McLeon's Bluff, 45; Maclin (Holland), 60, 61; Marshall, 43; Menard, 15, 96, 225, 233; Miller Mounds, 120; Moore, 155; Mound Bottom, 244; Mound Place, 157; Moundville, 7, 8, 12, 33, 106, 173, 176, 187, 188, 206 (n. 4), 229; Murdoch, 156, 157; Murphy Mound, 60; Natchez Fort, 230; Neeley's Ferry, 120, 123, 127; Nettle Ridge, 161; New Bethel, 156; Nickel, 159; Nodena, 70, 72 (search for), 69–117; Upper Nodena, 3, 73, 76–77, 98, 99, 102 (Table 6–2), 104, 107, 109 (Table 6–3), 111, 112 (Table 6–4), 113 (Tables 6–5, 6–6), 128, 129 (Table 7–1), 172; Middle Nodena, 3, 73, 77, 98, 99, 102 (Table 6–2), 107, 108 (Table 6–3), 111, 112 (Table 6–4), 113 (Tables 6–5, 6–6); Norfolk, 131, 140, 143, 167, 205; O'Quinn, 235; Old Town, 201; Oliver, 204, 205; Oneota, 24 (Fig. 2–2), 67, 220; Osotouy, 82, 207, 223; Owens, 149; Owl Creek, 193; Padden, 193; Parchman, 149; Parkin, 118–34, 172, 203; Pecan Point, 14, 70, 80, 128, 129 (Table 7–1) (*see also* Wilson-Joiner Cluster); Pemiscot Bayou Cluster, 78, 81–82; Stott and Potter, 81; Piachi (Durant Bend), 181; Plaquemine, 227; Point Remove (Tanico), 214, 219, 224; Portland, 232, 233; Pouncy, 157, 167; Powers Fort Complex, 8; 9; Rhodes (Brackenseik), 129 (Table 7–1), 157, 167; Rice, 235; Ring, 230; Riverside, 155; Rodgers, 57; Rolling Hills, 197; Rose Mound, 119, 120, 123, 127, 128, 129 (Table 7–1); Russell, 232; Salomon, 149; Sassafrass Ridge, 45 (and Table 3–1), 54; Sellars, 237; Snodgrass Village, 9; Shawnee Village, 161; Shell Bluff, 111; Sherman, 161; Snow's Bend, 188, 192, Stanley Mound, 119; Starkley, 155; Stone, 57; Sudan, 155; Sunflower Bend, 200, 201; Taylor's Shanty, 220; Thelma, 193; Tinsley Hill, 57; Toqua, 29; Trinity, 232; Trudeau, 172, 227, 232, 233, 234, 237, 238; Tunica, 172, 234; Turk, 45 (and Table 3–1); Turner, 159; Twist, 120; Vernon Paul, 120, 128, 129 (Table 7–1); Wallace, 15; Walls, 70, 129 (Table 7–1), 131, 140, 167, 201, 202; Wapanocca Lake Cluster, 78–80, 84, 128, 129 (Table 7–1); Bradley, 77, 78 (3CT9, 3CT43), 79–80 (as Pacaha the capital), 80, 83, 128, 129 (Table 7–1), 161; Bradley Ridge, 78; Banks, 75, 78 (and 3CT17), 96 (and 3CT139), 96, 161; West, 149; White, 45 (and Table 3–1); White Mound (Talipacana), 184, 186, 188; White Springs, 111; Wilford, 149; Wickliffe, 45 (and Table 3–1); Wiggins, 187; Wildy (*see* Pemiscot Bayou Cluster), 77, 78, 82; Wilson-Joiner Cluster, 80, 81, 84; Pecan Point, 80; Upper Nodena, 80 (and 3MS7, 3MS14); Notgrass, 80 (and 3MS17, 3MS73), 82; Middle Nodena, 80; Carson Lake, 80 (and 3MS60, 3MS61, 3MS65, 3MS69, 3MS71, 3MS80, 3MS91, MS57, 3MS94), 82; Woodlyn, 140, 143, 167; Wright's Bluff, 157, 232

Skeletal metrics and morphology, 104–5

Skeletons: human, 3, 16, 19–31, 61–62, 63, 73, 75, 98–117, 133, 215–16, 220; periosteum, 20, 28, 31; cranium, 20, 100 (and Table 6–1), 101, 103, 105, 106–8, 110, 230; long bones, 25 (Table 2–1), 28, 105, 108 (Table 6–3), 116; skeletal elements, 100 (Table 6–1), 108 (Table 6–3), (*see also* Dental pathology, Porotic hyperostosis); females, 18, 31, 73, 102 (Table 6–2), 104–5; children (subadults), 73, 75, 99, 101, 102 (Table 6–2), 103, 104–5; males, 73, 102 (Table 6–2); Caucasian, 73, 105–6; animal, 61

Slaves, 3, 11, 187

Social organization, 7, 9, 18, 27, 85, 86–87, 96–97, 123, 125, 132–33, 168, 169, 175, 176, 189, 213

Soil, 16, 48, 49 (Fig. 3–5), 73, 75, 83–84, 85, 95, 121, 123, 142 (Fig. 8–6), 143, 148 (Fig. 8–7), 150 (Fig. 8–9), 152 (Fig. 8–10), 153, 156 (Fig. 8–12), 158 (Fig. 8–13), 159, 160 (Fig. 8–14), 162 (Fig. 8–15), 163 (and Fig. 8–16), 226 (n. 3)

Soil Conservation Service, Osceola, 72

Southeastern Archaeological Conference, 70, 175

Spain, 13

Spanish: influence, 15, 81. *See also* de Soto, Hernando; de Soto entrada

Staphylococcus, 20, 107

Stature, 31

Stone, 12, 66, 67, 84–85, 86, 90, 92–93, 94; grinders, 111
Stratigraphy, 42
Stress, 17, 18, 19, 22, 25 (Fig. 2–1), 31, 32, 55, 115
Swanton, John R., 171

Technology, 18, 84, 86
Topography, 38, 57, 59, 140, 151, 206 (n. 2); levees, 44, 48, 60, 83–84, 85, 143, 201
Trade and exchange, 2, 6, 9–11, 13, 27, 67–68, 74, 77, 82, 92–93, 96–97, 165, 176, 177, 178, 216, 218, 219, 221, 223, 225
Tennessee Valley Authority, 244
Trees, 76, 84

United States De Soto Expedition Commission, 200–3, 205 (n. 1)
Universities: University of Alabama, 74, 99; University of Arkansas, 70, 74, 88, 96, 120, 121, 125, 219; University of Florida, 131; University of Illinois' Western Kentucky Research Project, 2, 38, 44, 46, 55, 56; University of Kansas, 211; University of Mississippi, 244; University of Missouri, 61–62, 63; Southeast Missouri State University, 62; Southwest Missouri State University, 62; University of Tennessee, 242
Urethral discharge, 109–10

Vacant Quarter Hypothesis, 42, 54–55, 59, 85, 170–80

Warfare, 2, 11, 73, 84–85 (weapons, warriors), 87, 128, 132–33, 175
Weanling diarrhea, 114
Williams, Stephen, 69, 70 (naming of Nodena Phase)